CANADIAN SCIENCE, TECHNOLOGY, AND INNOVATION POLICY

Canadian Science, Technology, and Innovation Policy

The Innovation Economy and Society Nexus

G. BRUCE DOERN, DAVID CASTLE, AND PETER W.B. PHILLIPS

McGill-Queen's University Press
Montreal & Kingston • London • Chicago

ISBN 978-0-7735-4723-0 (cloth)
ISBN 978-0-7735-4724-7 (paper)
ISBN 978-0-7735-9898-0 (ePDF)
ISBN 978-0-7735-9899-7 (ePUB)

Legal deposit second quarter 2016
Bibliothèque nationale du Québec

Printed in Canada on acid-free paper that is 100% ancient forest free (100% post-consumer recycled), processed chlorine free

This book has been published with the help of a grant from the Canadian Federation for the Humanities and Social Sciences, through the Awards to Scholarly Publications Program, using funds provided by the Social Sciences and Humanities Research Council of Canada.

McGill-Queen's University Press acknowledges the support of the Canada Council for the Arts for our publishing program. We also acknowledge the financial support of the Government of Canada through the Canada Book Fund for our publishing activities.

Library and Archives Canada Cataloguing in Publication

Doern, G. Bruce, 1942–, author
Canadian science, technology, and innovation policy: the innovation economy and society nexus/G. Bruce Doern, David Castle, and Peter W.B. Phillips.

Includes bibliographical references and index.
Issued in print and electronic formats.
ISBN 978-0-7735-4723-0 (cloth). – ISBN 978-0-7735-4724-7 (paper). – ISBN 978-0-7735-9898-0 (ePDF). – ISBN 978-0-7735-9899-7 (ePUB)

1. Science and state – Canada – History – 20th century. 2. Science and industry – Government policy – Canada – History – 20th century. 3. Technology and state – Canada – History – 20th century. 4. Technological innovations – Government policy – Canada – History – 20th century. I. Phillips, Peter W. B., author II. Castle, David, 1967–, author III. Title.

Q127.C3D63 2016 338.971'06 C2015-908747-3
C2015-908748-1

This book was typeset by Interscript in 10.5/13 Sabon.

Contents

Chart and Tables

TABLES

Abbreviations and Acronyms

AAAS	American Association for the Advancement of Science
AAFC	Agriculture and Agri-Food Canada
ABC	applied genomics for bioproducts and crops
ABIP	Agricultural Bioproducts Innovation Program
ACOA	Atlantic Canada Opportunities Agency
ACST	Advisory Council on Science and Technology
ADM	Assistant Deputy Minister
AECL	Atomic Energy of Canada Ltd
AHRA	Assisted Human Reproduction Act
AHRC	Assisted Human Reproduction Canada
AIF	Automotive Innovation Fund
AIT	Agreement on Internal Trade
ANAC	Aboriginal and Northern Affairs Canada
APF	Agriculture Policy Framework
ARDA	Agriculture and Rural Development Act
AUCC	Association of Universities and Colleges of Canada
AUTM	Association of University Technology Managers
BCNI	Business Council on National Issues
BERD	business expenditures on R&D as a percentage of GDP
BL-NCE	business-led Networks of Centres of Excellence
BRIC	Brazil, Russia, India, and China
BSE	bovine spongiform encephalopathy
CADTH	Canadian Agency for Drug Technologies in Health
CANARIE	Canadian Network for the Advancement of Research, Industry and Education
CANDU	Canada Deuterium Uranium (reactor)
CAP	Community Access Program

CARICOM	Caribbean Community and Common Market
CATA	Canadian Advanced Technology Association
CBAC	Canadian Biotechnology Advisory Committee
CCA	Council of Canadian Academies
CCC	Canadian Chamber of Commerce; Canola Council of Canada
CCE	Council of Chief Executives
CCS	carbon capture and storage
CDR	Common Drug Review
CDPQ	Centre de développement du porc du Québec
CDRM	Cabinet Directive on Regulatory Management
CDSR	Cabinet Directive on Streamlining Regulation
CEAA	Canadian Environmental Assessment Agency
CECR	Centres of Excellence for Commercialization and Research
CEF	Central Experimental Farm
CEO	chief executive officer
CEPA	Canadian Environmental Protection Act
CEPMed	Centre of Excellence in Personalized Medicine
CERN	Conseil Européen pour la Recherche Nucléaire
CESD	Commissioner of Environment and Sustainable Development
CETA	Canada-EU Trade Agreement
CFA	Canadian Federation of Agriculture
CFDS	Canada First Defence Strategy
CFERF	Canada First Excellence Research Fund
CFI	Canada Foundation for Innovation
CFIA	Canadian Food Inspection Agency
CFIB	Canadian Federation of Independent Business
CGC	Canadian Grain Commission
CGDN	Canadian Genetic Diseases Network
CIDA	Canadian International Development Agency
CIHR	Canadian Institutes of Health Research
CIPO	Canadian Intellectual Property Office
COPE	Committee on Publication Ethics
CORE	Centre of Regulatory Expertise
CRC	Canada Research Chair(s)
CRTC	Canadian Radio-television and Telecommunications Commission

CSA	Canadian Space Agency
CSR	corporate social responsibility
CSTA	Council of Science and Technology Advisors
CTCS	Canadian Trade Commissioner Service
CUDOS	communalism, universalism, arm's-length disinterestedness, and organized skepticism
CUFTA	Canada-US Free Trade Agreement
CURA	community-university research
CWB	Canadian Wheat Board
CWS	Canadian Wildlife Service
DFAIT	Department of Foreign Affairs and International Trade
DFATD	Department of Foreign Affairs and Trade Development
DFO	Department of Fisheries and Oceans
DNA	deoxyribonucleic acid
DPI	deep packet inspection
DREE	Department of Regional Economic Expansion
DRIE	Department of Regional Industrial Expansion
EACSR	External Advisory Committee on Smart Regulation
EC	Environment Canada
ECC	Economic Council of Canada
ELSA	ethical, legal, and social aspects (Europe)
ELSI	ethical, legal, and social implications (US)
EMR	Energy, Mines and Resources
EPA	Environmental Protection Agency (US)
ERDA	Economic and Regional Development Agreement
EU	European Union
FAO	Food and Agricultural Organization
FCC	Farm Credit Corporation; Federal Communications Commission (US)
FDA	Food and Drug Administration (US)
FIPA	Foreign Investment Promotion and Protection Agreement
FIRA	Foreign Investment Review Agency
GATT	General Agreement on Tariffs and Trade
GDP	gross domestic product
GE3LS	ethical, economic, environmental, legal, and social aspects of genomics
GERD	gross expenditure on R&D, as share of GDP
GFAPC	Growing Forward Agriculture Policy Framework

GHG	greenhouse gases
GM	genetically modified
GNP	gross national product
GRAS	generally regarded as safe
GRC	Global Research Council
GRDI	Genomics R&D Initiative
GST	Goods and Services Tax
HERD	higher education R&D; higher education expenditures on R&D (as share of GDP)
HGP	Human Genome Project
HQP	highly qualified personnel
HT	herbicide tolerant
ICSU	International Council for Science
ICT	information and communication technologies
IDRC	International Development Research Centre
IEA	International Energy Agency
IG	Institute of Genetics
IHAC	Information Highway Advisory Council
IMWIC	International Maize and Wheat Improvement Center (CIMMYT)
INDU	Standing Committee on Industry, Science and Technology
INSITE	Innovation Sustainability Technology
IP	intellectual property
IPPC	Intergovernmental Panel on Climate Change
IPR	intellectual property right
IR	international relations
IRAP	Industrial Research Assistance Program
IRB	Industrial and Regional Benefit
IRDI	Industrial Research and Development Internship
IRIC	Industrial Research and Innovation Council
ISP	Internet service provider
ISTC	Industry, Science and Technology Canada
ITAC	Information Technology Association of Canada
ITC	Department of Industry, Trade and Commerce
KBE	knowledge-based economy
LAV	light armoured vehicle
LDC	less developed country
LO	legitimate objective (in trade agreements)

LSARP	Large-Scale Applied Research Program
MC	Memorandum to Cabinet
MII	Agri-Food R&D Matching Investment Initiative
MIT	Massachusetts Institute of Technology
MOSST	Ministry of State for Science and Technology
MP	member of Parliament
MRC	Medical Research Council of Canada
MSED	Ministry of State for Economic Development
MSERD	Ministry of State for Economic and Regional Development
MSPRA	Ministry of State for Privatization and Regulatory Affairs
NABST	National Advisory Board on Science and Technology
NAFTA	North American Free Trade Agreement
NATO	nodalities, authority, treasury, and organization (Hood's taxonomy)
NBAC	National Biotechnology Advisory Committee, replaced by CBAC
NBS	National Biotechnology Strategy
NCE	Networks of Centres of Excellence
NEB	National Energy Board
NEP	National Energy Program (1980–84)
NGO	non-governmental organization
NHDRP	National Health and Development Research Program
NIH	National Institutes of Health (US)
NOC	Notice of Compliance
Nortel	formerly Northern Telecom
NPE	non-practising entity
NPM	new public management
NRC	National Research Council of Canada
NRCan	Natural Resources Canada
NREN	National Research and Education Network (US)
NRTEE	National Round Table on the Environment and the Economy
NSA	National Security Agency (US)
NSERC	Natural Sciences and Engineering Research Council of Canada
NSF	National Science Foundation (US)
NSI	National System of Innovation

NWPTA	New West Partnership Trade Agreement
OAG	Office of the Auditor General
OECD	Organization for Economic Cooperation and Development
OGI	Ontario Genome Institute
OIPC	Office of Intellectual Property and Commercialization
OIT	Office of Industrial Technology
ONSA	Office of the National Science Advisor
OOGP	Open Operating Grant Program (of the CIHR)
OPCC	Office of the Privacy Commissioner of Canada
OPEC	Organization of the Petroleum Exporting Countries
OTA	Office of Technology Assessment (US)
P3	public-private partnership
P4	public-private-producer partnership
PBI	Plant Biotechnology Institute
PBO	Parliamentary Budget Officer
PBR	plant breeders' rights
PCO	Privy Council Office
PHAC	Public Health Agency of Canada
PIPSC	Professional Institute of the Public Service of Canada
PLF	Progressive Licensing Framework
PM	personalized medicine
PMO	Prime Minister's Office
PNT	plants with novel traits
PPM	production and processing methods
PSO	Parliamentary Science Officer
R&D	research and development
RAE	research assessment exercise
RAF	risk analysis framework; research excellence framework
RAS	Regulatory Affairs Sector
RCGO	Royal Commission on Government Organization (Glassco Commission)
RCMP	Royal Canadian Mounted Police
RCNRT	Royal Commission on New Reproductive Technologies (Baird Commission)
RIA	regulatory impact analysis
RIAS	Regulatory Impact Assessment Statement
RIM	Research in Motion
RR	Roundup Ready™ (wheat)

RSA	related science activities
RSC	Royal Society of Canada
SAGE	*Science Advice for Government Effectiveness*
SAGIT	Sectoral Advisory Group on International Trade
SARS	severe acute respiratory syndrome
SBDAs	science-based departments and agencies
SCC	Science Council of Canada; Supreme Court of Canada
SCN	Stem Cell Network
SD	sustainable development
SDTC	Sustainable Development Technology Canada
SFT	Speech from the Throne
SME	small and medium-sized enterprise
SOA	special operating agency
SPD	Science Policy Directorate
SPI	science-policy interfaces
SPS	sanitary and phytosanitary
SR&ED	Scientific Research and Experimental Development (Tax Credit)
SRTC	Scientific Research Tax Credit
SSHRC	Social Sciences and Humanities Research Council of Canada
S&T	science and technology
STEM	science, technology, engineering, and medicine
STI	science, technology, and innovation
STIC	Science, Technology and Innovation Council
TBS	Treasury Board Secretariat
TBT	technical barriers to trade
TCPS	Tri-Council Policy Statement: Ethical Conduct for Research Involving Humans
TILMA	Trade, Investment and Labour Mobility Agreement
TK	traditional knowledge
TPD	therapeutic products development
TPP	Therapeutic Products Programme; Trans-Pacific Partnership Free Trade Agreement
TRIPS	Trade-Related Aspects of Intellectual Property Rights
TRIUMF	Tri-Universities Meson Facility (now involves many universities, not just three)
TTO	technology transfer office
UK	United Kingdom

UNDP	United Nations Development Programme
UNEP	United Nations Environmental Programme
UNFCC	United Nations Framework Convention on Climate Change
UPA	Union des producteurs agricoles
UPOV	International Union for the Protection of New Varieties of Plants
URL	uniform resource locator
"WEO"	World Environmental Organization
WHO	World Health Organization
WIPO	World Intellectual Property Organization
WTO	World Trade Organization
WTO-TRIPS	World Trade Organization Trade-Related Intellectual Property Rights
WWF	World Wildlife Fund

Preface

This book is the product of the authors' work on science and technology (S&T) and innovation policy in Canada and internationally over the past five decades. During this extensive period of teaching, reading, discussion, research, and interviews, we owe numerous debts of thanks, gratitude, and learning to many individuals and to many agencies and institutions involved directly and indirectly with the story of Canadian S&T and innovation policy, governance, and democracy.

We are also grateful to many S&T and innovation policy academics and practitioners from across Canada and internationally. We have drawn on their scholarly research that we cite and debate throughout the book. In particular, we would like to thank colleagues whose research, ideas or comments on drafts of the book manuscript have informed this book. Jeff Kinder, Paul DuFour, Michael Prince, Richard Schultz, Christopher Stoney, and John Coleman deserve special recognition and thanks.

A continuing intellectual and personal set of thanks are owed to colleagues and staff at our home academic institutions, the School of Public Policy and Administration at Carleton University, and the Politics Department, University of Exeter in the United Kingdom, the University of Victoria, Edinburgh University, and the Johnson-Shoyama Graduate School of Public Policy, University of Saskatchewan.

We also thank the publisher's two anonymous peer reviewers for constructive and important comments and suggested revisions to the initial manuscript submitted. These have helped strengthen the final product.

G. Bruce Doern, David Castle, and Peter W.B. Phillips
November 2015

CANADIAN SCIENCE, TECHNOLOGY, AND INNOVATION POLICY

Introduction and Analytical Framework

The title of this book evokes the volume's two central and different contributions. The first major contribution of this book is that it provides by far the most extensive and in-depth account and explanation of Canada's changing science, technology, and innovation (STI) policy and related governance system over the past half century. This is done via an analysis of eight STI policy domains in Canada covering the four prime ministerial eras from Pierre Elliott Trudeau to Stephen Harper and through an examination of thirty-one policy histories. All of this integrated empirical coverage is set in the context of important international forces and imperatives and intricate realms of public and private governance and political, economic, and social power and agenda setting.

In the course of this needed and different analytical journey, we also present new critical analyses of related developments such as significantly changed concepts of peer review and the emergence of Big Data in the digital age and Internet information economy and society. The different ways in which federal versus provincial STI policies have had an impact on both levels of government are also examined, including some aspects of federal policy that reflect the influence and power of the provinces and of universities. We examine numerous realms of technology across Canada in universities, business, and government and in ever more complex modes of internationally and nationally networked and partnered research and development (R&D) and various efforts to assess new technologies or seemingly succumb to them. These include biotechnology, genomics, and the Internet but also earlier technologies such as nuclear reactors, satellite technology, and evolving computer technologies.

The second closely related and needed new and major contribution is centred on how we show and argue that scholars, students, and practitioners in the field of Canadian STI policy need an enhanced understanding of the *innovation economy and society nexus* in operation across this fifty-year story. We examine whether an S&T-centred innovation economy and society has been established in Canada and whether the STI framework adequately captures the way in which STI policy is being put into practice in Canada over the past half-century. An innovation economy and society is one that aspires to and achieves some kind of moving and interacting balance between STI directed at commercial, private, market objectives and STI deployed to achieve social objectives, including delivering public goods and supporting values related to redistribution, fairness, and community and citizen empowerment, often judged and based in part on the extent of democratic engagement in governance and decision making.

This book is mainly for academics and students interested in the developments, strengths, and weaknesses of Canadian STI policy, and its related governance and democracy. We have also kept in mind governmental, business, and STI professional readers who follow STI policy issues on a regular basis and those citizens and interest group participants who benefit from, or may be disadvantaged or in some cases harmed by STI policy developments and shortfalls.

There is almost always a creative tension between the pursuit of scientific excellence for its own sake and the understanding that "good enough" science is often at the root of major technological and industrial changes; first-to-market frequently trumps best-in-market. In this book we investigate the transformations and change and also forms of inertia that have occurred in Canadian STI policy. We want to understand whether, and to what extent, the current concerns raised about Canadian STI policy are new, or are embedded and long-standing, and whether they can be changed and reformed to achieve other more defensible goals in private markets and in the closely related public social realm.

Three principal questions are addressed in this book: (1) How, why, and to what extent has Canadian STI policy, governance, and democracy changed in the past fifty years? (2) Where has STI policy resided in Canadian federal political and policy agendas across the past fifty years? (3) How have concepts of science-based governance prevailed or been changed across the past several decades?

ANALYTICAL FRAMEWORK

Our analytical framework involves three components. First, we develop and deploy the concept of an *innovation economy and society nexus* in relation to STI policy in Canada. Second, we identify and analyze eight

S&T and innovation policy *domains*. Third, we examine three *elements* within each domain to help understand drivers of change and inertia: (1) policy ideas, discourse, and agendas; (2) economic and social power; and (3) time, temporal realities, and conflicts. Chart 1 provides a visual map of the conceptual and empirical journey involved. Before turning to the framework, we offer five key definitions.

STI policy is defined by the Organization for Economic Cooperation and Development as policies "on the contribution of science, technology and industry to well-being and economic growth" (OECD 2015a, 1). This is a very broad inclusive definition that must be augmented by other key definitions as they have emerged historically or as terms are variously used in policy and political discourse.

Thus, *S&T policy* refers to policies aimed at the funding, conduct, and dissemination of basic and applied scientific and technological research in the natural, engineering, and medical-health sciences (sometimes called STEM), increasingly supplemented by the social sciences. But S&T policy is also frequently used interchangeably with concepts such as science policy, R&D policy, and commercialization policy. S&T policy has historically been seen as the *linear continuum* of activity, with basic research seen as leading to applied research and then to "development" and commercialization. S&T policy also promotes and governs the use of scientific and technical knowledge in public policy and regulation ("science in policy") where governments draw on their internal S&T capabilities, or the capacities of others, to carry out their responsibilities under laws, rules, and international agreements, especially in public interest socio-economic areas such as the environment, health, consumer, and safety policy and regulation (Doern and Kinder 2007).

Innovation policy usually refers to stated efforts to foster the use of S&T to produce new and competitive products and new production processes, and the innovative organizational approaches and management practices that support these activities. The OECD defines innovation as diverse activities that go "far beyond the confines of research labs to users, suppliers and consumers everywhere – in government, business and non-profit organizations, across borders, across sectors and across institutions" (OECD 2015b, 1). The role of science in innovation is, however, vigorously contested. Some authors assert that historically firms and economies have been and can be highly innovative without any specific scientific or technological invention (Mokyr 2002). Moreover, innovation policy is complicated because it is increasingly embedded in complex innovation systems or creative milieu (Wolfe and Lucas 2003;

Etzkowitz and Leydesdorff 1998; Florida 2002). Innovation is usually characterized as a *non-linear* networked interactive process involving often unpredictable recombinations of research and action by key individuals, universities, businesses, and government. Innovation policy places considerable emphasis on systems, communities, or networks of researchers using an interdisciplinary approach (Gibbons et al. 1994). The overall innovation and innovation policy paradigm has also increasingly been cast in relation to the existence of national and local-city-regional networks of innovation and "clusters" (Porter 1990; Phillips et al. 2012). They are again seen as being the product of complex non-linear interactions among universities, corporations, governments, capital markets, systems of regulation, and informed consumerism (Freeman and Soete 1997; de la Mothe and Paquet 1998). Innovation in the current pervasive Internet and social media age has also been cast as *permissionless innovation* based on the argument that information-centred Internet innovation does not require the approval of authorities (and much of it is to a great extent beyond the reach of any local, regional, or national authority). Indeed, some assert that this new form of innovation was deliberately designed to function this way (Thierer 2014; Leiner et al. 2014).

The definitional realm for innovation involves complexities regarding product innovation, process innovation, marketing innovation, and organizational innovation as set out in the *Oslo Manual Guidelines for Collecting and Interpreting Innovation Data* (OECD 2015c). But in the politics of innovation policy, the definitional realm for innovation also involves, as we will see in its use in agenda-setting Throne Speeches and Budget Speeches and across the eight STI domains examined in this book, discourse that may simply be used to suggest or seek "improvement" or "reform" or "change" or "progress."

Governance is a concept that emerged in the literature on politics, policy, and public administration over the past thirty years or so (Aucoin 1997, 2008; Rhodes 1997). In one sense, governance can be expressed simply as an effort to recognize more explicitly that governing involves more than government, more than the state, and more than public policy pronounced and implemented by the state and its bureaucracies (Pal 2012). Governance implies the state playing a role characterized at times more by steering than rowing (Osborne and Gaebler 1992) and by more explicit efforts to improve service delivery. But governance also justifies the continued need for a strong state and state-led capacities (Bell and Hindmoor 2009). In Canada it has equally been present in varied forms in Canada's provincial governments and related urban and local government policy systems (Atkinson et al. 2013).

Governance also refers to the deployment of a much more varied array of policy instruments, fully exploiting the nodalities, authority, treasury, and organization (acronym: NATO) options in Hood's taxonomy (1986). This includes the traditional tools of spending, taxation, and regulation but also softer more exhortative instruments of governing through guidelines and codes of behaviour designed to influence and change the behaviour of different kinds of entities, clients, and beneficiaries (Jordana and Levi-Faur 2004). Regulatory governance theory has always recognized that state-centred regulatory bodies almost always do more than regulate. These bodies are typically multifunctional in nature. In addition to regulation, they are planning entities; they exhort, persuade, and provide information; they adjudicate; and they network (Schultz and Alexandroff 1985; Doern, Prince, and Schultz 2014).

Democracy we define broadly to include all the main values, criteria, and arenas in play in the Canadian political system. These include the following: elected, representative, Cabinet-based parliamentary democracy; federalism; interest group pluralism; engaging civil society; and direct democracy, including the use of burgeoning Internet-based social networks (Pal 2013; Bickerton and Gagnon 2009; Williams 2009; Rainie and Wellman 2012). These in turn involve various types of engagement and consultation that involve S&T advice and controversy (Castle and Culver 2006).

The Innovation Economy and Society Nexus

Economies and societies are always linked in some fashion in both human and policy terms. Often, societies are analytically cast as playing perpetual catch-up in the face of powerful economic and industrial forces. Thus, examining Canada as an innovation economy and society nexus necessarily involves investigating several further themes and issues in each of the S&T and innovation policy domains. We initially and advisedly cast the innovation economy and society nexus as an analytical *heuristic*.

The innovation economy and society is, in simple terms, a two-category analytical world, but the relationships comprising the nexus between the two are complex and replete with boundary overlaps and changing discourse regarding how to describe and understand it. The innovation descriptor emerged as a policy term in the late 1980s and

CHART 1 Analyzing Canadian science, technology, and innovation policy

CONTEXT	ANALYTICAL FRAMEWORK	8 S&T AND INNOVATION POLICY DOMAINS
CONCEPTUAL FOUNDATIONS (chapter 1) • S&T and innovation policy ideas, paradigms, and discourse • S&T and innovation governance, power, and democracy • Economic and social innovation CANADIAN S&T AND INNOVATION POLICY AND AGENDAS IN LIBERAL AND CONSERVATIVE PRIME MINISTERIAL ERAS (chapter 2) CANADA-US AND INTERNATIONAL STI POLICY AND INSTITUTIONS (chapter 3)	INNOVATION ECONOMY AND SOCIETY NEXUS "HEURISTIC" Aspires to and achieves some kind of moving and interacting balance between: • S&T directed at commercial, private property, and market innovation objectives and • S&T directed at social objectives, including values related to public goods, sustainable development, fairness, and full democratic engagement S&T AND INNOVATION POLICY DOMAINS • As middle-level interacting systems of ideas, values, policies, laws, rules, processes, and institutions • As vertical and horizontal in direction and scope • With economic and social intent and intended, unintended, or ill-considered social and economic impacts DOMAIN ELEMENTS • Policy ideas, discourse, and agendas • Economic and social power • Time, temporal realities, and conflicts	CHANGE AND INERTIA ACROSS 40–50 YEARS OF S&T AND INNOVATION POLICY AND IMPLEMENTATION IN THE DOMAIN OF: • Macro S&T and innovation policy (chapter 4) • Government S&T departments and agencies (chapter 5) • Granting, universities, and levered money (chapter 6) • Industrial S&T and Innovation (chapter 7) • Intellectual property, invention, and Innovation (chapter 8) • Agriculture, food, biosciences, and biotechnology (chapter 9) • Genomics, life sciences, and technology (chapter 10) • Internet, communications, and social media (chapter 11)

extends to the present day, but its presence can be found in earlier eras when policy discourse and content was different (as we see below). The entire field of STI policy is often categorized in other sets of two-category worlds, including science and technology, research and development, linear and non-linear innovation processes, social and natural sciences, and disciplinary and interdisciplinary research.

The notion of what empirically should be considered economic versus social is complex. Research, innovation, technology, and policy each simultaneously relates to realms and nexus features such as industrial sectors and subsectors, natural resource industries, health, safety, the environment, education, research ethics, and impacts on populations, places, and regions, and also subsets of individuals and communities that are more particular and perhaps harder to define and find.

In an initial way, the following list previews some of these innovation economy and society nexus themes and issues, some aspects of which emerge in the literature review in chapter 1 and others that emerge in each of the later empirical domain chapters and are assembled for final analysis in chapter 12.

- How is the innovation economy and society different from earlier eras of economic and social policy?
- What might a "balanced" Canadian innovation economy and society look like, and can "balance" be empirically observed and assessed overall?
- Do current macro democratic processes and agenda-setting practices make room for social concerns and values or are they marginalized in such processes?
- Does an innovation economy and society need more integrated assessments and monitoring – especially pre-market and post-market audits of technologies and processes, on top of the product reviews?
- Are universities, academic research projects, and researchers alert to the concerns of an innovation economy and society, or are they being transformed by new, blended funding models into agents of a business-led innovation economy?
- Would sustainable development ideas, in combination with precautionary approaches to decision making, help to address the values and linkages of the innovation economy and society nexus?
- Are more diverse kinds of policy and political discourse (S&T, innovation, and otherwise) needed to attend to concerns of smaller social groupings that might either benefit from or be harmed by innovation?

- To what degree is each of the eight S&T and innovation policy domains also a de facto nexus?
- How do the temporal features of public policy decision making influence the innovation economy and society nexus?

Policy and Governance Domains

Policy and governance *domains* are integrated complex realms of policy ideas, institutions, interests, instruments, and rules (Doern, Maslove, and Prince 2013). The following are the eight S&T and innovation policy domains mapped and empirically examined in this book:

1 Macro S&T and innovation policy (chapter 4)
2 Government S&T departments and agencies (chapter 5)
3 Granting, universities, and levered money (chapter 6)
4 Industrial S&T and innovation (chapter 7)
5 Intellectual property, invention, and innovation (chapter 8)
6 Agriculture, food, biosciences, and biotechnology (chapter 9)
7 Genomics, life sciences, and technology (chapter 10)
8 Internet, communications, and social media (chapter 11)

The first five are broad domains encompassing particular realms of S&T and innovation policy and governance, and the last three deal somewhat more with transformative technologies, where governance and democratic systems seek to assess their pros and cons, albeit while playing a game of constant catch-up with forces that are to a greater extent more external and global than in the other earlier established domains.

As an analytical category the notion of a *domain* is often interchangeably used with terms such as "jurisdiction" (as in the federal, provincial, or international constitutional jurisdiction) or as a "regime" (as in analyses of risk regimes or of individual policy areas such as the food safety or the drug approval system; see, e.g., Hood, Rothstein, and Baldwin 2001; Harris and Milkis 1989). We map and examine domains as relatively complex S&T and innovation policy and governance spheres, levels, and temporal periods where different policy, taxation, spending, regulation making, and compliance challenges are faced, debated, reframed, or ignored.

Because of political-economic debate, academic discourse, and institutional politics, there are bound to be both agreements and disagreements about the precise boundaries of domains. Indeed, boundary issues and overlaps are often a key feature driving the analysis of domain challenges, debates, and outcomes. Some domains have been present and recognized in academic analysis for very long periods (e.g., the industrial S&T and innovation policy domain) and others (such as the Internet, communications, and social media domain) are somewhat more recent but certainly not brand new. New domains are contending for analytical and practical recognition in academic S&T and innovation policy analysis as well as other more specific policy field literature. These include domains centred on the "new underclass" or on particular or combined technologies such as nanotechnology or S&T in risk assessment regarding lifestyle choices and activities.

Policy Domain Elements

Elements refer to basic important features that help us understand some of the basic causes of both S&T and innovation policy change and inertia, as well as providing insights into the overall innovation economy and society nexus. The three main elements deployed in and discussed at the end of each empirical chapter are:

1 S&T and innovation policy ideas, discourse, and agendas
2 S&T and innovation economic and social power
3 Time, temporal realities, and conflicts.

The three elements that we map and deploy in the eight domain chapters are essential features for understanding domain change and inertia. But each of the three elements poses different challenges regarding qualitative and quantitative evidence and the time periods being covered.

The mixes of dominant and contending *S&T and innovation policy ideas, discourse, and agendas* are found in historical, but still relevant, advocacy of policy. The varied discourses and ideas show up in the changing content of Speeches from the Throne, Budget Speeches, and numerous ministerial speeches and talking points. Ideas and preferred discourse are also revealed in opposition political party, interest group, think-tank, and academic papers, and in the diverse kinds of sound-bite discourse developed by the electronic media, Internet bloggers, and social networks that are cited by political parties in partisan discourse.

Our eight domains also contain entrenched and shifting systems of *S&T and innovation economic and social power*. Within the state, power can grow, ebb, or wane based on minority versus majority governments and the styles and performance of ministers and prime ministers. In short, style often trumps substance. Firms and industry associations largely project private economic power. Both Canadian and global businesses are important in this field. For the most part, private power works in tandem with the Department of Finance and Industry Canada, and also with specific line ministries or special operating agencies with industry-specific mandates and capacity. Consumer, medical-health (professions and patient-disease groups), and environmental interests also exercise power and influence, albeit usually not on as sustained a basis as business. All of these actors are networked with each other and the S&T and innovation knowledge communities; the way they are organized and how they function in the diverse kinds of science-policy interface adds both richness and complexity to the analysis.

Varied and changing kinds of *time, temporal realities, and conflicts* underpin the structure and dynamics of the different domains. These range from diverse temporal decision processes, cycles, and realities, including the following: intermittent crises or disasters; annual or multi-year business, political, or electoral cycles; and long-term structural, life-cycle, generational, or intergenerational demographic changes. Indeed, S&T issues frequently involve having to develop strategies and communicate knowledge about vast epochs of time such as in debates about climate change, nuclear waste management, carbon capture and storage, and global biodiversity.

THE EMPIRICAL ANALYSIS OF POLICY AND GOVERNANCE HISTORIES

Across the eight domains, we examine thirty-one illustrative policy and governance histories (see Table 0.1) that draw empirically on numerous reports and studies of Canadian and international policy and regulation by academics, think tanks, and government, complemented by our own research. This work is underpinned as well by a large number of interviews conducted by the authors with S&T and innovation policy officials and players across the eight domains. The historical content of the empirical chapters as a whole broadly covers a forty- to fifty-year period, sometimes in a basic chronological manner but often moving back and forth between the four prime ministerial eras in an effort to show the complex interplay of ideas, power, and temporal realities and conflict. In addition, the contextual analysis in chapter 3 of seven Canada-US and international S&T and innovation policy developments, forces, and institutions adds to the weight and depth of our empirical story overall.

Table 0.1 Policy and governance histories examined in the eight domains

S&T and innovation policy	*Policy and governance histories domains*
Macro S&T and Innovation Policy (chapter 4)	1963–2015: Prime ministers, central agencies, and macro S&T policy statements and strategies 1968–2015: Budgets, budget processes, and annual S&T policy 1980–2015: Regulation, regulatory processes, and S&T and related science activities (RSAs) 1990–2015: Policies and guidance regarding science advice and the role of scientists 1963–2015: Public service renewal and S&T
Government S&T Departments and Agencies (chapter 5)	1945–2015: National Research Council (NRC), nation building, and struggle for a stronger business focus 1980–2015: Health Canada and drug and health product regulation 1971–2015: Environment Canada and its S&T strategies
Granting, Universities, and Levered Money (chapter 6)	1960–1980: Foundational granting councils – Medical Research Council (MRC), National Sciences and Engineering Research Council (NSERC), and Social Sciences and Humanities Research Council (SSHRC) 1989–2015: Networks of Centres of Excellence (NCE) 2000–2015: Canadian Institutes of Health Research (CIHR) 1997–2015: Canada Foundation for Innovation (CFI)
Industrial S&T and Innovation (chapter 7)	1945–1970s: Dirigisme, made-in-Canada, and made-for-Canada industrial policy Mid-1985–2015: Creating the climate for innovative firms to compete in global markets 1986–2015: Scientific Research and Experimental Development Tax Credit (SR&ED) policy 1990–2015: Clusters, business, and local-city-regional innovation systems
Intellectual Property (IP), Invention, and Innovation (chapter 8)	1980s to late 1990s: IP and patent policy emergence in the federal and global trade and innovation policy agenda 1980–2006: Fight to extend property rights to living matter 1983–2015: Finding the right balance in patents 1995–2015: Finding better ways to generate benefits from IP
Agriculture, Food, Biosciences, and Biotechnology (chapter 9)	1960s–2005: Global innovation leadership in the development of canola 1995–2015: Groping for new models of research management and the shift to partnerships 1993–2015: Anti-globalization, new models for social engagement, and genetically modified (GM) wheat
Genomics, Life Sciences, and Technology (chapter 10)	Late 1980s–2015: Reproductive technologies, Assisted Human Reproduction Canada, and federalism imperatives Early 1990s–2003: International Human Genome Project and Canadian research support for the changing life sciences 2000–2015: Creation of Genome Canada and its research centres Mid-1990s–2015: Emergence of personalized medicine as a socio-economic Genomics discursive arena and "product" realm

Table 0.1 Policy and governance histories examined in the eight domains (*Continued*)

Internet, Communications, and Social Media (chapter 11)	1970s–1990s: Recognizing and reacting to the Internet as a transformative international technology 1994–2002: Developing a Canadian strategy for the Information Highway and for high-speed broadband access 1998–2015: Social media and economic and social innovation – Google, Facebook, and Twitter and Canadian and information and communication technologies (ICT) impacts and development 2010–2015: Internet neutrality and privacy – national and global governance pressures, shocks, challenges, and responses

MAIN ARGUMENTS

Throughout the analysis as a whole we advance five arguments that underpin our answers to the three research questions. Our first overall argument is that scholars and policy practitioners need a more explicit understanding of the *innovation economy and society nexus*. It is important to differentiate the new world from the more familiar post–Second World War system of capitalism and the Keynesian social welfare state. By focusing on this nexus, we are also more likely to dispense with other unhelpful framing devices such as push-pull descriptions of innovation or simple market versus state accounts of economic development.

Second, we argue that STI policy development, implementation, and review is typically a low priority field of activity in relation to larger and more compelling macro-economic fiscal policy and micro-economic and social priorities. Moreover, STI policy has for the most part remained on the periphery of public opinion in Canada. These conditions help explain why Canadian innovation performance lags despite its high priority in several official S&T policy statements, discourse, and in some fiscal plans.

Third, we argue that the dominant position and the historical image of natural resources as solely extractive activities generates strains in the STI policy debate. In contrast to many other jurisdictions, where STI policy envelops and builds upon the core economic strengths of the country, in Canada STI policy to a great extent isolates the core sectors and instead targets greenfield areas that, if successful, would diversify our economy and lessen our dependence on global commodity markets. This fundamentally harms the continuous development of both economic and social innovation. In spite of this neglect (some might say because of it), the primary and resource sectors have in important ways from the beginning been S&T based but have become even more increasingly focused on the application of S&T, and many have become global innovation leaders.

A fourth argument is that science-based regulation has never been a complete enough descriptor or framework for de facto regulatory and risk-benefit governance. Science has always been and still is only one of the inputs to, but not the sole determinant of, regulatory and policy decisions. In the past that was not viewed as a major issue, as STI-informed regulatory processes generally flew under the radar, allowing experts in science and risk management to quietly do their jobs without much external input or oversight. Now risk regulators are in the spotlight – scholars, industry, foundations, civil society, and other governments are keenly interested in the choices being made. Technical expertise appears to be increasingly submerged (some would say subsumed) under political and partisan attacks from all sides. Discourse about science, technology, and innovation has often gone viral, with few norms or limits to debate.

Fifth, and finally, we argue that Canada's formal technology assessment processes are stressed when presented with transformative technologies, such as biotechnology, genomics, digital media, and the Internet. Our systems are designed to regulate and assess risks and benefits that flow from products, rather than to address the underlying technologies or production processes. But in some realms what constitutes a "product" is itself highly debatable and controversial, especially in the life sciences. Although that approach underpins the global consensus on risk regulation, it is under almost universal attack in Canada and abroad – some people want to see and consider the broader transformative effects of whole classes of technology. The absence of a place to direct those new demands for engagement with science and technology challenges governments everywhere.

Part I of the book examines Canadian S&T and innovation policy, governance, and democracy conceptually and in a historical context. It first examines key parts of the foundational literature. It then examines key prime ministerial eras and maps the changing interests and power structure that underpin S&T and innovation policy and governance agenda setting. Key international developments and pressures are also identified and mapped. Part II provides an extensive complementary empirical analysis of the eight key STI policy and governance domains, including their basic historical features and the manner in which the three domain elements help us to understand change and inertia, and also to probe the nature of Canada regarding the innovation economy and society nexus. Our conclusions and main arguments then follow in chapter 12. We end with a short Postscript about the new Justin Trudeau Liberal government.

PART ONE

Conceptual Foundations and Historical Context

1

Conceptual Foundations

This chapter examines the conceptual foundations that, along with our empirical research, led to the development of the analytical framework of this book set out in the introduction. This conceptual review and the resulting framework draw on a range of academic, governmental, and other studies that are historical and contemporary as well as Canadian and comparative. The literature emerges from diverse social science disciplines and fields including political science, economics, business, risk management, sociology, public policy, S&T studies, geography, governance, and public administration.

This literature reveals an ever-broadening sense of complexity in the nature of S&T and innovation policy, governance, and democracy in Canada as it seeks to influence, interpret, benefit from, and at times, partly resist or reshape a vast amount of international diffusion pressures. The S&T and innovation domain analysis in Part II of the book draws on an additional subset of policy relevant literature, including work on intellectual property (IP), biotechnology, genomics, the life sciences, the Internet, and social media.

CONCEPTUAL FOUNDATIONS

Our perspective on the foundational theoretical and applied literature on S&T and innovation policy, governance, and democracy is organized into three central streams: ideas, paradigms, and discourses; power and democracy; and the innovation economy and society. We explore each stream briefly, commenting on ways in which they are linked and overlap as well as how they can be used both with regard to our analytical framework and the discussion of our central themes.

S&T and Innovation Policy Ideas, Paradigms, and Discourses

This stream of literature refers to broad interpretations of stages in policy ideas and related paradigms and the changing discourse used in the STI community and in political dialogue. The presentation is broadly chronological, but it must be stressed that most of the paradigms are still evolving as various authors debate and characterize the S&T and innovation policy ideas involved.

Science policy as an overt policy field in Canada dates to the mid- to late 1960s, when it was typically expressed as "policies for science" but also "science in policy" (Doern 1972). Policies aimed at science development and support are, however, easily traceable back to the overall effort during the Second World War and immediately afterwards to build and improve Canada's scientific capacities. At the end of the war, the US government had commissioned Vannevar Bush to provide advice on the construction of post-war science (Bush 1945). His report set the stage and provided an initial framework for sustained partnership and engagement between government and the academy (Price 1965). In Canada post-war science policy discourses led to the further development of the National Research Council of Canada (NRC) and the birth of Canada's nuclear science, directed both to medical research and power generation, including the Atomic Energy of Canada Ltd (AECL) CANDU reactor (Thistle 1965; Bothwell 1988).

One primary goal of science policy was to get more industrial research done by industry because Canada's record was well below that of other Western countries, especially the United States (Dufour and Gingras 1988; Enros 1991; Doern 1972). When federal science policies were first seriously debated Canada, in the mid-1960s, the federal government was the main funder and performer of R&D, and industry's role was much more modest as a share of the economy. These early figures were the polar opposite of the situation in most of Canada's main competitor OECD countries (Dominion Bureau of Statistics, 1970). In the 1970s, this focus on increasing industrial research continued, and a federal "make or buy" policy made explicit that "buying" or procuring R&D was to be the rule for government departments *except* when the research was needed to support policy and regulatory tasks central to the role of government (Skinner 1978;Treasury Board Secretariat 1973; Ministry of State for Science and Technology 1975). In spite of fifty years of focused effort to spur private initiative, industrial R&D in Canada remains inadequate.

S&T policy was often conceived, after the Second World War, as *a linear continuum where stages of implied causality and progression follow a straight path*. S&T was anchored in basic or curiosity-driven research that led to R&D where more applied research and prototype developmental production systems were tested, and ultimately this led to commercialization, sales, and market development. A prevailing, if often tacit assumption has long been that the volume of commercializable outputs from this continuum is proportionate to the inputs, hence the long-standing debate about the levels of investment required to grow and sustain an innovation economy. The continuum typically did not refer to patenting or intellectual property or financial capital as phases or typical parts of the discourse. The continuum traces its routes to the US post-war S&T policy articulated by Bush in 1945 (Crow 1994). In Canada, however, this conception came to see different core institutions as lead players in different parts of the continuum, with universities as the home base of basic research (supplemented with some work in a few corporations); business and government in applied research and R&D; and business as dominant in commercialization.

S&T policy as a linear continuum is broadly cast in the theoretical literature as having been supplanted by innovation policy and theory about innovation systems (see more below), but this is not entirely true politically or in terms of how S&T issues and programs are communicated and debated every day by different governments, universities, and science-based agencies. They often characterize their roles as public and private science or public-interest science (Atkinson-Grosjean 2006; Porter and Phillips 2007). Both S&T policy and innovation policy have always had to weave their way through continuing debates and actions about *industrial policy* and *regional policy* aimed at selected sectors and regions of the economy. Overt industrial policy, and the proverbial sin of governments entering into the activity of "picking winners" (mostly poorly), has been in decline but is not entirely expunged rhetorically or practically.

This is in part because there are some corporations nationally and regionally that are quite skilled at "picking governments" either in the name of controlling foreign investment, needing various kinds of adjustment assistance, or simply buying "time" to become competitive. The biggest deviation occurred in the 2008–10 period, with the federal and provincial bailout of the Canadian auto industry, albeit those funds came with some provisions regarding the need for auto industry production and product innovation (Waddell 2010). Mazzacato (2013) has made the case for the existence and role of the "entrepreneurial state" as a particular key feature of S&T and innovation in key firms largely via continuous and patient government funding in industries and technologies such as the Internet, biotechnology, and nanotechnology.

Innovation and innovation policy theory and discourse have been in the political-economic ascendancy for the past twenty-five years. At the centre of this conceptual interpretation is *innovation as non-linear and networked interactions*, often involving *collaborative, networked science*, where research and technological change actually occur and are achieved in diverse and even unpredictable ways. The concepts of national innovation systems, local-regional innovation systems, and clusters emerged internationally and nationally to capture views about the non-linear nature of research and innovation (Foray 1999; Niosi 2000; Freeman and Soete 1997; de la Mothe and Paquet 1998; Nelson 1993).

Along with concepts such as the knowledge-based economy (KBE), the linked ideas of innovation, networks, clusters, and applied commercialization often imply that one cannot actually deliver determinative policies but rather governments can only realize strategies for recognizing and fostering networks of interaction. In a sense therefore they raise a very old issue about whether research itself can ever be fully "planned" and whether observed successes or failures can be attributed to any given innovation or S&T policy (Guston 2000).

A further important strand of innovation theory centres on the concept of *permissionless innovation*, the basic freedom to experiment with new technologies and business models (Thierer 2014; Naughton 2012) without needing permission from authorities. This concept was seen to emerge directly from the development of the Internet as an information network of networks. Vinton Cerf and Robert Kahn, both engineers and the key designers of the Internet, sought to ensure that two principles were pivotal for the Internet: *no central control* and *Internet neutrality*, the idea that the network should not be optimized for any particular application or user. These two principles combined to produce a system of permissionless innovation that released the massive growth of the Internet, the World Wide Web, and literally millions of new sites, businesses, social media, and systems of social production (discussed further in chapter 11). In addition, we are hypothesized to be on the cusp of a new era of production, coined the "Internet of things."

It has also been argued that science and knowledge production has changed in more fundamental ways. Gibbons et al. (1994) illustrated the value of approaching science and policy issues from a higher, more abstract level by describing the social transformation within and among knowledge-producing institutions. This and related work differentiated between so-called Mode 1 and Mode 2 systems of knowledge production, with Mode 1 roughly referring to familiar academic, hierarchical, and traditional peer-review approaches, and Mode 2 highlighting emerging networked, flexible, and interdisciplinary approaches with much broader notions of peer review (Gibbons 2000). These authors pointed to the evolution of advanced science systems toward greater heterogeneity, pluralism, and fuzziness of boundaries (such as between public and private, between knowledge producers and knowledge users, and between natural and social sciences). In particular, the authors describe a shift in the modes of funding and institutional design of knowledge-producing institutions, with greater emphasis on and new approaches to ensuring public accountability for these institutions.

Knowledge as a concept and a process takes on more complex meanings and involves changing authority relationships in and among the sciences (Whitley, Glazer, and Enwall 2010). It may initially be equated by many with "research" and the dissemination of research to various users. Knowledge can also be seen more broadly as "knowledge transfer," where knowledge consists of services rendered and transferred through teaching, research, and service of other kinds. Policy is also increasingly focusing on the importance of "tacit knowledge," knowledge that goes beyond theoretical and basic causal knowledge to include experiential knowledge about how to do things and how to think about things in ways that are imparted through both classroom teaching and also interpersonal mentoring and coaching (Wolfe 2009; Pavitt 1991; Polanyi 1967). Some scholars offer more differentiated typologies, defining knowledge by what it embodies: know-what recipes, know-why basic principles, know-how, know-who, know-where, and know-when, which embody much of what is tacit indigenous knowledge (Crookshanks and Phillips 2013).

This mix of ideas and discourse involves considerable overlap among concepts, and in ways is several steps removed from operational policy as expressed in laws, agency mandates, rules, and guidelines. But in examining these kinds of discourses, it is important to understand the particular S&T and innovation policy domains examined in Part II of this book. We also see this below in the related literature centred on the shift from macro-economic stabilization policy to micro-economic growth policies that involve a complex mixture of both the innovation economy and innovation society.

Related science activities (RSA) refers to activities that complement and extend R&D by contributing to the generation, dissemination, and application of scientific and technological knowledge (Godin 2004; Djellal et al. 2003). Much of this type of knowledge generated in government is vital to the effective regulation of different types and sources of risk. Many key federal S&T departments and agencies, including Environment Canada and Health Canada, have tasks and mandates that are dominated more by RSA than R&D, constituting up to 80 percent of their science budgets and typically lodged in the brains and knowledge possessed by their front-line S&T staff (Doern and Kinder 2007). Crucially, they also engage and share knowledge with their counterparts in other countries' regulatory agencies and with international agencies in a changing and more complex regulatory landscape.

There are at least three textbook definitions of *regulation* that offer concentric circles of ever widening scope: first, regulation in the narrowest sense involves promulgation of rules by government accompanied by mechanisms for monitoring and enforcement; a somewhat larger definition incorporates any direct state interventions in the economy, which would include both rules and other incentive structures put into place; and, most widely, regulation could encompass all mechanisms of social control or influence affecting all aspects of behaviour from whatever source, intentional or not, which opens the field to all impacts of commission or omission (Atkinson et al. 2013). Each of the three conceptions of regulation in this typology can be parsed according to the relative contributions of statutory versus non-statutory mechanisms, but they are somewhat less than wholly clear because they include the consequences of actions, inaction, and collateral effects, often too tightly centred on the state.

British scholar Julia Black (2002, 20) argues that one could perhaps more constructively investigate regulation as the "sustained and focused attempt to alter the behaviour of others according to defined standards or purposes with the intention of producing a broadly identified outcome or outcomes, which may involve mechanisms of standard-setting, information-gathering and behavior modification." Black's de-centred definition adds particular value as it captures the changing nature of regulation, which has recently expanded as its definitions, actors, forms, domain, and instruments have evolved. This decentring has flowed from the way our world is becoming more complex, fragmented, and interdependent and from the reality that the distinction between public and private matters is increasingly blurred (Coglianese 2010). In such a world, regulation is often co-produced, not formally sanctioned by the state, and seldom is it confined to any specific territory. Co-produced regulation has been so increasingly present in the past two decades that it now underpins theories of regulatory capitalism as distinct from earlier periods of regulatory liberalism and deregulation (Braithwaite 2005; Levi-Faur 2005).

We see these diverse regulatory and RSA trends in our later domain chapter policy and governance histories on health, drug, food, biotechnology, genomics, and environmental regulatory and policy governance. There we can see direct regulatory science, monitoring, and service roles broadening. The system is expanding to include both *pre-market* product assessments and approvals or rejections and also *post-market* and life-cycle monitoring (Doern, Prince, and Schultz 2014). RSA is a particularly unique, but often less acknowledged feature of government science and public goods provision in these diverse and changing regulatory landscapes.

Peer review as an idea and process is at the heart of science and research and relations in the natural sciences, engineering, health, and social sciences. Peer review is anchored first and foremost in the way academic journals determine which research papers are published and which are rejected. Peer review is central to the functioning of academic disciplines and interdisciplinary research; it is central also to universities and their public view of themselves as knowledge institutions. Peer review is held as being important in government science labs, with individual science employees and research teams seeking to conduct and publish their work. The granting councils of the federal government (see chapter 6) also adhere to this idea but are affected as well by the significant growth of disciplines and research areas within which peer-review committees and processes are anchored (Murphy 2007; Atkinson-Grosjean 2006).

There are also global journal quality league tables and research citation rankings that increasingly drive the research funding system, and largely define what is high-quality research and a high-quality research university. In the United Kingdom (and Australia) this has also centred on periodic research assessment exercises: the RAE, or research assessment exercise, now rebranded as the REF, research excellence framework, conducted on a department-by-department and discipline-by-discipline basis within and among universities, with rankings published nationally and on university websites. State funding and student recruitment are increasingly driven by these rankings. Canada does not have its own national ranking system. In addition to being evaluated in a range of private rankings (such as the Leiden, QS, Shanghai Indexes, THES rankings, and various professional school rankings), *Maclean's* magazine's rankings of Canadian universities as both research and teaching institutions has largely defined the public imagination of the relative quality and impact of the major Canadian universities.

So-called *grey literature* also warrants some emphasis. Huffine (2010) makes the case for the value of grey literature to scholarly research in the digital age. He defines grey literature as "a body of materials that cannot be found easily through conventional channels such as publishers" (ibid., 1). Grey literature therefore includes government research, non-profit reports, and think-tank assessments. Peer review enters the equation of how to view grey literature. Peer review of a sort may be practised by think tanks and by government departments and agencies, but there is a grey tinge to the effort. It is hard for readers and users of this work to tell what kind of peer review or simply "review" the work has been subjected to. Some work would pass the scholarly norms of originality and unique contribution, but much of the grey literature is summative and normative, which adds considerable value but is seldom original or unique. The world of private and increasingly academic consulting also has these kinds of greyness.

In academic and also governmental and private research, ideas regarding *merit review* are increasingly present. These refer, for example, to reviews regarding whether to fund capital-intensive and infrastructure facilities and technologies. The concept of *formal leveraging* and broader merit analysis was introduced formally with the formation of the Canada Foundation for Innovation (CFI) whose central purpose was to foster new research infrastructure (Lopreite and Murphy 2009). Peer review was not enough for these decisions for two reasons. The management of universities or research hospitals had to be involved and had to be assessed as to their capacity and claims of merit on their own and in comparison with other competing project applications. This was less formally undertaken in earlier periods when Canadian research funding had to deal with so-called Big Science centred on large structures; the difference was that the earlier merit review was more politically charged, being bound up in classical distributional regional and locational politics (Doern 1972).

Second, research partnerships were now increasingly involved in part because all projects now had to contribute 60 percent of the capital cost, which usually meant private sector money was needed to meet the full leveraging targets. There are also aspects of merit review and peer review as de facto combined ideas in the funding of new or renewed centres of excellence as networked forms of research emerged in the late 1980s and beyond (Atkinson-Grosjean 2006). In part this was because applications for funding had to be assessed regarding their proposed *impacts* for specific research users. Similar review features became a part of particular thematic or priority funding programs in each of Canada's main granting councils and underpin the extensive bioscience investment programming offered by Genome Canada in which predictions about the socio-economic benefits of the research are required in research funding proposals (Doern and Stoney 2009).

The notion of *value for money* of research and related program and performance evaluation increasingly drives government research and is to an increasing extent becoming the norm in universities, especially in relation to emerging technologies (Marchant, Abbott, and Allenby 2013; Boden et al. 2004). This is a broader public finance idea that draws on the conceptions of the new public management (NPM; Aucoin 1997). Although commonly cited as a driver for science investments, is not clear whether value for money evaluation is actually practised at times of fiscal restraint, when fiscal deficits and real or perceived crises trigger major cuts to research budgets, to research staff, or to both, as happened in the mid-1990s federal Program Review and in the federal budget cuts made after 2008 (Shepherd 2013; Swimmer 1996). Investments and reinvestments, in contrast, almost always are justified based on their value for money.

Intellectual property, invention, and creativity are now inextricably linked in the minds of scholars and policy-makers. Until the early 1990s, there was no obvious sustained discourse about intellectual property in mainstream S&T policy debate. This was true despite the fact that Canada's laws on intellectual property had been in place for decades. These laws and those agencies mandated to manage patents, trademarks, copyright, and industrial design are grounded on the premise that private exclusionary rights are necessary for promoting creators, inventors, and firms to invest in private invention and creativity (Doern and Sharaput 2000; Doern 1999b; Maybee and Mitchell 1985). Patents are the main focus in this book's domain analysis in chapter 8 but in terms of innovation policy and discourse overall the other IP rights loom large as well. Trade secrets drive the chemical industry and plant breeders' rights (PBR); and pedigrees are the currency in the agricultural sector, and copyright is vital to software development. But IP rights such as copyright has also been subject to historic and continuing transatlantic policy and value differences as to whether intellectual works belong more to the creator or to society, more an evolved human right or simply a cultural product (Baldwin 2014). Trademarks are vital to most products directed at end users, but they also have clashed with the onset of the Internet and domain name creation and other designations.

Until quite recently the international criteria for assessing national S&T performance tended to focus on measures such as the gross expenditure on R&D, which measures the spending as a share of GDP (called GERD). But this was very much an input-oriented kind of criterion and said nothing about outputs and outcomes and performance related to commercialization. Intellectual property and patents emerged as more overt policy levers when the federal government, in concert with innovation ideas being practised by other OECD countries, began to compare itself and respond to its relative rate of patenting in the international league tables; Canada wanted its firms to patent more and took steps to encourage this practice (Doern and Sharaput 2000).

Patents of exclusive use are granted to new products, technologies, business processes, or compositions of matter that are shown to involve an inventive step that is novel, useful (effective), and non-obvious. A key trade-off was that patent recipients would have the right to exclude others from using their patented claims for up to twenty years (since 1989), but the underlying knowledge that led to that inventive step would be disclosed in the patent and would be made public for others to see and potentially improve the invention with further related or new patents. As a result, national and international patent agencies moved much more to centre stage globally compared with their previous existence (Doern 1999b), even though research showed that patents are largely a reflection of inputs and activity and not a measure of value (Jaffe and Trajtenberg 2002).

At about the same time as patents were becoming an instrument of choice, there was some diffusion from the United States to Canada of the norms and instruments to commercialize academic science. The Bayh-Dole Act, 1980, in the United States fully assigned rights to inventions developed using federal government funds to the universities and incentivized them to aggressively develop capacity to exploit inventions. University technology transfer has become the norm in the United States since then. This had some influence in Canada, but patenting at Canadian universities had no such direct federal legislative entry point and faced many practical constraints (Madgett and Stoney 2009; Bird 2009).

There is little doubt, however, that over the past two decades the idea of patents as *protection* has become dominant and now is core to international trade rules (Drahos and Mayne 2002; Sell 1998). It has also been reflected in a monumental increase in the volume of patents being applied for and granted nationally; many patents are pursued in a wide range of countries, particularly the United States, the United Kingdom, Europe, and Japan, coordinated through the Patent Cooperation Treaty, making them effectively global. But concern about various *dissemination* notions of the overall patent IP agenda has also begun to emerge.

Scholars and NGOs have promoted changes to support concepts of common property (e.g., seeds as the common heritage of humankind); to forbid certain kinds of activities, products, and processes from being patented (e.g., humans); and to facilitate patent pooling and open-source practices to support competition and the larger public interest in innovation systems (Dreyfus, First, and Zimmerman 2010; Ventose 2011; Castle et al., 2010; Duttfield 2008).

Others have argued that "patent failure" is increasing because of the inability to establish property boundaries, especially in complex communications equipment and the biosciences. Golden Rice became a high-profile example when it was discovered that it used approximately seventy discrete intellectual properties owned by more than thirty institutions; this was further complicated because the rights were applied differentially in differing countries, with no more than forty-five claimed in any country and the assignees were changing each year, as the global biotechnology industry was restructuring (Kryder et al. 2000). More recently, the telecommunications industry has been beset by "patent trolls," non-practising entities (NPEs) that acquire patents (often from bankrupt firms) and use them to extract (some assert extort) licensing fees for their intellectual property. There is a prevailing view that all of these challenges greatly harm innovation (Moir 2013; Bessen and Meurer 2008).

Notions of creativity have also moved beyond patents and into more direct research on the nature of creativity in fine arts and design and in relation to culture defined broadly (Martin and Christensen 2013). Notions of culture are also being more critically explored (Thomas and Chan 2013). For innovation in some sectors, such as the textile and related "fashion" industries or clusters, creativity takes on different meanings and poses new challenges (Galvin 2012). The same is true for the role of design and taste in the global and Canadian wine industry (Hira 2013) and for most creative entrepreneurial milieu (Florida 2002).

Transformative and disruptive technologies refer to periods of time and development where economies and societies are remade by foundational technological changes. This concept is closely linked to Schumpeter's concept of *creative destruction* (Schumpeter 1934, 1954). The related concept of disruptive technologies has found its way into analytical use both in describing interacting technologies such as digital technology, nanotechnology, and the neurosciences but also even smaller kinds of disruption that create new information and data industries and products (Zussman 2014). The latter are found in on-line businesses that can grow large fast and may well be considered a threat by some to competition

(*Economist* 2014b). As chapter 7 will show, Industry Canada has recently taken to describing disruptive technologies as a key presence in the Canadian economy in both a positive and negative sense but also an unavoidable one (Industry Canada 2014b).

Past examples of technological transformations include the introduction of steam power, electricity, moveable type, and the internal combustion engine (Phillips 2007). Given the profound effect of some of these technologies (both positive and negative), a debate began in the 1970s about how and whether governments can practise rational and transparent forms of technology assessment and engage in technology foresight processes to help guide developments. Several new or recent technologies have emerged and, equally important, are interacting, producing myriad applications that generate both intended and unanticipated consequences for economic affairs, social interactions, and notions of human identity. These include general purpose technologies such as the Internet, information and communications technology, biotechnology, and nanotechnologies, all of which have links to and potential for generating other sectoral technologies, products, and processes in different industries and social realms (Mazzacato 2013). The analyses of such technologies also compare them historically in terms of social resistance strategies and how they change the technologies' trajectories and shape. This is a central feature of Martin Bauer's analysis of "atoms, bytes and genes" (2015).

The pace of technological change after the Second World War brought forth institutions and processes that would enable a society and polity to assess new technologies, their benefits, and their potential adverse impacts before they emerge in the form of new products and production processes (Nordman 2004). Such issues were raised regarding the peaceful uses of nuclear energy in the 1950s and 1960s, and, in the 1970s, regarding technologies such as supersonic aircraft. These and other issues prompted the United States to establish the Congressional Office of Technology Assessment in 1972. The OTA published numerous technology assessments but was abolished in 1995 following the Republican victory in the mid-term elections (Mooney 2005; Nye 2006). Its demise was due to sharp partisan politics and policies by a conservative-dominated Congress that some saw as "anti-science" as well as being opposed to government intervention of the kind that is implied in technology assessment processes. Canada has never had as formal an effort to undertake prospective technological assessments; nevertheless, it tried in its own way to build capacity, which, as in the United States, has now been downgraded. Technology assessment institutions and issues continue to be examined through international bodies, for instance, the International Association for Technology Assessment and Forecast Institutions, various university institutes, and through diverse literatures and journals.

Phillips (2007) stresses the inherent complexity, multi-stakeholder shared relations, and competing nature of modern transformative technology governance and shows the highly linked nature of innovation, ranging from knowledge, invention, the gestation of an invention, and finally, production, marketing, and consumption. Our final three chapters examine S&T and innovation domains that have involved diverse debates about the extent to which they should be subject to such broader-based governance and democratic engagement processes, at the level of law and regulation making, at the level of particular products and processes, or at the level of individuals and groups that are actually or potentially harmed by such technologically centred and fast-changing features (O'Doherty and Einsiedel 2013).

S&T and Innovation Policy Governance, Power, and Democracy

Policy governance and power over all policies, including S&T and innovation policy, is centred in Canada in the *prime minister, Cabinet, and Parliament*, with Canada's nominal minister of science but one voice in a Cabinet of thirty-five to forty ministers and a single member in a 338-member House of Commons. From the outset in the early 1970s, science policy literature has shown that ministers of science are almost always junior ministers on the periphery of power (Dufour 2014a). Moreover, the thirty-four variously titled science ministers over forty-two years were each in office on average for less than a year. The first Cabinet-level minister headed the Ministry of State for Science and Technology (MOSST), but with a very small staff and no control over S&T budgets. The minister was differentially reliant on the use of exhortation and persuasion amid other ministers and the public service bureaucracy (Doern 1972). Science ministers, of course, do foster relationships and engagement with different changing parts of the S&T community, but for the most part they have been minor, low-profile, stand-alone players in the Cabinet governance and power structure.

The first reason for minimalist power and influence is that several non-science ministers were, and continue to be, de facto S&T ministers because their departmental and agency structures and mandates are significantly S&T-based. Indeed, during the Mulroney era, MOSST disappeared and

S&T found its way into the name and mandate of the then new Department of Industry, Science and Technology which later in 1996 morphed into Industry Canada, the current lead micro-economic department (Doern 1996). S&T was thus easily moveable but also retrievable. In the present Cabinet structure, at a minimum ministers and departments such as Industry Canada, Environment Canada, Natural Resources Canada, Fisheries and Oceans Canada, Health Canada, and Agriculture and Agri-Food Canada all have significant S&T policy mandates vis-à-vis their sectors and industries (Clancy 2011; Doern and Kinder 2007; Doern and Reed 2000).

The structure of Cabinet power resides under increasing amounts and forms of prime ministerial oversight and control. This has been true for some time but especially since the late 1990s under the Chrétien-Martin and Harper governments (Savoie 2010, 2013; Martin 2010; Doern, Maslove, and Prince 2013). In significant ways, prime ministers act as de facto S&T and innovation policy ministers, variously through actions that support, discourage, or sideline it on national and international agendas. Closely aligned in this portfolio and power equation is the minister of finance who exerts varying degrees of control vis-à-vis macro-economic fiscal, economic, and budgetary policy. As we will see in chapter 2, Paul Martin as Jean Chrétien's minister of finance was pivotal as a de facto S&T and innovation minister both for ill and for good (Kinder 2010; Doern and Stoney 2009).

Although executive governance is our overall focus initially, the House of Commons as the elected body of representative democracy is also a part of the story (Aucoin, Jarvis, and Turnbull 2011; Fletcher and Blais 2012; Malloy 2004). Opposition parties and their science, technology, and innovation policy agendas (or lack thereof) will be examined in later chapters, as will the roles of parliamentary committees and parliamentary watchdog agencies. Committees at present include the Standing Committee on Industry, Science and Technology (INDU) but also other subject matter committees where S&T and innovation issues often emerge, including in explicit particular social and social policy contexts and fields. Canada's provinces are also relevant here, particularly at the executive level.

The provinces have jurisdiction over education – Canada's universities are created by provincial statute. They also invest in S&T and regulate in many fields with a significant impact on S&T, including natural resources, health care, and agriculture. There is, however, little sustained provincial focus or strategic organization vis-à-vis S&T and innovation; most efforts are embedded in provincial departments and regional or sectoral strategies with little or no legislative power or engagement. There has, however, been reasonable provincial interest in encouraging cities as innovation clusters in a number of defined policy fields. Our domain chapters also show in different ways how universities exert independent influence. This includes chapter 6 on the granting system domain where key university presidents were instrumental in lobbying for the establishment in 1997 of the Canada Foundation for Innovation as a federally funded foundation for research infrastructure funding. The CFI was established as a far more arm's-length entity in that it was not even subject to audits by the Auditor General of Canada (Lopreite and Murphy 2009).

Claims about *democratic governance* in any policy field, including S&T and innovation policy, have to contend with at least five arenas and kinds of democracy and various notions of democratic deficit and imperfections (Lenard and Simeon 2013; Doern and Prince 2012; Dryzek and Dunleavy 2009; Pal 2013). The five arenas and related but different criteria of democracy include: (1) *representative democracy*, in the Canadian case Cabinet-parliamentary democracy, where the public interest is claimed to be reflected through elections and majoritarian decision making; (2) *federalized democracy*, where politics and policy divide constitutionally between national and provincial governments, yielding both conflict and cooperative joint action and various kinds of bilateral and multilateral federalist bargains and diverse views of democratic action; (3) *interest group pluralism*, where democracy emerges through the continuous interplay of interest groups of numerous kinds (especially business groups and consumer interests) involved in lobbying, engagement, and consultation with government; (4) *civil society democracy*, where those representing the weak and marginalized in society, especially environmental, human rights, Indigenous, and women's groups, take up the language of equality seekers and rights holders in general, often using the opportunities created by the Charter of Rights and Freedoms; and (5) *direct democracy*, where individual citizens, voters and non-voters alike, express their views and have influence through their own individual actions, either in polling, focus groups, and on-line social networks or as citizens and as consumers functioning in the marketplace. Individual scientists see themselves as involved in each of these types of democracy, variously as aides to effective decision making or as critics or whistleblowers within government and private firms (see more below).

S&T and innovation policy agenda setting needs to be seen in the context of broader conceptual theories of agenda setting. Theories of agenda setting are anchored in the complex interactions between politicians and policy-makers, the media, and the public. More specifically, agendas arise from the issue attention cycle where different players including voters, citizens, the mass media, and governments themselves engage in framing and reframing problems and opportunities (Eissler, Russell, and Jones 2014; Wolfe, Jones, and Baumgartner 2013; Howlett 2011; Kingdon 1995). The dynamic nature of the system and our collective inability to sustain focus and interest on issues leads to "punctuated equilibria," where most policy stays stable for long periods while, in a punctuated fashion, some policies change quickly and dramatically (Baumgartner and Jones 1993; Jones and Baumgartner 2005, 2012).

To this initial conceptual discussion one needs to consider what public opinion research adds to the picture of S&T agenda setting. Polling by Environics has repeatedly asked a sample of Canadians what, in their opinion, is the most important problem facing Canada. The responses show some interesting results for the period from 1987 to 2011, spanning the last half of the Mulroney era and all of the Chrétien-Martin and Harper years until 2011 (Canadian Opinion Research Archive 2013). In these twenty-five polls, not once is S&T or innovation mentioned in response to this question, nor are they even offered up as possible concerns in a list that otherwise contains twenty to thirty likely or expected issues devised as prompts by the pollsters. Topics such as the economy and jobs are listed as possible responses, as are health care and national unity. Those, of course, tend to garner the highest responses from those polled, varying only by year. Other issues listed include the environment and the BSE/mad cow crisis, which have an S&T component but are seldom framed as such.

A further separate type of question asked was: "Do you think Canada is spending too much, just the right amount, or should be spending more on 'new technology research'?" This question was asked each year from 1984 to 2001. The "too much" proportions hovered around the 10 percent range on average over the period. The "just right" proportions in both the Mulroney and Chrétien-Martin years until 2000 ranged between 35 and 40 percent but increased in 2001 to almost 50 percent. The "should spend more" proportions stood at around 40 to 45 percent for the Mulroney years and then came down to the mid-30 percent range during most of Chrétien years (Canadian Opinion Research Archive 2013).

These kinds of polling data both suggest a very oddly mixed picture and raise some concerns about the types of question posed or, as seen above, not posed. It is not clear that Canadians would know that much about whether spending was too much, too little, or just right, so the opinions disclosed might be more a reflection about whether a government was "handling S&T" appropriately. Authors such as Geoffrey Hale (2002) have argued more generally about the elite nature of S&T policy matters and the difficulties governments face on how to speak about these issues and how to foster a sense of broader social inclusion. We address these kinds of overall agenda-setting dynamics and puzzles in our discussion of Speeches from the Throne in chapter 2 and Budget Speeches in chapter 4.

S&T and innovation policy functions in a Canadian capitalist economy in *which business and corporate power* and policy access is exercised through diverse lobby groups and direct corporate access to ministers and to officials in the Department of Finance. They also lobby sectoral, science-based departments whose ministers and arm's-length agencies make expenditure, tax, and regulatory decisions that differentially affect S&T and innovation policy in often very different competitive situations. Most of the lobby groups are sector based, but a small set are more cross-sectoral or framework oriented, including the Council of Chief Executives (CCE) and the Canadian Federation of Independent Business (CFIB; Bricker and Ibbitson 2013; Hale 2006).

Regional lobbying is also at play through the power of varied natural resource groups (augmented by provincial government support) in areas such as agriculture, fishing, forestry, energy, and mining. As the foundation of Canada's staples economy, the enabling policies that address sectoral concerns often find themselves at odds with an economy oriented toward S&T and innovation (Hessing and Howlett 2005; Campbell, Pal, and Howlett 2004; Coleman 1988). Staples theory centred on export dependence of natural resources has also featured in Canada's debate about capitalism, both historically and in the present day (Kellogg 2015; Howlett and Rayner 2006; Howlett, Netherton, and Ramish 1999; Innis 1956). Federal regional economic development agencies are also a part of the policy system, mobilizing regional interests to lobby for innovation and cluster formation in their mandates and funding activities (Bradford and Wolfe 2012).

The power of business is also ultimately anchored in the decisions made regionally, nationally, and internationally to spend or not spend on new capital projects and new product and service development, whether innovative, science based, or otherwise. Business corporations, in particular, have both direct and indirect political and social power (Wilks 2013). This centres on their capacity to invest in infrastructure and to expand or close plants, and also via business ethics embedded in the strategic plans and relations among shareholders and investors.

Analysis of varieties of capitalism is highly relevant overall and, specifically, to the STI story (Hall and Soskice 2001). Within Canada provinces exhibit different kinds of capitalism, ranging from more state-centred varieties historically in Quebec and Saskatchewan to market-based varieties in Alberta and Ontario. Also, the dominant role of foreign owners, initially British, then American, and now more global, especially in particular strategic or regionally important industries, has been a central theme in Canada (Clarkson 2009; Panitch 1977).

Canadian capitalism includes a number of key firms known to be progressive and innovative, initially exploiting our telecommunications and transportation strengths and more recently focused on new green environmental technologies as alternatives to the carbon economy (Macdonald 2011). They include such historically key high-tech firms as Nortel, RIM-Blackberry, and Bombardier, but also others that have had varying trajectories of rise and fall and stability (Reguly 2013; Yakabuski 2013).

In addition to the concepts of the entrepreneurial state discussed above, the notion of *regulatory capitalism* has also emerged conceptually and empirically (Braithwaite 2005, 2008; Levi-Faur 2005). It is highly relevant to understanding some aspects of innovation policy and governance centred on modern regulation and supply chains and production networks as it differentiates regulatory capitalism from the neo-liberal capitalism era of the 1970s and 1980s. Regulatory capitalism seeks to capture the fact that regulation is growing markedly but is less a feature of state rule and enforcement in the name of public good and much more a system of co-regulation between the state, business interests, and firms but also other non-state interests and networks in support of innovation (Doern, Prince, and Schultz 2014; Ugur 2013; Grabosky 1995).

S&T and innovation policy and governance also functions in middle levels, and often micro-levels, of at least forty federal agencies, funders, and laboratories. These include the four main granting councils examined as a domain in chapter 6 but also the larger array of departments and agencies that are examined as a domain in chapter 5 and selectively in chapters 7 to 11. Conceptually speaking, the literature at this level looks to realities and contradictions centred on concepts such as principal-agent theories in the context of agency, bureaucracy, and hierarchies, but also in relationship to the growing natural or mandated, forced, and/or induced networking achieved or attempted through statutory requirements, guidelines, or leveraging requirements (Doern and Stoney 2009). This literature often emphasizes the growing and complex managerialism extant within, and around, these agencies as they support or restrain S&T development and innovation in complex industrial, resource, and social settings, often complicated by diverse spatial contexts (Guston 2000). Indeed, S&T and innovation policy is frequently or even normally hyphenated into worlds of S&T and innovation-related environment policy, energy policy, oceans policy, and numerous other policies (Kinder 2010).

The STI policy governance structure has also included an array of broad-based but also changing science policy advisory bodies. The National Research Council of Canada (NRC) had such a role since it was created in 1916. But a central much broader role was played by the Trudeau Liberals' Science Council of Canada, established as a complement to the Economic Council of Canada, created in 1963 by the Pearson Liberal government (Doern and Phidd 1983). Later, the National Advisory Board on Science and Technology (NABST) was established by the Mulroney Conservatives, in part because they thought the Science Council had supported initiatives that were state-led and interventionist rather than market-oriented or -focused. The expert-based National Biotechnology Advisory Committee (NBAC, 1992–98) is also relevant here. It was replaced in 1999 by a more diverse and, therefore, more representative Canadian Biotechnology Advisory Council (CBAC). Both Liberal government advisory bodies focused mainly on the area of biotechnology and related genomics research issues (Peekhaus 2013; Doern and Prince 2012).

The Science, Technology and Innovation Council (STIC) replaced a number of expert-based advisory groups, now representing a cross-section of interests. Created in 2007 by the Harper Conservatives, it has published biannual *State of the Nation* reports. During that period the Royal Society of Canada (RSC) also has attempted to fill the gap, but found the going difficult. In 2005 the Royal Society and a range of other scholarly groups created the Council of Canadian Academies (CCA), to an extent modelled on the US National Academies of Science. After a long period of gestation, it was finally accepted that such a body might help fill the kind of arm's-length analytical gap in Canada that was partly performed in the US policy system by the American Association for the Advancement of Science (AAAS) and other bodies (Dufour 2012; Papadopoulos 2009). The CCA has produced reports based on expert panels on numerous subjects including the state of industrial R&D and more specific issues such as fracking in the oil and gas industry.

Although such S&T and innovation advisory bodies, like all policy advisory bodies in numerous policy fields, can do good work and contribute to debate, they all suffer from a dilemma about how their overall role can and should be judged. The odds are high that many, or most of their policy recommendations will not be implemented by governments that are receiving advice and political pressure from a wide variety of sources and interests. Thus, a score-card approach would likely show a very low overall batting average, with one or two specific proposals every few years finding its way to implementation. The S&T and innovation advisory and performance reports are interesting in and of themselves, as later chapters will show, and offer evidence that can be used to judge policy performance since they reveal trends in how S&T and innovation paradigms and discourse are evolving (Phillips and Castle 2010, 2012, 2013).

There has long been a literature on *scientists in politics* and the related science-policy interface – in short, the many types of relational situations and dynamics in which scientists and policy-makers find themselves (Schooler 1971; Doern 1972). In some respects, scientists and technologists see themselves as non-political in the sense that they do not seek power or in the sense that they advise and make recommendations based on what they see as rational empirical evidence, facts, and generally accepted theories. But both theory and case studies reveal the inherently normative values scientists bring to their interpretations of the evidence (Barrett and Abergel 2002). Some public sector scientists see a public role for their work; however, the lion's share of the attention goes to a handful of notable public scientists and activists, such as David Suzuki, who lobby from outside government for attention and action. They are the go-to commentators and are sought by the media to comment publicly on contentious issues. Many of these work for, or are involved with, public interest groups, including environmental and health institutions and other NGOs. Regardless of the source of the science input, Schooler (1971) was right from the outset in stressing the highly varied (high, medium, and low) levels of influence by scientists across policy fields and departments of government.

The science-policy interface has been shown to function in situations of policy change when new laws, regulations, or guidelines are being promulgated, frequently after considerable types and forms of consultation and engagement with various publics (Leiss 2000). But these kinds of interface occur in any given policy field only every few years. Much more frequent, even day-to-day, interaction occurs during the risk-benefit and safety assessments of numerous new products, drugs, and processes at the pre-market stage and, increasingly, in post-market monitoring that involves a wider range of scientists, knowledge professions, patients, consumers, and citizens. This is the world of the previously mentioned "related science activities." It is also a diverse transactional world of numerous situations regarding ministers and officials asking clear or general questions and scientists and other evidence suppliers providing answers that can also be clear or vague and conditional (Saner 2014). Deeper analyses by historians of science regarding Big Science and policy controversies such as tobacco smoke, DDT, acid rain, and climate change are also important here, including the book *Merchants of Doubt*, by Naomi Oreskes and Erik Conway (2010).

One key factor is the overriding principle that government scientists are public servants who serve their political masters, Cabinet ministers, and parliamentarians. The higher up one goes in the civil service and among Cabinet ministers the fewer scientists there are, especially natural and engineering scientists, though the social sciences and management and lawyers are well represented. Furthermore, the higher up one proceeds the more likely it is that advice of all kinds is verbal and not written or even electronic. Rather it consists of policy actors verbally summing up other people's summaries of other people's summaries in time-constrained situations, events, and contexts.

As civil servants, government scientists have not been given unlimited scope for saying what they wish publicly or to talk with whomever they wish. By tradition, however, they have been give more liberty to speak publicly than most civil servants and, therefore, to function in the larger public interest. This tradition has been under pressure in most countries for quite a while. In Canada, the tradition was severely challenged, some say broken, in the Harper era by a government and prime minister that have sought explicitly to silence (some say muzzle) the independent voice of its scientists (O'Hara and Dufour 2014). The most overt control is managed through the new dictates of communications policy that requires all officials speaking in public to get pre-approval for their texts and messages. Some see this as an effort to control the public agenda; Savoie (2013) asserts this is part of a larger strategy of continuous attack-politics and campaigning strategies.

This has led to significant media criticism and public campaigns about the "death of evidence" and to public rallies across Canada, sponsored by the Professional Institute of the Public Service of Canada (PIPSC 2013). Rallies in 2013 exhorted the public to "stand up for science"

(Kilpatrick 2013). *New York Times* editorial board member Verlyn Klinkenborg (2013) has likened what he calls "the black arts of muzzling science in Canada" as far more entrenched and disturbing than those practised in the George W. Bush presidential years. Evidence for Democracy and Simon Fraser University published a study in 2014 that sought to grade several federal science departments (Magnuson-Ford and Gibbs 2014). Using several criteria regarding communication and openness, it was generally critical of Harper-era practices in most departments but also gave more positive grades to the Department of National Defence (ibid., 5).

The role of science in policy came to the fore in the late 1990s in the wake of public controversies over the safety of blood supplies, declining fish stocks, aspects of nuclear reactor safety, and biotechnology (Enros 2013; Public Policy Forum 1998; Doern 1999a; Helliwell and Smith 1999). These kinds of concerns are global and international with several countries and jurisdictions having to deal with similar problems and challenges. Controversies related to BSE, foot and mouth disease, stem cell research, and general debates about the use of the precautionary principle in regulation have been framed and largely debated globally. The Chrétien government attempted to respond to these challenges by forming an external advisory group, called the Council of Science and Technology Advisors (CSTA), to provide strategic advice on internal federal government S&T issues to the Cabinet Committee for the Economic Union. In May 1999 it published its report, *Science Advice for Government Effectiveness* (or SAGE report), which provided a set of principles (see chapter 4) to drive science advice (CSTA 1999).

Innovation Economy and Society

Conceptually, the discussion of the innovation economy and society has been influenced by a basic *shift from macro-economic stabilization policy* that after the Second World War had dominated macro and fiscal policy (the Keynesian era) to *theories and policies informed by drivers of micro-economic-centred growth*. We refer to the broad features of this shift here but related strands of it also inform our analysis in later individual S&T and innovation domain chapters. There are four competing paradigms that attempt to explain the incidence of innovation in relation to growth. In essence, theorists and policy-makers in the modern, knowledge-based economy are currently trying to debate and possibly balance competing views of how society can optimize the creation and use of new knowledge while promoting and fostering growth. This task has engaged scholars, scientists, policy-makers, inventors, entrepreneurs, capitalists, consultants, and NGOs in a heated and by no means simple debate (with both positivist and normative aspects) about which model is best.

Growth theorists and development economists abstract from the focus on innovation, asserting that optimal investment will be derived in well-structured economies, while a mix of business economists, economic geographers, and economic sociologists posit that innovation is the result of a combination of "special people" doing "special things" in "special places" (Leadbeater 2008, 2012). The special people can be experts or amateurs.

Neo-classical economics, encompassing traditional growth theorists and many development economists, has tended to abstract from the explicit role of institutions, place, and systems, asserting that the optimal amount and distribution of savings and investment, and by implication productivity growth, will be forthcoming in economies and societies that are, in World Bank terms, "well founded." Such economies and societies have a mix of stable economic foundations (e.g., fiscal and monetary systems that generally deliver low and stable debt-to-GDP ratios and inflation), policies that generate competitive markets (e.g., effective private property rights, liberal international trade, anti-trust, and stable and equitable tax and regulatory systems), and good social foundations (e.g., effective and efficient education, health, justice, welfare, and physical infrastructure). The logic is that the magic of Adam Smith's (1776) "invisible hand" of self-interest will lead individuals and organizations to deliver optimal amounts of savings and investments, or where there are demonstrable market failures, selected, limited, and targeted government policy interventions can correct the course.

Three interrelated theories of development, innovation, and growth are often conflated into the innovation debate. The search for deeper meaning was at least partly triggered by the advent of the "new'" economics of growth (Romer 1990; Lucas 1988), trade, and geography (Krugman 1996). The new economics showed that endogenous growth could be exploited to generate economic advantage; however, it did not explicitly examine the growth processes.

The earliest and most popular theory of development, innovation, and growth among policy-makers is that the pursuit of policies that create "special places" will generate all these positive outcomes. Marshall (1890) and Porter (1990) both noticed the presence of "agglomerations" of firms and industries and attempted to explain them in the context of what are now called economies of scale. Firms co-locating in regions are likely to gain through competition and various traded interdependencies such as extensive, supportive, forward and backward linkages and "thick" labour markets (i.e., where both demand for and supply of skilled labour is high).

In the 1980s the "special processes approach" to innovation gained ascendancy. Variously described as national systems of innovation (Lundvall 1992), regional systems of innovation (Wolfe and Gertler 2004), and the triple-helix model (Etzkowitz and Leydesdorff 1998), the special processes approach posits that economies of scope generate untraded interdependencies between people and firms that coexist in innovation systems. These structural determinants and dynamic interactions happen outside formal transactions – they emerge organically and are a collective product of local culture and context.

At about the turn of the millennium, a group of business theorists and economic geographers began to explore the role of the individual in the context of the innovation process (e.g., Florida 2002). A set of interrelated theories and policy prescriptions assume that "creative" individuals are at the core of the innovation process. A mix of sociology, psychology, and economic theories assert that highly creative individuals can be nurtured and "unleashed" in communities that invest in technology and talent and celebrate tolerance to diversity and change (e.g., Florida's three Ts).

The institutional and social pathways of change are not fully delimited in this literature, but the central tendency of those promoting this approach is to generate high-quality, creative human capital through education and investments in research and networking programs and to invest in talent attraction through nurturing the local quality of life. Firms that are attracted by this type of milieu tend to be pursuing knowledge-based strategies, seeking to be on the leading edge of technology, product, and organizational innovation. Also relevant here is Peter Swann's (2014) analysis of *common innovation*, cast as the contribution of ordinary people to establishing the wealth of nations; in it the ordinary citizen is a central player in diverse locales and situations.

In the light of this conceptual mix, international bodies such as the OECD suggest that the objective of policy should not be innovation for innovation's sake; rather, its objective should be the application of innovation for the betterment of individuals and society at large. The 2010 innovation strategy put forth by the OECD, *The OECD Innovation Strategy: Getting a Head Start on Tomorrow*, recognizes that there is no one-size-fits-all model that governments can implement. In that document, the OECD suggests five key priorities for government action that can, however, help to underpin innovation-led development. These are discussed further in chapter 7's analysis of the industrial S&T and innovation domain, but they span core areas of education and curriculum development, labour market policy, securing better balances between patent protection and diffusion of inventions, and also better data on innovation programs over the short, medium, and long terms.

The innovation economy as revealed in the literature streams above is a clearer concept, relatively speaking, than notions of the *innovation society*. Conceptions of the innovation society gain expression both in the academic and policy literature but also via research institutes seeking to capture what the concept means to them (Hawkins 2012). An interesting starting point in this regard is found in the EU-sponsored program coordination forum, Innovation Sustainability Technology (INSITE). Its mandate and name is premised on a view that "in the Innovation Society citizens are just consumers, hungry to satisfy 'needs' they didn't know they had; firms work to create those needs and the products that satisfy them – that is, they *innovate* or they *die* – governments make sure that nothing – from social justice to cultural enrichment – gets in innovation's way" (INSITE 2013, 1; original emphasis). It argues that the "only way society can respond to this is by changing the way in which it monitors, evaluates and engages in the processes through which it transforms its own organization" (ibid.). Of course, there are other conceptions about what consumers and consumer-citizens and interests are like, but there is a certain grain of truth to the above notion of real and unreal consumer needs (Wilks and Doern 2007). Drawing the line to determine how autonomous consumers will be is a highly contentious political issue in Canada and worldwide.

The literature parses this conflict in two ways. Some assert the challenge has deep root causes that need exploration. Others see it simply as a mismatch of government will or purpose to the reality of new purposeless innovation. Analysis by Andersson, Tornberg, and Tornberg (2013) addresses such challenges by asking whether social systems are "complex or worse?" They argue that social systems have been recalcitrant in part because "societal systems combine at least two methodologically troublesome qualities… complexity and complicatedness … and that these two qualities may have not been systematically pursued" (ibid., 1). Complicatedness as a paired quality is not particularly well defined, but it is advanced as a necessary concept to deal with social systems.

Rayner's analysis (2012) agrees with the complexity features of social innovation but argues that to make sense of that complexity both individuals and institutions need to develop "simplified, self-contained versions of that world," versions that inevitably constitute what he calls "uncomfortable knowledge" (ibid., 108). This is in part because these newer versions involve the expunging of other existing versions even though they are not entirely disproved or wrong.

Bouchard's (2013) analysis of innovation in Quebec links innovation to the social economy, a strong tradition in Quebecois social and socio-economic literature, seeing it as a good and viable laboratory for studying and understanding positive and progressive social innovation. Mook, Quarter, and Ryan (2012) also see market-based social economy firms as businesses with a difference and that they indeed bear comparison with firms that are influenced by concepts of social responsibility as they seek a new, viable balance between the economic and the social.

The Internet as a defining and increasingly dominant and enabling technology has become a focal point for much of the analysis. Beyond its foundational impact on the production of information and things, it is also key for the systems of e-commerce and e-governance. In the context of its birth and massive growth, the Internet and social media giants such as Google, Facebook, and Twitter have become vehicles and centres of social production and the networked society (Castells 2011). For many public and private interests, individuals, families, charities, and communities, the Internet has greatly reduced the costs of communication and joint action. It has also fostered new avenues for direct democracy by individual citizens, including via social networks, though the democratic promise of the latter is still more in the potential it has than in the practice (Rainie and Wellman 2012; Borins and Brown 2008). Moreover, in commercial terms, Internet development promotes arguments that the "wealth of networks" is the new driver of technology and profit and of democratic participation (Benkler 2006).

Often characterized as created without permission, the Internet offers a number of valuable insights into international policy diffusion, including its original establishment in 1983 (Naughton 2012; Dahlgren 2013). The Internet and related broadcasting technologies are being increasingly examined and criticized as avenues for pedophiles, sexual predators, and bullies; there are many examples of local and global responses (Caron and Cohen 2013). Similarly, Internet-based technologies and marketing centred on Facebook, Twitter, and dynamic social networks are producing deep concerns about privacy and its regulation, particularly the access to data that target particular consumers and citizens (Etzioni 2012; Office of the Privacy Commissioner 2011). In this regard and in other senses the Internet raises concerns about human rights (Jorgensen 2013).

Each of these different initial conceptions and analytical entry points shows that analysts of social innovation look for and promote discourses in diverse arenas of national, international, and local/regional governance and among diverse scientific, technological, and cultural communities. This greatly complicates the search for complete or convincing innovation impacts that can be measured and assessed (Council of Canadian Academies 2013c).

The longer term literature on social issues and policy has tended to reside in core literature on the Canadian welfare state (Rice and Prince 2013), where the focus has been on the interplay of socio-political forces and the development of the welfare state since the Second World War and up to the 1960s, with a particular focus on policy, programming, and related tax and spending initiatives such as employment insurance, medicare, pensions, and greatly expanded education funding, including university funding. Thus, social policy evoked a well-understood set of values, ideas, and discourse.

The innovation society does not yet have a similar clarity or public resonance, but this has not precluded problems and struggles from entering the social policy and politics literature. The literature has responded to problems emanating from earlier industrialization and more recent globalization impacts. Traditionally, these range from worker occupational health and safety hazards such as asbestos, the treatment of the disabled, the notion of the social economy, and reproductive technologies and their impacts on women and families (Doern and Prince 2012). Transformative technologies have added some new concerns, including knowledge and strategies related to nutrition, obesity, and genetic testing, to name only three (Castle and Ries 2009). Typically, the actual or guessed-at populations and subpopulations of society involved in an issue are smaller and disbursed in ways that make it difficult to mobilize them politically in what are increasingly complex and diverse agenda-setting contexts and dynamics.

Conceptions of a *risk society* and related notions of safety, uncertainty, risk, and risk-benefit are now fundamental drivers of S&T policy (Beck 1992). Safety as a public value and standard has been conflated into a focus on absolute quality – at least partly because the origins of many S&T-centred regulatory mandates arose from evidence regarding the deaths of human beings and animals and from related notions of the sanctity of life. The thalidomide crisis in the 1960s, for example, led to explicit changes that placed the safety of drugs at the heart of health product regulation, anchored by the need for objective science-based regulation and extensive pre-market testing. This focus has diffused across the science-policy space.

The challenge is that risks relate to potential harms and adverse impacts on individuals, families, groups, communities, and subpopulations, which immediately imply some sense of uncertainty. Risks can be voluntary or involuntary and are so diverse, extensive, and ever-present that the literature often speaks explicitly of the "risk society" (Beck 1992; Hood, Rothstein, and Baldwin 2001). Such risk challenges are associated one way or another with a trilogy of key concepts and decision stages, specifically: risk assessment, risk management, and risk communication.

Risk assessment refers to the determination of a quantitative or qualitative value of risk that is related to a specific identifiable hazard. Analytically it involves calculations of the size or magnitude of a potential loss (of life or other harm) and also the probability of its occurrence. *Risk management* is defined as the decision of allowable risk and the taking of steps to either eliminate or to reduce such risks as much as is reasonably possible by introducing control measures ranging from outright bans right through to provisions for proscribed use (Leiss 2000; Doern 2007). *Risk communication* involves any number of processes of disseminating information about risks and uncertainty. It also entails communicating about risk avoidance, prevention, and management to citizens, communities, and firms on a continuous targeted basis.

There is some political and institutional slippage between the primacy and absolutist nature of safety discourse and policy and regulatory mandates in reality. Safety concerns are certainly not deleted from mandate statements or from the concerns of front-line regulatory assessors or inspectors. Larger notions of risk-benefit are in evidence and have worked their way into the logical structure of mandates and organizational cultures in different regulatory bodies and realms (Doern, Prince, and Schultz 2014). Some of this was driven by business pressure as reflected in smart regulation and some driven by innovation pressures and the need in some sectors for greater regulatory speed in obtaining regulatory product approvals.

The risk society is also bound up in conceptual and practical discussions of the so-called *social licence* to undertake development, such as building energy pipelines or other complex infrastructure (Cleland 2014). In earlier eras more conventional but broad notions of the public interest and political pluralism and engagement were used as a basis for advocacy and discourse. Now, social licence has become more diffuse and challenging. The concept of *sustainable development* (SD), in particular, refers to the notion that policies should be anchored in preventative and intergenerational ideas about environmental harms and benefits, most comprehensively articulated by the Brundtland Commission (1987).

Sustainable development is an idea and paradigm whose intent is to ensure that all areas of policy and governance should ensure that the environment and ecosystems are left in at least as good a state for the next generation as they were for the current generation; social protection and innovation are explicitly taken into account (Boutros et al. 2010; Lafferty and Meadowcroft 2000). Cast somewhat more loosely, SD policies are often seen by governments as those that take into consideration in a balanced way the economic, social, and environmental effects of policies, the so-called "triple bottom line" (Toner and Frey 2004). For example, in the process of practising sustainable development, harms to endangered species have been identified and policy solutions have been extended, albeit with very mixed results. But in essence, social aspects often are only one aspect of the triple bottom line and only have to be considered in decision making without necessarily becoming determinative (Kinder 2010).

The *precautionary principle* is an environmental decision-making idea drawn from debates and concerns about protecting biodiversity that has been extended into other environmental and health issues and into general regulatory policy. It is an approach that allows policy-makers, regulators, and scientists to take provisional risk-management measures when an assessment points to the potential of harmful environmental, human, and health effects but there is a lack of scientific certainty (Vogel 2012; Goklany 2001, 2007; Saner 2007). The European Union has been particularly strong in formally endorsing this idea in EU law in areas such as food policy, in part due to direct pressure and actions by the European Parliament. At times the principle has become a substantial hurdle, as proponents of new products are forced to address all imaginable (and some might say impossible) hazards to gain regulatory approval.

A softer interpretation of precaution is adopted and practised in Canada, as it is in most countries. Canada's caution arises, in part, because of the pressure to sustain open and competitive international markets for our exports. Nonetheless, notions of precaution are always embedded in the underlying policy discourse in the environmental, social, and health realms as new products and production processes emerge (Doern and

Prince 2012). Indeed, Stirling (2013, 2) argues that "precaution reminds us that innovation is not a forced one-track race to the future. Instead – like biological evolution – technical progress entails constantly branching paths ... the notion of exclusively science-based decisions under uncertainty is an oxymoron."

Theories of complexity, as already noted above, have been developed and debated within the natural and engineering sciences and their subdisciplines and have begun to enter considerations about society and social innovation. In this immediate context, we refer to it in the context of theory for and about public policy in the twenty-first century. Authors such as Geyer and Rihani (2010) and Morcol (2012) offer complexity theory as an antidote to the policy-makers and policy scholars who treat policy and governance systems as being orderly, predictable, and controllable. Pal (2013) argues that complexity theory is needed to understand policy- and decision -making in the face of predictable, yet chaotic, surprises, such as accidents, floods, terrorism, epidemics in public health, and breakdowns in networked infrastructure. It has also been deployed by Norberg and Cumming (2012) to analyze sustainable development in diverse policy fields, and by Walby (2007) to examine the nature of what she refers to in social policy terms as "multiple intersecting social inequalities."

Concepts of complexity are a part of the literature on public policy and governance without complexity theory being centrally or overtly used. The *governmentality* literature (Hunt 1999), for example, is inspired by the writings of Michel Foucault on "the government of one's self" in particular economic and social affairs. The governmentality literature offers a set of concepts for examining changes in the role of the state and for tracing changes in state-society relationships in contemporary politics (Crawford 2006). Other theory is centred on concepts of *rescaling*, which argues that globalization is causing a rescaling of collective action problems such that they are no longer very well matched in the scale of the nation state or subunits of the nation state. As Cerny (1995, 34) explains, "The more that the scale of goods and assets produced, exchanged, and/or used in a particular economic sector or activity diverges from the structural scale of the national state – both from above (the global scale) and from below (the local scale) and the more that those divergences feed back into each other in complex ways, then the more that the authority, legitimacy, policy-making capacity, and policy implementation effectiveness of states will be challenged from both without and within." Not surprisingly, this kind of theory has usefully found its way into diverse policy fields where science and innovation intersect, including environmental policy, climate change, and urban and local government (Brenner 2009; Okereke, Bulkeley, and Schroeder 2009; Bulkeley 2010).

Last but not least, it is useful to reiterate the basic conceptual fact that innovation policy (economic and social) and related features of network-based clusters have been defined from the outset as being non-linear. They are thus complex whether or not one embeds and investigates them in the context of a theory of complexity. Although not yet fully explored, complexity offers the concepts of emergent properties, the effect of scale on generating these emergent properties, and the power log effects on outcomes from the forces of complexity. This is probably the leading edge for scholarly theorizing and modelling (McPhee Knowles and Phillips 2014).

Many of the above kinds of literature implicitly deal with the nature of *time* without necessarily being as explicit as they could be about different temporal periods in economic and social innovation policy. Concepts of time in public policy making and governance capture temporal periods that include immediate short-term crises (real and manufactured); diverse temporal notions of what "capital" means regarding investment in physical infrastructure, human and natural/environmental capital in different intergenerational periods, and even epochs of time; decision cycles related to business and political-electoral cycles; and fiscal budgets that are annual and cyclical or structural in nature. But the literature on policy, governance and time has quite early roots in policy and social analysis. Edelman's work (1964, 1977, 1988) deals with both crises and the political use of symbolic politics and discourse to advance and skew issues and policy critics. More recently, Doern, Prince, and Maslove (2013) have utilized notions of temporal budgeting in the Canadian fiscal and budgeting system, showing how aspects of innovation policy often were downplayed because of these dynamics. Auld (1985) was one of the first academics in Canada to advance the need for real capital budgets to pay for different kinds of capital that had long life cycles and to renew and improve the asset rather than simply fund annual spending. Work by Gaudreault and Lemire (2003) shows continuing problems with tracking and investing in physical capital in Canada, both federally and provincially. The recent infrastructure-centred "stimulus" programs have been justified based on controversial claims that such activity is helping community, environmental, and social innovation, with little or no consideration to the life-cycle needs for these investments (Bennett 2012).

As we show in chapters 5 and 6, there seems to be a historical disconnect between S&T and innovation policies, incentives, programs, and budgeting; money seldom maps well with sustained priorities. There is up-front funding to incentivize change but no sustained operating or recapitalization funds available. This has been a critical concern in many of the Networks of Centres of Excellence (NCE) and networks of the Canadian Institutes of Health Research (CIHR), laboratories of the Canada Foundation for Innovation (CFI) and National Research Council Canada (NRC), and other strategic priorities started or encouraged by the federal government.

Individual authors and policy think tanks have also increasingly broadened the notion of kinds of capital and hence temporal concepts to be researched. Thus, Riddell (2008) has argued for explicit attention to *human capital*, the federal Policy Research Initiative (2006) attempted to define and advocate the fostering of *social capital*, and the Pembina Institute (2011) promoted the concept of *natural capital* as supplied by the natural environment and ecosystems. A further interwoven strand of literature centres on demographic changes and varied kinds of intergenerational policy and investment spending and conflicts over regional, national, and global priorities (Eberstadt 2010; Kotikoff 2002; Foot 1990). In Canada, the federal Parliamentary Budget Officer has aggressively sought to ensure that such temporal full costing is carried out and made public, but this has had only limited impacts.

Concepts of policy impairment, inadequacy, and failure have been part of public policy theory and analysis for decades and are especially relevant to the innovation economy and society debate. Earlier theories and empirical analyses have shown how various kinds of impairment, caused by systematic indoctrination, tend to reinforce the current social and political order but also to reduce the underlying intelligence needed for democratic problem solving (Lindblom and Woodhouse 1993). Some assert this was an underlying cause of the regulatory failures and slow response to the 2008–13 banking crisis, both at the beginning with a failure of macro prudential regulation and most recently with only weak efforts at reregulation after the crisis (Baker 2013; Rixen 2013). Impairment can involve human limitations, imperfect or incomplete information, pressing time or resource constraints, and, crucially, inequalities of power and authority. Notions of policy inadequacy and failure and of the limits of purely rational policy-making and implementation have been a part of Lindblom's earlier theories of incrementalism, but also related theories of goal displacement, studies of policy crises and failure, and policy evaluation literature. Perfectly functioning policies and programs are a rarity due to mandate conflicts, resource weaknesses, and limits and disagreements regarding causal analysis (Atkinson 2011; Pal 2011).

CONCLUSIONS

Drawn from three quite broad strands of literature, we have set out the conceptual foundations underpinning our analysis of S&T and innovation policy in Canada. As noted, we have set this in the international context. Key aspects of ideas, paradigms and discourses, governance, power, and democracy inform our thinking and research about the innovation economy and society. These foundations have informed the development of our analytical framework and are drawn upon again in our later empirical chapters along with other literature sources more specifically relevant to understanding change and inertia in eight S&T and innovation policy and governance domains.

2

Canadian S&T and Innovation Policy and Agendas in Liberal and Conservative Prime Ministerial Eras

Our historical overview considers Canadian S&T and innovation policy in Liberal and Conservative prime ministerial eras covering the past five decades. Canadian S&T and innovation policy, democracy, and governance are also a part of this history, as are national, federal-provincial, and international policies and institutions, each influencing the other's timing and content. The more detailed international story is told in chapter 3, and we return to many of these and other policy agenda issues in more detail in the policy domain chapters in Part II of this book.

This chapter surveys the major federal S&T and innovation policies (and related laws, statements, plans, strategies, and changes) from prime ministers Pierre Trudeau to Stephen Harper including, where appropriate, important features of the partisan, political party platforms and agendas. These partisan dimensions show both continuities and key differences across Liberal and Conservative governments and prime ministerial eras. The nature of federalism and federal-provincial agendas are also a part of the story, because S&T and innovation policy often enters the field or is blocked by overlaps between policy fields and their diverse federalist dynamics. Thus, S&T in agriculture, food, resources, higher education, trade, the Arctic, transportation, health care, cities, or the Internet involve different kinds of federalism and constitutional or shared jurisdiction. Also relevant are changing kinds of views and concepts of overall federalism and decision making within and across prime ministerial eras. These include concepts and eras of executive federalism, bilateral and multilateral federalism, fiscal federalism, constitutional federalism, free trade federalism, open federalism, and one-off federalism (McGrane 2013; Atkinson et al. 2013; Gibbins, Maioni, and Stein 2006).

We examine, in turn, the Trudeau Liberals from 1968 to 1984, the Mulroney Conservatives in the period from 1984 to 1993, the Chrétien and then Martin Liberals in the period from 1993 to 2006, and finally the Harper Conservatives from 2006 to 2015. The prime ministerial eras reveal aspects of and evidence regarding S&T and innovation policy, but such policies can only be fully understood by locating them within and across the broader economic and social agendas in each era. For this initial sense of federal S&T agendas, we examined briefly each Speech from the Throne (SFT) across the prime ministerial eras.

Speeches from the Throne are the quintessential agenda-setting expressions of priorities and values by the federal government as developed by prime ministers as political party leaders. SFTs do not occur annually but rather at approximately eighteen-month intervals on average as new sessions of Parliament begin. They are bound up in the inherent nature and dynamics of agenda setting and issue management in Canada (Pal 2013; Soroka 2002; Doern and Phidd 1992) and in the continuous changing art of "patching and packaging" complex ends and means in policy formation and political communication (Howlett and Rayner 2013). SFTs tend to be thematic narratives centred on new or changed laws and ways of expressing agendas and priorities. We are interested in how S&T and innovation policy is presented, the discourse used, and how it ranks or seems to rank in federal government priorities. STI policy overall may be a small or larger section of an SFT or may simply be subsumed in a section dealing with another named policy field or agenda item. Also of interest is when and why S&T and innovation policy is ignored in these agenda-setting occasions.

Indeed, the chapter's account needs to be seen in the context of broader conceptual theories of agenda setting as previewed in chapter 1. In the first instance, it is important to keep in mind that the long and variable times between action on science, technology, and innovation and the outcomes in terms of products, sales, and jobs. This complicates framing and addressing policy in this area as the innovation cycle never effectively fits easily in the context of planning for Speeches from the Throne, annual budgets, or even government electoral mandates. This accounts for two observed phenomena. In such policy spaces we are more likely to see "punctuated equilibria," where policy stays stable for long periods before, in a punctuated fashion, some policies change quickly and dramatically in response to a pressing threat or opportunity (Jones and Baumgartner 2005, 2012). The long, variable, and uncertain timing of innovation also affects attention spans of different players, including voters and citizens, and perhaps more so the mass media and government communication efforts that seek to frame and reframe issues (Eissler, Russell, and Jones 2014; Wolfe, Jones, and Baumgartner 2013).

Here we summarize and comment on an aggregate view of references to S&T and innovation policy in the thirty-two Throne Speeches from 1968 to 2013. The economic and social agendas are also revealed in policy and reform agenda strategies related to overall macro-economic policy, embodied in measures relating to taxation, regulation, and public spending and also in political exhortations. The realities and complexities of these policy instruments are a necessary part of our historical account of each prime ministerial era both here and in later chapters. These instruments range widely across systems of hard rules; soft guidance; tax systems involving deductions, credits, and tax expenditures; spending attached to levered rules and provisions; diverse forms of exhortation and nudging; and diverse complex agreements and contracting (Doern, Maslove, and Prince 2013; Howlett 2011).

Budget Speeches are also crucial for S&T and innovation policy agenda-setting purposes and will be examined in chapter 4 as part of our analysis of the macro S&T and innovation policy domain. They almost always are introduced annually, typically in February or March of each year. They may have themes but their core role is to guide the macroeconomy; in that sense Budget Speeches contain numerous specific tax and expenditure measures, new, altered, or cancelled (Doern 2010). As shown in chapter 4, these measures can be large, medium, or small in size, not to mention gesture-like and applicable for diverse time periods (e.g., small amounts for a couple of years; very big amounts but for time commitments of seven to ten years and thus likely well beyond the current government's tenure in office).

Among our findings regarding a review of all thirty-two Speeches from the Throne is the general strong proclivity of all prime ministers to stress economic and fiscal priorities, albeit punctuated with messages related to national unity, social policy, and employment.

PRIME MINISTERIAL AND POLITICAL PARTY ERAS: S&T AND INNOVATION POLICIES, IDEAS, AND REFORMS

The Trudeau Liberals

The Trudeau Liberals governed through majority governments in 1968–72, 1974–79, and 1980–84 and had a minority government interregnum from 1972 to 1974. Under its initial "Just Society" and rational decision-making ethos, Trudeau focused on distancing his government from the seeming chaotic decision making of the previous Pearson Liberal minority government by pushing reforms for greater public participation (Doern and Phidd 1983).

Somewhat counterintuitively it was the chaotic Pearson Liberal government that had given birth to the early formal notions of a federal science policy that was intended to be broader and more explicit than developed earlier by the National Research Council of Canada (NRC; Doern 1972). Developments in the Pearson era included C.J. Mackenzie's 1964 *Report to Prime Minister on Government Science* (The Mackenzie Report), which led to the establishment of the Science Secretariat as a part of the Privy Council Office (PCO). In 1966 the government formed the Science Council of Canada (SCC), an external advisory body. But consideration of structure did not end there. The Senate Special Committee on Science Policy (the Lamontagne Committee) began work in 1968 and held hearings in 1968–69 and 1975–77 before publishing four volumes on options, the last in 1977.

When Pierre Elliott Trudeau came to office as prime minister in 1968 he established a Cabinet Committee on Science and Technology and lent considerable initial support to the development of more coherent federal science policy, defined as both "policy for science" but also "science in policy." The focus overall, however, as emphasized by the Lamontagne Report, was to get more R&D done by industry and to shift the focus away from government dominance of S&T. The Liberals also established in 1971 the Minister of State for Science and Technology (MOSST), but it was a decidedly junior ministry without much clout vis-à-vis the established federal departments with science mandates (Aucoin and French 1974). Economist Donald McFetridge concluded that "the decade of the 1970s was characterized by the emergence and development of a number of agencies designed to direct, coordinate, and evaluate federal scientific activity. They have presided over a decline, in real dollars, in federal scientific activity and support" (McFetridge 1981, 255).

There were many reports by the Science Council of Canada about the state of S&T in Canada, but the Trudeau Liberal government, in its expression of priorities in Speeches from the Throne, worked S&T into its narrative only briefly and without much emphasis. In 1970 it spoke of a "great wealth of untapped and uncoordinated scientific talent and experience not now being adequately utilized" (SFT 1970, 3). In 1973 the Liberals said that they would undertake "a review of research and development incentive programs to improve their effectiveness in promoting innovation in the private sector of the economy, including examination of patenting, licensing and other policies related to technology research and development" (SFT 1973, 9). In 1974, building on the early work of the Ministry of State for Science and Technology, the Liberals pronounced that "the objective of the Government's science policy is the rational generation and acquisition of scientific knowledge and the planned use of science and technology in support of national goals" (SFT 1974, 4).

In the early and mid-1970s the Trudeau Liberals were forced into a policy agenda centred on wage and price controls as Canada (and the global economy) faced high and rising inflation that was not amenable to normal central bank monetary or contra-cyclical fiscal policy. Following the interregnum of the Clark Conservative government in 1979–80, when the Tories promised to introduce measures to expand R&D, the Trudeau Liberals returned to power with a government agenda that was both broadly aggressive and interventionist but not about S&T. It aggressively launched initiatives that ranged from repatriation and the Canadian Charter of Rights and Freedoms on the constitutional front and on the economic front the National Energy Program (NEP) as a partial response to a global energy crisis in the wake of a tripling of oil prices and a resource-centred mega-projects industrial strategy (Doern and Toner 1985). The National Energy Program sought to revive the economy by mobilizing support for an array of planned natural resource projects that were queuing up for investment and construction across all regions of Canada. This initiative was not dissimilar from the resource and energy pipeline infrastructure projects promoted in the 2008–15 period under the Harper Conservatives (see below).

Buried in the nether regions of the NEP's stated priorities were some further mentions of S&T. In 1980 the Trudeau Liberals asserted that "Canada's capabilities in science and technology lie at the heart of our competitiveness," and the government committed itself "to increase Canada's overall expenditures on research and development to 1.5 percent of the value of the Gross National Product" (SFT 1980, 15). The Trudeau Liberals final Speech from the Throne had a brief section on the need for "Competitive World-Class Industries" (SFT 1983, 7–8). It went on to introduce plans for the "increased use of micro-electronic technology," new tax incentives for R&D, and a new Office of Industrial Innovation. One of the policies rushed into effect in 1983 was the Scientific Research Tax Credit (SRTC). It provided a mechanism for R&D firms in a non-taxable position to transfer their unused tax incentives to outside investors. This precipitated a massive secondary market in credits that ultimately cost the federal treasury more than $2 billion in the first ten months of the program (Doern 1987, 1996; Bernstein 1986).

The Trudeau S&T and innovation policy agenda was decidedly structural and governance focused, much like Trudeau himself, the self-styled rational policy-maker. This was a marked contrast to the incrementalism of the Pearson era. STI overall gained some attention in the early Trudeau years, but for most of the time after 1974 it had to find its way through, in, and around the much higher-priority signature themes and crises of the Liberal years.

The Mulroney Conservatives

The Mulroney Conservative majority government spanning 1984 to 1993 offered an intermittently expressed S&T and innovation agenda including some measures subsumed within the Conservatives' 1990 $3 billion Green Plan, an ambitious federal environmental initiative (Gollner and Salee 1988; Conway 1992). The realities of other policy and fiscal challenges in the Mulroney era left some room for S&T and innovation matters but only episodically. First, for its entire period in office, the Mulroney Conservative government had to deal with large fiscal deficits, the largest in Canadian history. This meant that funds for new initiatives, S&T or otherwise, were scarce for most of this nine-year period. This was the period when post-war macro-economic stabilization through demand management ended as the prime approach to macro fiscal policy. The shift in the Thatcher-Reagan era toward a micro-economic policy of growth, deregulation, privatization, and supply-side structural fundamentalism was profound (Doern, Maslove, and Prince 2013).

One of the early initiatives carried out in concert with Mulroney's deficit reduction focus was to end the ill-structured Liberal SRTC program, which was bleeding the treasury and raising concerns about possible tax corruption. The SRTC was ended in 1985 and replaced in 1986 with the Scientific Research and Experimental Development Tax Credit (SR&ED), which was intended to be especially favourable to smaller Canadian R&D firms without taxable revenue by enabling them to receive refundable tax credit payments for outlays for "experimental development" (Doern 1987; see also chapter 7 herein). During the 1984 election campaign the Conservatives promised to be much better supporters of R&D for business than the Trudeau Liberals. The promise included the expansion of R&D support to 2.5 percent of GNP, about double its then current low level, which was well below other Western countries. Another change early in the Mulroney era came in the government's efforts through new appointments and new directions at the National Research Council to make the NRC research institutes more commercially focused and competitive (Doern and Levesque 2002).

The Mulroney government's policy battles in its first term involved energy and trade. New energy accords replaced the interventionist Liberal National Energy Program that was much reviled in Western Canada, especially. Free trade through the Canada-US Free Trade Agreement (CUFTA) absorbed the rest of its energies and provided the platform for the 1988 election, which ultimately was virtually a referendum on free trade (Doern and Tomlin 1991). Later the government negotiated the North American Free Trade Agreement (NAFTA), which extended the scope and added Mexico to the deal (Cameron and Tomlin 2000). Free trade had indirect S&T policy and political links because trade disputes and dispute resolution centred in part on defining *science-based* and *objective science* as essential in determining whether new or revised rules by member countries were pursuing legitimate public objectives (which were allowed) or constituted new forms of trade protectionism (Trebilcock 2011). In the context of the emerging trade regime that defined prohibited, suspect, and permitted measures related to production and trade in goods (especially agricultural and natural resource products), S&T investments were for the most part viewed as "green" and generally exempt from dispute or countervailing trade measures.

The second-term Mulroney agenda centred on restructuring the Canadian economy, through tax reform, introduction of the Goods and Services Tax (GST), and deregulating fields such as telecommunications, airlines, and railways. The Mulroney agenda also was ultimately caught up in constitutional wrangling leading to the Meech Lake and Charlottetown Accords, which led to abject failure and to an eventual crushing election defeat in the 1993 general election. Given these big agenda preoccupations, it is hardly surprising that S&T and innovation only infrequently gained the attention of the Mulroney Conservative government over its nine years in office.

The first Mulroney Speech from the Throne, in 1984, found room for a brief mention that the government would "enhance innovation," especially among small and medium-sized enterprises (SMEs). In the large economic renewal section of its next Speech from the Throne, the Conservative government proposed a "four-point program to ensure that support for science, technology and education is more clearly focused in the national interest" (SFT 1986, 12). This program included the formation of what became in 1987 the National Advisory Board on

Science and Technology (NABST), a new federal S&T strategy, a national conference on technology and innovation, and the establishment of the Canadian Space Agency.

The NABST brought a more market-based and business-focused approach to S&T advice, with private sector members dominating most of its committees (Kinder 2010). But the Conservatives were already in receipt of the Wright Report (MOSST 1984), prepared by the Task Force on Federal Policies and Programs for Technology Development, which had focused on the performance of federal laboratories. The Task Force, headed by Douglas Wright, president of the University of Waterloo, Canada's then-new "technology" university, sharpened the focus on federal government laboratories and their weak relations with industry. Although not directly attributed to the report, over the coming years some of the NRC centres refashioned themselves as institutes and engaged more in partnerships with industry, for example, the Plant Biotechnology Institute (PBI) at NRC, Saskatoon, which was created in 1989.

At the institutional level, in 1987 the Mulroney government formed Industry, Science and Technology Canada (ISTC), which amalgamated MOSST with the industry and regional development departments and ministries to give emphasis to S&T with an avowed industry angle. ISTC was already structured to recognize not just traditional industrial sectors but also crosscutting enabling technologies, especially the Information Highway (Doern 1990).

The 1989 Speech from the Throne cited the Mulroney Conservative government's support for the early NABST work and its suggestions to create "a network of centres of excellence and programs for the development of strategic technologies such as biotechnology, advanced industrial materials and information technology" (SFT 1989, 6). The Networks of Centres of Excellence (NCE) program had already been launched in 1988, a program that Atkinson-Grosjean (2006, xiii) has described as "the most dramatic change in Canadian science policy since the National Research Council was established in 1916."

The 1991 Speech from the Throne, the last of the Mulroney era, was dominated by national unity issues and discussions about barriers to internal trade, the concern that eventually led to federal-provincial negotiations and the 1994 Agreement on Internal Trade (AIT; Doern and MacDonald 1999). The Mulroney Conservative Party was decimated in the 1993 election, which was won by the Liberals.

Despite its often episodic record on science matters, the Mulroney government can claim to have had a substantive impact on S&T and innovation policy. S&T policy structures were changed, and some S&T programs and institutions were pushed in more commercial directions, including some early changes to how government labs were to be managed and networked.

The Chrétien and Martin Liberals

The three Liberal majority governments of Jean Chrétien from 1991 to 2003 (with Paul Martin as minister of finance) and the Paul Martin minority government from 2003 to 2006 offered an expansive S&T and innovation economy agenda that was first unveiled in the 1993 election campaign Red Book agenda (Liberal Party of Canada 1993). Some of these found their way into Chrétien's first Speech from the Throne, including a proposed Canada Investment Fund to help "innovative leading-edge technology firms obtain the long-term capital they need" (SFT 1994, 2). Measures were also to be developed regarding technology partnerships between universities, research institutions, and the private sector, and additionally, a Canadian strategy for the Information Highway was unveiled.

But these initiatives succumbed along with many other plans as the government focused on the large, now fifteen-year-old, federal fiscal deficit and accumulated debt (Pal 1998; Swimmer 1996; Hale 2000, 2001). Major expenditure cuts flowing from Program Review included reductions in science funding and staff reductions ranging from 20 to 30 percent in key federal science-based departments (Kinder 2010; Doern and Kinder 2007).

The Liberals did not highlight these cuts in their 1996 Speech from the Throne but S&T was given considerable attention in other ways. The key message was the need for better "effectiveness and focus in the federal science and technology effort" (SFT 1996, 2), including proposals to support the aerospace industry, environmental technologies, and biotechnology. The Speech also sought to provide a predictable policy and regulatory framework for the Information Highway and to provide enhanced access to the Information Highway via programs such as SchoolNet and the Community Access Program (CAP) in rural areas (see chapter 11).

Following re-election in June 1997, the Liberals' next Speech from the Throne focused on their promise to balance the budget by 1998–99. Near the end of the Speech there was discussion about how the government would invest in "knowledge and creativity" through its endowment of the new Canada Foundation for Innovation (CFI) as part of a "leading edge national system of innovation." The Speech also promised to "make the information and knowledge infrastructure accessible to all Canadians by 2000" (SFT 1997).

Earlier, in concert with Program Review, the Liberals had abolished Industry, Science and Technology Canada, and replaced it with Industry Canada. Thus, the S&T moniker disappeared from its name but not from its mandate (Doern 1996). Its ministers thereafter were quite prepared to speak the discourse of innovation- and knowledge-based growth and especially to ride the policy coattails of the popular and self-evidently transformative "Information Highway" (as the Internet was called in Liberal policy discourse).

The 1999 Speech from the Throne was couched in the language of the new imperatives and opportunities of the global knowledge-based economy (KBE). It announced a range of measures to enhance Canada's "knowledge infrastructure," including the formation of the Canadian Institutes of Health Research (CIHR) (see chapter 6) as the replacement body for the Medical Research Council of Canada (MRC). The new CIHR consisted of thirteen networked institutes, each with its own scientific director, focused on disciplinary and interdisciplinary research related to different diseases and health populations (Murphy 2007).

Following their 27 November 2000 re-election victory, the Chrétien Liberals were basking in an era of fiscal surpluses and economic growth (Pal 2000). The 2001 Speech from the Throne opened with a commitment to build "a world-leading economy driven by innovation, ideas and talent" (SFT 2001). A stand-alone "Innovation" section promised that Canada would be in the "top five countries for research and development performance by 2010" (ibid.). There was a section entitled "Skills and Learning," and in one entitled "Connecting Canadians" there is mention of the creation of a National Broadband Task Force (ibid.). The final Chrétien government's Speech from the Throne, in 2002, continued these innovation themes, but this time in the form of "smart regulation" for a knowledge economy and promises to develop an improved intellectual property (IP) framework that would relate to issues such as new life forms (SFT 2002) – which, as we show in chapter 8 and 10, did not emerge in any coherent way.

Paul Martin became prime minister and Liberal Party leader after a bitter battle with Jean Chrétien within the Liberal Party. His first Speech from the Throne (2004), delivered shortly before an election was called, offered parliamentary reform, measures to rebuild trust in the wake of the Chrétien government sponsorship scandal, and a new deal for cities and communities. S&T matters were raised near the end of the Speech, but mostly to proudly reiterate the $13 billion increase in S&T spending since 1997 (Martin claimed credit as this happened under his tenure as minister of finance, albeit with full support from Jean Chrétien). But the key shift in message was that "now we must do much more to ensure that our knowledge investment is converted to commercial success" (SPF 2004). Martin announced the establishment of new a National Science Advisor (NSA), who would report to him and would engage in consultations to forge a "truly national science agenda." Arthur Carty, moved from president of the National Research Council to take on the assignment. Martin's second Speech from the Throne (SFT 2004), shortly after being re-elected as a minority government, had only limited reference to S&T matters; the main focus was on the need to foster capabilities in key enabling technologies such as biotechnology, information and communications, and advanced materials.

Overall, the Chrétien-Martin Liberal era saw serious early cuts in federal S&T funding under Program Review. But it embraced and exploited political value in fashioning a part of the government's economic agenda under the combined, but changing, rubric of innovation strategies, knowledge infrastructures, and strategic investment in network-centred activities such as those inherent in the Canada Foundation for Innovation and the Canadian Institutes of Health Research. The Liberal governments of this period spoke of "enabling technologies" and the rapid deployment of science and technology for "smart regulation," but mostly focused on the Information Highway. As the Liberals left office in 2006, they could point to significant new S&T spending, indeed the highest proportion of GDP ever spent on S&T (about 2.3 percent of GDP), but they left the unfinished business of increasing commercial focus and socio-economic impact. Nevertheless, given the long lags in impact from such investments, one could reasonably attribute a significant portion of the more recent improvements in scientific research and patenting to the policy and investment decisions of this earlier period.

The Harper Conservatives

The Harper minority governments from 2006 to 2011, and its majority government between May 2011 and October 2015, centred their agenda on dealing with the recession and fiscal deficits and, after 2011, on energy and natural resources, which reside at the centre of the Conservatives' Alberta and Western Canadian political base. The initial agenda, revealed in the 2006 Speech from the Throne, was anchored in a five-point program of accountability (as a direct response to the scandals that plagued the Liberals); the 2006 Throne Speech also promised reductions of the Goods and Services Tax and to tackle crime. No S&T or innovation policy announcements or views were offered at this early stage.

Eighteen months later, the 2007 Speech from the Throne led with commitments to strengthen Canada's sovereignty. This included a promise to "build a world-class Arctic research station that will be on the cutting edge of Arctic issues, including environmental science and resource development" (SFT 2007, 2), and in addition "our Government will complete comprehensive mapping of Canada's Arctic seabed" (ibid.). This is a clear example of science embedded in policy. The economic priorities centred on the government's "Advantage Canada" package of initiatives, but also on Canada's new Science and Technology Strategy, which would emphasize the greater need to mobilize S&T to commercial purposes, echoing the departing message of the Martin era (Treasury Board Secretariat 2007) and Castle and Phillips (2010). Also included was a commitment to improve the "protection of cultural and intellectual property rights in Canada, including copyright reform" (SFT 2007, 3). Even more prescient about future Tory policy was a commitment that the government would "stand up for Canada's traditional industries" – the government had already referred to Canada as a global "energy super power," with a firm eye on the burgeoning Alberta oil sands (Macdonald 2011).

When the global banking and economic crisis began in 2008, the Harper minority government's agenda shifted focus to the fast-growing fiscal deficit and the development of the economic stimulus program, marketed under the banner of the Conservatives' Economic Action Plan (SFT 2008). This agenda shift sought to maintain economic growth, deal with the recession, and tackle specific issues such as an Ontario auto industry bailout and regulatory streamlining via initiatives to reduce red tape. But a later section of that Throne Speech was entitled "Expanding Investment and Trade," and it offered promises to ensure that advances in S&T were central to improved competitiveness. A prime commitment was the government vow to invest "in world-class research facilities" (SFT 2008, 3). The government called a snap election but failed to secure a majority.

The extremely brief 2009 Speech from the Throne after the election contained no mention of S&T or innovation. But in the 2010 Speech from the Throne the stimulus measures of the Conservatives' Economic Action Plan were central, including plans to return to budgetary balance through future cuts that would eventually impact in then (and largely still) unknown ways on federal science budgets and staffing. S&T matters were subsumed under the "Building the Jobs and Industries of the Future" section of the Speech. Here the Harper government stressed that Canada had to "combine the best of our intellectual and natural resources to create jobs, growth and opportunity" (SFT 2010). This would be done through a further bolstering of the federal S&T Strategy, launching a digital economy strategy, developing space-based technology largely in support of Arctic sovereignty, supporting the development of clean energy technologies (especially carbon capture and storage), and a re-promising of the High Arctic Research Station.

The 2011 Speech from the Throne, developed as the agenda for the new majority government, saw the Harper Conservatives feature "jobs and growth" in the context of its Economic Action Plan and deficit reduction strategy. S&T and innovation policy was not on offer. The Conservatives, instead, developed aggressively and openly a more explicit policy of what they called "responsible resource development" that supported resource exports and oil and gas pipelines and that was coupled with policies that weakened environmental assessment policies and processes (Toner and McKee 2014; Clarke et al. 2013). This full package, part of a huge omnibus budget bill, was rammed through Parliament in 2012. Apart from the contested process, there is some disagreement over whether the new measures streamlined or weakened oversight. Undoubtedly, the federal government withdrew from a range of regulatory activities. In almost all cases, however, the federal changes either worked to vacate spaces where provinces had the constitutional lead (e.g., environmental assessments for "developments" that operated and have impacts exclusively in a single province) or set specific timelines for project review processes without fundamentally changing the steps or standards for decision making.

The Harper government's 2013 Speech from the Throne continued the jobs and opportunities theme but also included a promise to introduce a law requiring balanced budgets in normal economic times. Prominence was again given to resource development but subject to the proviso that the Conservatives would "enshrine the polluter-pay system into law" (SFT 2013, 5). In the growth section of the Throne Speech, the government stressed that it had "invested more than 9 billion dollars to support science, technology and innovative companies" and that it would "release an updated Science, Technology and Innovation Strategy" (ibid., 6). The government also would continue "making targeted investments in science and innovation chains from laboratory to market" (6). Later in the

Speech, the Conservatives indicated that the long-promised Canadian High Arctic Research Station had been established and that it would be opened in time for the 150th anniversary of Confederation in 2017. This Arctic research initiative was extended when it was announced on 1 June 2015 that a new organization called Polar Knowledge Canada (POLAR) was established that will "provide a world-class hub for science and technology research in Cambridge Bay, Nunavut called the Canadian High Arctic Research Station" (Polar Knowledge Canada 2015).

By no means "announced" as an S&T priority, but increasingly evident in 2012–13 in the context of an energy-resource policy, the Harper government exerted greater control over the public engagement of scientists. One interpretation is that this involved "muzzling" scientists employed by the federal government (O'Hara and Dufour 2014). Although this has been the effect in many cases, the long-standing privilege for scientists to independently comment on policy matters was an anomaly in executive government: other professions, such as lawyers, economists, and statisticians have always had limited liberty to contribute to public debate on matters related to their government roles; this will be discussed in more detail in later chapters.

Since 2006 the Harper Conservatives have certainly been conscious in an overall sense of the importance of S&T policy, but they have tended to cast it under discourse and policy related to Arctic sovereignty, standing up for Canada's traditional resource industries, and what they increasingly assert is responsible resource development, including clean energy technologies. And, as noted above, the 2013 Speech from the Throne spoke only in linear terms of science and innovation "chains" from laboratory to market.

SPEECHES FROM THE THRONE AND NATIONAL S&T AND INNOVATION POLICY AGENDAS

Our second way of tracking and assessing federal S&T agendas through the four main prime ministerial eras is through the agenda-setting dynamics, discourse, reframing, and issue management process, as underpinned by chapter 1's discussion of agenda-setting theory and practice.

Speeches from the Throne: Summary Picture

The place or mention of an S&T and innovation policy agenda in the history of Speeches from the Throne varied significantly in the thirty-two

Throne Speeches since Trudeau came to power in 1968 and extending to the 2013 Harper Throne Speech (see Table 2.1).

It must be emphasized that our reading of all the thirty-two SFTs confirms the general proclivity of all prime ministers to stress economic and fiscal priorities ahead of S&T ones. Table 2.1 conveys in a different, more consolidated way the basic story of the four prime ministerial eras traced above. The Throne Speeches story for the Trudeau era shows an initial low-key S&T policy presence in 1970, 1972, and 1973 and again in four later SFTs, including 1980 and 1984; but it was not present at all in five SFTs, most of them in the mid- to late 1970s. In the Mulroney era, two SFTs had no S&T content while in three it received middle-level attention, especially in 1986 and 1989.

The Chrétien-Martin era overall had the strongest overt S&T and innovation policy mentions and/or promises, with references showing up in all of the era's Speeches from the Throne. Its general discourse was also caught up in innovation policy language and in frequent reference to the Information Highway. In terms of performance, however, assessments in the literature have cast the Chrétien-Martin governments' efforts, outputs, and outcomes as highly problematical and limited (Wolfe 2002; de la Mothe 2003; Kinder 2003). In contrast, half of the Harper era's six Throne Speeches had no S&T mentions, and while the other half made considerable references, most of them were couched in the context of other policy priorities such as Arctic research and clean energy technologies. Similar, very mixed report cards on the Harper S&T and innovation performance record have emerged (Toner 2008; Phillips and Castle 2010, 2012). All governments have talked about enacting policies to enhance the commercial impacts of technologies, and most have sought to deliver a digital strategy, but none have had any measurable success in those areas.

It is worth stressing overall that these Throne Speech references to S&T and innovation policy across the four prime ministerial eras refer mainly to promises and plans. They tell us little about performance on specific measures or S&T overall, although each prime ministerial era confessed publicly several times to the need to do more about industrial R&D and commercialization. In terms of this book's later S&T and innovation policy domain analysis, the SFT analysis overall shows two or three mentions of intellectual property plans and reforms, including copyright, and some mention of the granting councils and their planned reform, but often without much explicit mention of Canada's university research system or, indeed, of basic research and its place in the changing Canadian economy and society.

Table 2.1 S&T and innovation policy as expressed priority in Speeches from the Throne (SFTs) in the Trudeau, Mulroney, Chrétien-Martin, and Harper eras

TRUDEAU SPEECHES FROM THE THRONE (12), 1968–1983
- No S&T mentions in 5 SFTs (1968, 1974, 1976, 1977, 1978)
- References in 7 SFTs, including to the innovative economy, R&D incentive programs, science policy, industrial R&D, new granting councils, technology employment, microelectronic technology, Office of Industrial Technology

MULRONEY SPEECHES FROM THE THRONE (5), 1984–1993
- No S&T mentions in 2 SFTs (1988, 1990)
- Varied, middle-level ranking/mentions in 3 SFTs (1984, 1986, 1989) regarding more focused S&T and education, establishment of NABST; new S&T strategy; national conference on science and technology; Canadian Space Agency, Network of Centres of Excellence, development of strategic technologies

CHRÉTIEN-MARTIN SPEECHES FROM THE THRONE (8), 1994–2005
- Considerable middle-level mention in all of the SFTs
- Varied expressions of S&T policy as Canadian technology network; technology partnerships; strategy for Information Highway; critical enabling technologies, SchoolNet and Community Access Program; investing in knowledge and creativity; formation and funding of CFI and CIHR and leading-edge national system of innovation; information and knowledge infrastructure; National Broadband Task Force; "smart regulation"; $13 billion S&T spending since 1997; new Office of the National Science Advisor (ONSA)

HARPER SPEECHES FROM THE THRONE (7), 2006–2015
- No S&T mention in 3 SFTs (2006, 2009, 2011)
- Varied middle-level and low-level mentions in other four SFTs (2007, 2008, 2010, 2013) expressed as building world-class High Arctic Research Station; new S&T strategy; protecting cultural and intellectual property rights, including copyright; investing in world-class research facilities; industries of the future; combine best of our intellectual and natural resources; space-based technology; clean energy technologies; $9 billion in S&T spending since 2006; and science and innovation chains from laboratory to market

Thus, not surprisingly, given the nature of Throne Speeches as agenda evidence, we need to look elsewhere for other triggers for or arenas of S&T and innovation policy, often in designated policy fields. Indeed, S&T and innovation policy and natural science and social science *in* policy are part of numerous policy fields in several science-based departments, ranging from foreign policy to health, environment, consumer affairs, transportation, food, and fisheries and oceans, to name only a few. Policies are announced in many other ways and in the context of other events, in between Speeches from the Throne. This often involves regional or sectoral ministers announcing S&T and innovation–related measures in relation to their more specific ministerial agendas, albeit often only when new or changed framework policies are ready for release and for a public defence.

In the multi-domain context of S&T and innovation we take note of Doern and Prince's (2012) summary of where biotechnology as a technology and science stood in terms of agenda setting and profile among prime ministers and ministers of finance, as revealed through Speeches from the Throne but also Budget Speeches in the 1980 to 2010 period. They show that biotechnology received mainly low priority mentions in only seven of twenty Throne Speeches and only six of thirty-four Budget Speeches and that these were almost all in the Chrétien-Martin era (Doern and Prince 2012, 98). This analysis also shows the types of political language used in these agenda speeches regarding biotechnology: "One type is generic or high level, calling biotech strategic or enabling, which taps into dominant societal beliefs about science and technology. The second and more common type is discourse specific to one or other realm or industrial sector, such as life science, medical research, agriculture, bio-fuels, or pharmaceuticals. Both types of political language are designed to support and promote biotechnology through either the use of an abstract yet positively evocative word, for example, 'enabling,' or of an identifiable economic application or research entity" (ibid.).

We return to these agenda-setting discourses regarding biotechnology and genomics and the life sciences in our domain analysis in chapters 9 and 10. But it is worth noting that both generic and more specific types of policy discourse have been present in the overall S&T and innovation policy agendas across all of the four prime ministerial eras.

CONCLUSIONS

The chapter has examined federal S&T and innovation policy agendas and evolution in the main federal prime ministerial and governing political party and partisan eras, the Trudeau Liberals from 1968 to 1984, the Mulroney Conservatives from 1984 to 1993, the Chrétien and then Martin Liberals from 1993 to 2006, and finally the Harper Conservatives from 2006 to 2015. In broad terms, the Throne Speech discourse has shifted from S&T per se to more frequent use of the innovation policy and even innovation network concepts, with a clear shift to presenting Canada as needing to become more innovative. The notions of S&T policy as being linear in nature or innovation policy as being non-linear do not get much mention by Throne Speech writers, either directly or as background explanations or theoretical drivers of change.

The story reveals an S&T and innovation policy presence in the agenda but typically at lower and occasionally middle levels of ranking and expression within and across broader economic and social agendas present in each era. But this story is not the whole story. In a government of thirty-five to forty ministers there are at least thirty-five to forty policy fields. Thus, a study of other federal policy fields such as transportation, fisheries, environment, and food, for example, would undoubtedly show that STI is relevant and thus has greater staying power in agenda formation and in Canadian political economic life (and possibly even in the attention spans and top-of-mind concerns of Canadians). Our domain chapters explore this further.

As noted in chapter 1, public opinion on S&T and innovation policy as revealed in polls over this almost fifty-year period conveys a similar low or masked priority. It is far from clear whether this is because of lack of public interest or because, at some level, pollsters have decided that S&T is not of much interest to Canadians and thus pollsters don't frame questions that would reveal this more nuanced awareness and interest in the topic.

S&T and innovation policies and agendas increasingly come from sources and developments outside Canada. We now turn in chapter 3 to a discussion of some of the key international determinants of Canadian S&T and innovation policies and agendas in the past fifty years.

3

Canada-US and International STI Policy and Institutions

The changing nature of international STI policy and Canada's role in shaping it and responding to it is an essential context for understanding our analyses of the eight S&T and innovation domains in later chapters. We have, of course, already seen some of the impact of these international dimensions in chapters 1 and 2, including the global shift from demand management macro fiscal policy to micro-economic growth theory and policy. We need, however, a clearer sense of the basic contours of the evolution of international STI policy and institutions, particularly across the past five decades. Our focus is on the policy and institutional story, but we also refer to the nature of international relations and power and to the governance structures that changed it and were produced by it. We need to see how Canada contributed to and sought to influence and shape international STI policy developments and institutions.

International S&T policy analysis overall has traversed a range of different and overlapping conceptual and theoretical vantage points (Dryzek 2013; Wijen et al. 2013; Volger 2013; Elliott 2004; Held et al. 1999). These include *international relations* (IR) theory, where the focus is on power and interpreting the international system through the practice of realism or "realpolitik," both in general and specifically in S&T and innovation policy matters; and *globalization* theory, which over more recent decades has posited that all policy is influenced by strong and complex forces including economic liberalism and new transformative technologies but also global public interest and societal, non-governmental forces. Canada's international story involves both the dominant Canada-US relationship and an array of global engagements. The former includes the changing role of the United States as a neighbour, as a trading partner, as an investor, and as a global superpower increasingly functioning under a

Table 3.1 Canadian researchers in a world and OECD context, 2011

	Researchers (n)	*% of World Total*	*% of OECD*
OECD	4,304,000	68.10	100.00
World (est.)	6,323,000	100.00	–
Canada	163,000	2.58	3.79
Federal and provincial	13,900	0.22	0.32

Sources: OECD, Total Researchers in Full-Time Equivalent, *Main Science and Technology Indicators* 2014, issue 2. Statistics Canada, Research and Development Personnel, 2012 (*Cansim Daily*, 16 Dec. 2014).

multipolar global power structure shaped by the European Union, as well as Brazil, Russia, India, and China (BRIC), and East Asian countries. In international governance terms, the emergence of multiple new points of power and influence is symbolized by the decline of the G8 and the rise to prominence of the G20 bloc of relatively wealthy or at least relatively large economies. International science, however, is not just a factor in global policy and power, it is also bound up in practical issues of policy design and implementation within nation states and among them (Howlett 2011).

From the outset it is essential to appreciate in the global context that S&T remains highly concentrated within the ambit of the OECD member countries, in which more than two-thirds of all research scientists in the world work. Collectively, they contribute about 90 percent of all publications and 99 percent of all patents and other working intellectual property (IP). Our estimates on the number of total researchers and federal researchers in Canada and the world (OECD, BRIC, and ten other leading countries) are shown in Table 3.1. These data put the Canadian component into stark context.

From a policy perspective, the international S&T "league tables" and rankings of Canadian performance are common reference points for the Canadian S&T and innovation debate. Canadian policy-makers continually lament Canada's weak showing on a range of measures, but particularly its weak performance on gross expenditure on R&D as a percentage of GDP (GERD). Canada has only episodically cracked the top half of the league of OECD countries. Other measures reveal some of the nuances of our situation. For example, the OECD's data on business expenditures on R&D as a percentage of GDP (BERD) showed Canada ranked at twenty-second of thirty-four OECD countries (OECD 2014, 1). Canada is ranked first among G7 nations in spending on R&D in higher education (HERD) relative to the size of the economy (ibid.). Not for the first time these kinds of data show comparative weakness and strength in these two dimensions of the S&T challenge (Industry Canada 2014b; Doern and Kinder 2007).

The analysis proceeds via an account of seven fairly representative benchmark examples in the evolution of an ever more complex international S&T and innovation policy and institutional system. We build on this initial international context and other international developments in the domain chapters in Part II of this book. First, the chapter highlights some core features of Canada-US S&T and innovation policy relations across the decades being covered. Second, we trace the basic nature of science as an international social system. Next we examine peer review and the changing structure of natural and social science disciplines and the role of grey literature. Then we look at the impacts and issues related to the massive growth of Internet and social media–centred *Big Data*. The fifth section looks at some core characteristics and developments with respect to the science-based aspects of international trade agreements. Sixth, we trace some of the core international scientific evidence in the climate change debate to illustrate the challenges of dealing with emerging science. Seventh, we look briefly at a few selected international agencies and their socio-economic STI policy mandates, influences, and challenges. Conclusions then follow.

CHANGING CANADA-US STI POLICY RELATIONS

Early historical examples of Canada-US S&T relations are not difficult to find. These include the initial mapping on both sides of the border and geological surveys to spur the development of the resource base of the two economies and societies (Vodden 1992). The geological and conservation science communities developed jointly and to this day share knowledge and research both among relevant public agencies and also among universities. The same has been true of farming and agriculture research and knowledge dissemination via systems of research stations, land grant–style universities, and agricultural extension workers located close to farm communities. This has evolved into even more intricate day-to-day research links in the Canada-US food regulatory system that now manages risks in the complex cross-border food chains spanning from farm gate to plate (Doern and Prince 2012; Phillips and Wolfe 2001).

US corporate structures and interests have been central to the Canadian economy as the United States is Canada's main export and import partner. Technology-related policy debates and actions were pivotal in the 1960s to 1980s when Canada's foreign-owned (mainly US-based) manufacturing sector was accurately described as a branch plant economy (Harris 1985; Blais 1986). Initially, this model was seen as a good way to have the latest production technologies easily and quickly brought into the Canadian economy. Over time, however, the branch plant approach was not seen as being structurally favourable for Canada; it inhibited efforts to develop Canadian technologies that could support longer production runs with greater export potential. In response, foreign investment policy regimes were increasingly tied to these developments, initially via the Foreign Investment Review Agency (FIRA), which screened foreign investment proposals for domestic economic impact. This morphed into Investment Canada in the 1980s, and it was tasked with encouraging foreign investment as one way to have an innovative economy in an increasingly freely trading world. Despite the liberalization in some sectors, a number of key sectors are still caught in the constraints of Canada-US policy, including the auto and auto parts industry and the aeronautics and airline industries.

US university research structures and debates have buffeted Canada, even though Canada's university system is structurally quite different. The United States has an extremely complex and diverse multilevel higher education system (Slaughter and Rhoades 2004). It consists of over four thousand public and private colleges and universities offering choice that ranges from elite research universities to two-year technical and community schools and four-year colleges. It is a dynamic and competitive system, not only because of the mixture of public and private institutions, but also because of the nimbleness of individual institutions (Teichler 2007). In a very real sense, the United States has fifty systems of higher education tied to American federalism and its fifty states, but even the state systems are not necessarily fully structured.

Canada created its university and later community college systems mainly through public finance and administration at the provincial level. For many years, the Canadian and US systems lived in splendid isolation. In recent years, however, more attention has been placed, albeit selectively, on US debates and reforms related to making universities a part of the innovation economy and restructuring them to be closer to the innovation process and to different modes of knowledge production (Jones, McCarney, and Skolnik 2005). Individual Canadian provinces have created new technology centres provincially, regionally, and locally. An early example of this was the establishment of the University of Waterloo as a mandated technology university based on the US model exemplified by the Massachusetts Institute of Technology.

We will see in chapter 6 other forms of international emulation in the Canadian university system, drawing on lessons from the United States, the United Kingdom, and Europe (Doern and Stoney 2009; Teichler 2007). American universities have also attracted tens of thousands of Canadian students, not to mention Canadian faculty, many of whom either migrate permanently or return to Canada with lessons learned from their experiences in the American system. Recent Canadian granting policies, as chapter 6 discusses, have sought to reverse the "brain drain" to the United States by trying to attract American faculty researchers to Canada and to repatriate Canadian faculty based in the United States, using private endowments and public investments in research chairs and centres as the inducement.

Other aspects of Canada-US realities drive the S&T and innovation policy effort. Energy and environmental science has been called on directly by the Canadian federal and Alberta provincial governments to provide evidence that can support a lobby directly in Washington and elsewhere in the United States to garner support for the Keystone XL Pipeline that will, if approved, transport mainly Alberta oil sands bitumen to refineries in the southern United States (Doern, Auld, and Stoney 2015). Alberta faced direct criticism and active campaigning from the All Risk, No Reward anti-Keystone coalition (Broder 2013). Meanwhile, and in concert, the Alberta government was proposing carbon capture and storage technologies and, because of US concerns, science-based environmental standards for the oil sands and related energy industry emissions in Alberta that would be tougher than advocated by the Harper government in Ottawa (Meadowcroft and Langhelle 2011; McCarthy 2013).

A second related energy example centres on what Aronczyk and Auld (2013) call "tactical repertoires" as they trace the "co-evolution of movements for and against the tar sands" functioning in both the United States and Canada in the past decade or more. In their story, a pro-climate change movement advocating major science-based emission reductions is met by a counter-movement "represented by an industry-friendly lobby group and backed by the Canadian government under the rubric of ethical oil" (ibid., 1). Playing out on both sides of the border, their analysis shows the development of the discourse and argument structure of both movements as they react to each other and try to get the right messaging out to governments and supportive interests regarding their preferred positions.

Last, but not least, there have been diverse examples of even more explicit Canada-US environmental agreements and disputes where the

United States was variously and at different stages the environmental hero or villain, with such relations always having a considerable S&T component. They highlight a series of challenges and achievements, including bilateral efforts such as various Great Lakes pollution-monitoring agreements as well as more multilateral examples such as the processes that led to identifying and then reducing the impacts of acid rain and dealing with complex forestry, biodiversity, conservation, and environmental and sustainable development (SD) challenges (Doern, Auld, and Stoney 2015; Hoberg 2002; Bernstein and Cashore 2002; Doern and Conway 1994).

Indeed, it is useful to keep in mind Hoberg's (2002) suggested strategic framework for understanding Canadian-American relations as we analyse the eight S&T domains in later chapters. Hoberg outlines six mechanisms of US influence on Canada-US relations that can and have occurred singly or in combination: the physical environment, US policies and actions through emulation (elite driven or activist driven), diplomacy, trade agreements, economic integration and harmonization pressures, and cross-border lobbying (170–3).

Cross-border lobbying involves for Canada more complex processes because of the nature of the US political system and its separation of powers in Washington, including the independent power of the US Congress, the appeal of litigation through the courts, and the realities of very short and always-looming electoral cycles. There is rarely one-stop shopping for any lobby in any democratic political system, but this is compounded in the US federal system with fifty state governments to contend with. Fourteen states, from Maine to Washington and north to Alaska, are at the front lines of sharing the vast common political border and the many common environmental, public health, food policy, and research concerns that emerge from both day-to-day management and emergencies.

SCIENCE AS AN INTERNATIONAL SOCIAL SYSTEM

It has often been observed and examined that science emerges from social contexts, often to explore and solve practical human opportunities, disasters, and problems (Johnson 2006). In several respects science emerged as an identifiable, discrete, social system well before there was a university-focused, state-centred, or business-oriented triple axis of S&T activity and support (Storer 1966). This historical fact is itself important in seeing and understanding the core innovation economy and society nexus that is central to this book as a whole. Science is forged at the international level through relations between and among individual scientists and then through their organizations, both broad based and discipline based. Indeed, Robert Merton identified a core set of values and norms that govern science, including communalism, universalism, arm's-length disinterestedness, and organized skepticism, which he shortened to CUDOS (Merton 1973). Over time, CUDOS was expanded to include originality as one of the driving norms.

These accounts of science as a social system are simultaneously both national and international. This was true even when communications systems and levels of interpersonal contact were much more rudimentary and difficult than they are today. For example, Margaret MacMillan's (2013) analysis of the 1900 Paris Exposition, in the historical context of the period preceding the First World War, stresses the marvels of an apparent age of rationality built up across Europe, with the exposition event showing scientific and technological discoveries from around the world (1–24). The 1851 Great Exhibition of London was a similar, even earlier example.

The institutions in Canada were influenced by particular systems of international relations, power, and dependency in the scientific realm. At the operational level, in English-speaking Canada, the land grant model was imported from the United States and the collegiate system transferred from the United Kingdom. Similarly, Canada's macro science-centred organizations, such as the Royal Society of Canada (RSC), were based on the British model but with "the important addition of literature and other elements found in the Institut de France" (RSC 2013, 1). However, to this day, as noted in chapter 1, Canada has found it difficult to create more broadly based and engaged models such as that of the American Association for the Advancement of Science (AAAS; Dufour 2012).

When one extends these initial earlier notions of international imperatives and influences to technology, invention, and creativity, the story takes on different meanings and trajectories. Thomas Kuhn's famous analysis and theory about the structure of scientific revolutions and the nature of related paradigm shifts drew on extensive international S&T and technological history (see Kuhn 1962; Kindi and Arabatzis 2012). Key technologies ranging from the telephone and telegraph, the automobile and its internal combustion engine, and railways and shipping profoundly influenced Canada and shaped its technological, social, and economic structure. Similarly, actual laws and concepts of intellectual property regarding patents and copyright introduced in other countries were gradually but differently shape-shifted to Canada. Patents and the political centrality of the small inventor often came from the United States, where the social value of invention is acknowledged and affirmed in the US Constitution, put there because a few of the US Fathers of the Constitution were themselves inventors. Although copyright laws emerged initially in Europe to manage and control the press, they migrated over time to the United Kingdom and elsewhere to be more about protecting publishers and authors, such as the popular Charles Dickens, from having their works copied and stolen by others (in Dickens' case, by American newspapers).

Canada chose the British-American concept of property rights, where inventors had moral creator rights grounded as economic assets that could be fully assigned to others to exploit (Doern 1999a). The continental system, now embodied in some of the EU measures, grants inventors' inalienable moral rights to their work that are perpetual, inalienable, and imprescriptible. In short, even if the right to exploit the creation is assigned to others, the creator has the power to direct how the creation might be used and presented.

Although extremely brief, this initial discussion and illustrations of the importance of science as an international social system and of international technologies and inventions and their impacts is important for the rest of this chapter, for our understanding of the eight changing S&T and innovation domains, and for appreciating the conceptual and empirical notions of the innovation economy and society nexus.

PEER REVIEW, THE STRUCTURE OF NATURAL AND SOCIAL SCIENCE DISCIPLINES, AND GREY LITERATURE

It is but a small step in international terms to move from science as an international social system to a consideration of peer review as its social and ethical set of rules and processes for assessing the validity and contribution of new research. As we saw in chapter 1, this is bound up in the changing structure of academic disciplines, by recent forms of review, including merit review of science infrastructure projects, and by management review of programs for the support and funding of science (see chapter 6 for further detailed discussion). Partnerships, leveraging, and the imperative to translate knowledge to use through commercialization or other transfer mechanisms has massively changed the system. Here we survey peer review and its changing international structure and legitimacy, both historically and in recent years.

At its core, peer review is the recognized process by which researchers submit to academic and professional journals their research reports that set out their findings and methodologies and delimit their contributions to the advancement of knowledge in a given discipline and subject area. Shapin and Shaffer (1985) trace this practice back to the English Restoration and the creation of the Royal Society; they assert that the model of focusing on witnessed science that is dispassionately disseminated, advocated to normalize civil discourse in the sciences, was the foundation of the scientific revolution that followed. Although not universal, the norm is for peer reviewers to assess articles for potential publication without knowing the name of the author (some high-impact journals use editor reviews or other non-blinded review structures). Decisions are then made by journal editors to accept or not the submitted research for publication. Peer review is, in this sense, a social construct for the community of scientists.

In most respects peer review is taken as a given in international science. However, it has, in fact, had a more complex history and has undergone criticism in different ways (Nielsen 2011; Phillips 2007; van den Hoonaard 2013a, 2013b; van den Hoonaard and Hamilton 2016). Although peer review can be traced to the sixteenth century, early science was by no means universally peer reviewed. Scientists in the age of alchemy and alchemists often kept their research secret as a defensive measure against fraud and theft. Peer review was partly set up to share knowledge, give credit, and to prevent fraud (Lorch 2013). Early journals, as a part of their development trajectories, often had to spend more time making sure that articles were submitted in the first place rather than just managing the peer-review process. In his career overall, the brilliant Albert Einstein rarely published in peer-reviewed journals. Well-known journals often did not have peer review; the journal *Nature* only adopted peer review in the 1960s.

The international science and peer-review system has undergone criticism on a number of fronts (Bohannon 2013; Harley 2013; van den Hoonaard and Hamilton 2016). One systemic challenge is that many of the big-name journals are operated by for-profit publishers rather than by learned societies. The competition between journals puts all editors under pressure to be the first to publish novel research that seems valid and useful. Similarly, authors and researchers are keen to have their work published early and in the "best" journals as the articles can then be used as PR occasions for securing their always-needed next research grant or to support tenure, promotion, or merit cases. In their mutual haste, errors can be and are made. There is evidence of this leading to both errors of commission and omission.

There have always been examples cited about how peer review harms innovation because it is discovered later how some breakthrough research was rejected by a journal or journals, in part, because "peers" did not recognize or understand its contribution and applicability. At the other end of the spectrum, there is evidence of several instances of outright research fraud via fabricated data that got published in haste by journals. Both are probably not insurmountable problems, as few papers are excluded from publication at least partly because of the burgeoning academic press while post-publication replication is an effective and efficient way to weed out flawed and fraudulent work.

One concern, however, is that many of the articles published today do not provide adequate disclosure of methods and data to enable the *replication* of research findings and results and hence the independent validation of research claims (*Economist* 2013b; Collins 1985). The Internet has enabled the rapid development of post-publication review and commentary by scientists and bloggers in varied social media, and some journals now require formal deposit of data, algorithms, models, and samples, made easier by the limitless and inexpensive storage and transmittal capacity of the Web (Lorch 2013). PubPeer is now an Internet site that allows anyone to comment on any published research, a practice that science journals are often reluctant to allow themselves. PubMed, a large database of research on medicine, now allows a comment feature, but other journals have found unmoderated commentary can be problematic as many of the posts are inflammatory or disputatious.

These problems in peer review are significant, but the attempted solutions and reforms pose a different set of problems (Stone and Jasny 2013; van den Hoonaard and Hamilton 2016). The extent of concerns can be partly seen in the establishment in 1997 of the Committee on Publication Ethics (COPE) by a small group of medical journal editors in the United Kingdom; it now has over nine thousand members globally from all academic fields (COPE, 2014).

International peer-review challenges and issues are complicated by the growing number of new journals, including interdisciplinary titles in fields such as the health sciences and medical research. Interdisciplinary research can mean research by teams of natural and health scientists and social scientists and hence a more complex mix of peers in the review process and need for balance in the criteria for assessment and publication (Barry 2013; van den Hoonaard 2016).

These peer-review and science communication issues are inextricably linked to *grey literature*, which is defined as the growing volume of research and analysis found outside the peer-reviewed journals and academic book publishers. The distinguishing feature is that there is lesser reliance on assessment by peers; sometimes there is no review, sometimes there is full review, and at times the review is done by sponsors or those with interest in the research, which on the face of it offends Merton's CUDOS norms (Huffine 2010; Lawrence 2012). Grey literature includes research by arm's-length think tanks, including many known for their political and ideological leanings, and by government departments and research agencies, international agencies (see more below), consulting firms (including academics engaged in consulting on contractual projects), and various bloggers and social media.

The volume of this research literature is massive and growing at both the international level and in Canada. Issues of reputation, and perceived or real scientific validity, are bound up in the growing grey literature realm. Generally viewed as a lower form of publishing by academic scholars and scientists, the grey literature can, nevertheless, at times perform an important knowledge translation function, making available in a policy- or market-ready form the often highly specialized and disconnected peer-reviewed knowledge.

BIG DATA

Big Data refers to the production, collection, storage, retrieval, control, and economic and social mining of massive amounts of information held by governments and industry, increasingly mediated through the Internet. Big Data is a major international phenomenon, predominantly shaped by the United States but with global reach. Canada's economy, social system, and its S&T and innovation architecture has (as chapter 11 shows in considerable detail) been massively influenced by the Internet, developed initially in the United States and then popularized and extended by US-based social media firms such as Google, Facebook, and Twitter (Dahlgren 2013; Mayer-Schönberger and Cukier 2013; Davenport, Barth, and Bean 2012; Coleman and Blumler 2009). This new medium has changed modes and strategies of innovation, production, communication, marketing, and democracy.

Big Data raises concerns about privacy and the limits of regulation in socio-economic innovation realms (McDonald 2014; Culnan, McHugh, and Zubillaga 2010). Indeed, the Internet as an "infosphere" has been declared to be historically the "4th revolution" (Floridi 2014) following in the wake of earlier historic transformative technologies. Big Data warrants inclusion in this contextual chapter on Canada-US and international S&T and innovation influences because, while it has garnered attention, most of the analysis has only made a cursory attempt to look at its definitional contours, the public discourses used, and the possible policy responses and difficulties.

In a recent speech, senior Bank of England official Andrew Haldane (2013, 2) pointed out that "almost all of the information is new. Over 90% of it has become available in the past 10 years. Nearly half has become available in the past 3 years. This rate of expansion is unlikely to slow." Citing Davenport et al. (2012), Haldane continued, "Google alone processes 25 million gigabytes of data each day." Haldane argued that such a data and information explosion, in concert with overall global economic and social integration, means that new forms of serious systemic risk will emerge. Haldane and other authors submit that both the world and individual nation states and their citizens will need serious science- and evidence-based public institutions to understand and deal with this new risk (Mayer-Schönberger and Cukier 2013; Lanier 2013).

Big Data also offers some potential for investigating questions that have been for decades the subject of small-scale inquiry (Kleinberg 2011). Big Data creates a present and fast-growing need for a different kind of scientist, a "data scientist," capable of scanning and analyzing the correlations embodied in these vast amounts of data (Davenport and Putil 2012), using a mix of algorithms and other techniques. While many are excited by this new venture, others raise concerns that this new approach could weaken causal research and theory building that is foundational to evidence-based policy-making. The key concern is that correlation does not necessarily equate to causation.

Skeptics and proponents alike are caught up in the Big Data debate. The definition itself is problematic. As noted above, the standard definition turns on its mass volumes, especially as collected by social media firms such as Google or Facebook. Richards (2014, 3) notes that the "reality about Big Data is that it is comprised of different forms of data: numbers, text, sound and images and it is moving quickly." Consequently, debates about Big Data centre on three features: "streaming data of different forms available from various social media sites, sensors and satellites; the technologies available to process the data; and the techniques available to use this data for descriptive, prescriptive, and predictive analytics" (ibid.).

Links are also drawn between Big Data and small science. As observed in the *Economist* (2014a, 66): "This may indeed be the unspoken secret of big data. The much-hyped term captures the idea that it has become possible to collect vastly more data than before, and to process it far more cheaply and quickly. The result, typically, is to identify the right combination of tweaks capable of bringing about marginal changes that, when multiplied by a huge number of instances, or allowed to work over a long time, produce a significant effect." These systems of "constant experimentalism and rapid implementation" (ibid.) have been seen to occur for Internet social media firms. Although much more difficult for mainly off-line firms this, too, is changing.

Consumers are in the spotlight – they are both the subject and object of most Big Data explorations. For example, in the US financial services industry various Big Data analytics are increasingly being developed and used to segment and target specific groups of consumers, with both positive benefits but also potential and actual adverse impacts for consumers who are most at risk of not being granted financial or insurance service eligibility (Chester and Mierzwinski 2014).

Most analysts of Big Data also immediately, albeit in different ways, acknowledge that it triggers complex issues of privacy (Jerome 2014). When seen as an international dynamic, Big Data and privacy issues are expressed with different kinds of discourse. For example, Lynch (2014, 2016) relates Big Data and privacy conceptually and metaphorically to the "pool of information." His concerns relate to Edward Snowden's whistle-blowing about surveillance by the US National Security Agency (NSA) on Americans and the citizens and governments of other countries (see further analysis in the chapter 11 account of the Internet, communications, and social media domain). Lynch drew attention to the fact that 90 percent of the data gathered related to Americans is incidental in nature. He argues that this much larger pool of information could "easily prove irresistible" for those with nefarious ends. He extends this logic by arguing "the pool of data is a pool of knowledge. Knowledge is power and power corrupts. As a consequence it is difficult to avoid drawing the inference that absolute knowledge might corrupt absolutely" (Lynch 2014, 2).

Craig Mundie, a former Microsoft chief research and strategy officer, calls for "privacy pragmatism" with a "focus on data use not data collection" (2014, 28). In Mundie's view the "widespread and perpetual collection and storage of personal data has become practically inevitable. Every day, people knowingly provide enormous amounts of data to a wide variety of organizations, including government agencies, Internet service providers, telecommunications companies and financial firms ... Indeed, there is hardly any part of one's personal life that does not emit some sort of 'data exhaust' as a byproduct" (ibid.). As a result, says Mundie, the era of big data "has rendered obsolete the current approach to protecting individual privacy and civil liberties" (29).

Mundie's suggested focus on regulating data use is based on the probably valid assumption that when "it comes to privacy, the data rarely matters, but the use always does" (2014, 34). User-centred solutions are based on the reality that "without a computer program, there is no use – and without use, there is no abuse" (ibid.). Mundie is convinced we should move to managing data use and not data collection, but he concludes that "moving from the current model to this new way of controlling privacy would require political will and popular support. It would require people to constantly re-evaluate what kinds of use of their personal data they consider acceptable" (37).

Diagnoses of the Big Data policy landscape have also produced what Preston (2014) calls the "death of privacy," in the sense that "our age has seen an almost complete conflation of the previously separate spheres of the private and the secret" (3). As Preston argues, "Google knows what you're looking for. Facebook knows what you like. Sharing is the norm, and secrecy is out" (1). He asks whether Google is "the de facto librarian of the Internet, given that it is estimated to handle 40% of all traffic? Is it something more than a librarian, since its algorithms carefully (and with increasing use of your personal data) select the sites it wants you to view? To what extent can Google be held responsible for the content it puts before us" (6). The concern, Preston stresses, is that Google is not an elected body yet it may have fundamental power to direct our wants and desires in ways that historically were mediated by democratically elected governments.

Finally, Big Data and related new technologies, such as 3D printed products and the "Internet of things," raises significant policy concerns about digital matters in international trade negotiations and trade agreements. As Donnan's review of digital trade shows, "innovation in some sectors is outstripping the trade rules designed to govern them" (2014, 7), in part, because the distinction between a good and a service vanishes with some products such as 3D printing, smart watches, and numerous apps that contain data-rich services embedded in them. Software lobbies in many high-income countries have begun to focus on the free cross-border flows of data, but other interests and countries want these included in newer explicit forms of "data protectionism."

Big Data overall has generated an array of internationally driven issues and prognostications about its core features, uses, and abuses. We return to some of these features with our analysis in chapter 11 of the Internet, communications, and social media domain.

S&T, LIBERALIZED MARKETS, AND FREE TRADE AGREEMENTS

Science and technology has always been a part of and been influenced by liberalized markets and the construction of international and regional free trade agreements such as the post–Second World War General Agreement on Tariffs and Trade (GATT), now the World Trade Organization (WTO), and a multitude of regional agreements, such as the North American Free Trade Agreement (NAFTA) signed in 1994 by Canada, the United States, and Mexico. Concepts of science-based governance have been entrenched in these structures, and the ability to foster industrial technology and related innovation policy has been variously restrained and supported by the embodied measures. Thus, S&T is interwoven in and around trade agreements, and diverse kinds of S&T are needed to manage dispute identification and resolution. There is an underlying assumption and increasingly universal view that the freer international trade is, the more innovation regarding products and processes will occur, although we have seen from previous chapters, that such innovation is not necessarily linear or predictable.

The central reality of trade agreements is that they seek to liberalize trade by controlling and restraining *governmental* behaviour. An initial early GATT focus was on tariffs (essentially a tax at the border) and their gradual reduction through several rounds of GATT negotiations (Trebilcock 2011). High tariffs had themselves been a key instrument used to protect firms and industries in Canada and other countries and a basis for national industrial and technology policy (see chapter 7). As tariffs came down, the focus shifted to controlling other governmental measures that were deemed to distort production and trade.

Domestic and export subsidies were the next obvious target. Countries have used subsidies in varied ways in the name of industrial policy, agricultural policy, defence policy, and regional policy. Such subsidies had both good and adverse and intended and unintended impacts on international trade and on technology support in general and on particular industries and their core technologies. As the global production system has become more interconnected, the conception of borders has necessarily become more complex. International trade policy has evolved from trade "at the border" (e.g., tariffs, quantitative limits, and entry rules) to trade "over the border" via national treatment of investments. Hence, trade policy now touches on and affects virtually all heretofore "domestic" policy fields. There are S&T and innovation aspects in virtually all of these fields.

None of the agreements is unambiguously liberal. All trade agreements in some way provide for derogations from the rigours of the rules for "legitimate objectives" (LOs), defined variously as health, safety, environment, and moral concerns where science- and evidence-based capacity is to be applied. In the international trade realm, such legitimate objectives are expressed as policy rights that states can use to legitimately pursue national sovereignty and the public interest. The qualifier in trade agreements is that when legitimate objectives are pursued, they must be carried out in the least trade-distorting manner possible by adhering to processes regarding sanitary and phytosanitary (SPS) rules and provisions related to technical barriers to trade (TBT). These processes generate disputes due to differences of opinion on the role of science, evidence, and knowledge in decision making, often wrapped up in the concept of precaution related to uncertainties in health, safety, and the environment (Trebilcock, Howse, and Eliason 2013; Heydon 2012; Coglianese 2010).

More and more countries are signing regional or bilateral trade agreements, each of which requires WTO approval but which could be somewhat different as long they are trade-liberalizing on an overall basis. This has certainly included Canada, with the Canada-United States Free Trade Agreement (CUFTA) and NAFTA agreements on the continent and bilateral trade and investment agreements with more than fifty other countries. Most recently the Harper government has invested heavily in negotiating new regional agreements, such as the Canada-EU Trade Agreement (CETA) in 2014 and the yet to be realized Trans-Pacific Partnership Free Trade Agreement (TPP).

A larger "trade and" agenda has emerged as a key feature of the expanding trade regime. This refers to the broadening content of agreements, which usually include "trade and" services, procurement, investment, and/or intellectual property, among others. These agreements also have tested the waters in such realms as competition policy. These issues were resolved in the overall Uruguay Round WTO agreement through separate related agreements on these subjects, with links to WTO institutions. Many of these changes increasingly are seen by critics as the products of US power and aggressive unilateralism, in short, raw hegemonic American political and economic power, especially vis-à-vis intellectual property (Ostry 1997; Sell 1998).

INTERNATIONAL SCIENCE AND CLIMATE CHANGE POLICY

Without doubt the biggest single recent international science-related example of science in policy is the climate change file (Giddens 2011; Helm 2012). We refer to climate change again in chapter 5, but here we capture some key features of its international science dynamics. The development of the Convention on Climate Change in 1992 and the negotiation of the Kyoto Protocol in 1997 followed years of international policy and economic analysis and extensive global scientific research to examine whether and how human emissions of greenhouse gases (GHG) affect the climate (Grubb, Vrolijk, and Brack 1999; Houghton 1997). The 1979 World Climate Conference created the World Climate Research Programme to help stimulate research, and in 1988 the Intergovernmental Panel on Climate Change was established. It is the IPCC's successive reports that provide the scientific and policy underpinning, first for the UN Framework Convention on Climate Change (UNFCC) and then for the Kyoto Protocol and the post-Kyoto era. The IPCC's first assessment report was published in 1990, the second in 1996, and the third in July 2001; others have followed since, including recent reports on the abrupt impacts of climate change and how to anticipate such surprises (see more below).

The IPCC is composed of 195 government-level nominees; however, it draws on the work of thousands of scientists from around the globe who draft the technical reports for the Panel. IPCC assessments reflect the strong overall scientific consensus about the man-made contribution to climate change that more and more governments are beginning to accept, though many have not yet acted upon with firm commitments. The Kyoto Protocol was adopted in 1997, with targets for reducing GHG emissions; however, the Protocol had not yet entered into force, and measures to realize the targets are not yet implemented. All parties knew that, at its core, the Kyoto Protocol was mainly regulatory and rule based in that targets for reductions were required. Accordingly, some countries, institutions, and interests have begun to change their behaviour in anticipation of eventual commitments. But policy solutions as a whole and the character of the policy mix, including specific regulatory and other provisions, will only emerge in more detail after prolonged and protracted negotiations among initially 178 countries (now almost 200) and with enormous pressure coming from national and international environmental NGOs and scientists the world over.

Most parties agreed that any final complex package had to meet not only the core test of reducing GHG emissions, but also of doing it in a way that is economically efficient – hence the need for flexible incentive-based regulations and other technological measures. It also had to be equitable, especially between developing and developed countries, but also *within* countries. Canada's strategies have proceeded in tandem with these international developments, initially through newly created domestic institutional processes. Many activities with research and technology orientations began in the early 1990s. After the 1997 Kyoto Protocol was negotiated, Prime Minister Chrétien and other first ministers directed the federal, provincial, and territorial ministers of energy and environment "to examine the impacts, costs, and benefits of implementing the Kyoto Protocol, as well as options for addressing climate change" (Environment Canada, 2000, 56). They were to do so under "one of Canada's guiding principles on climate change … that no region should be asked to assume an unreasonable share of the burden for GHG reduction efforts" (ibid., 19).

Thus was implanted at the political level the central importance of a guideline rule regarding equity considerations among Canada's regions, analogous to the larger "equity" debate in the global negotiations among countries. Such a guideline rule was undoubtedly needed in an overall national unity sense, but it was imperative for accommodating Alberta. Alberta was already rhetorically casting Kyoto as "another National Energy Program," alluding to the fiercely opposed interventionist energy policy of the early 1980s under Trudeau. And, of course, Alberta knew that as the province at the heart of the carbon-producing part, albeit not the core carbon-using part of the Canadian economy, it might bear the lion's share of any adjustment. This decision from First Ministers resulted in the establishment in 1998 of the National Climate Change Process (the National Process).

A further science-related issue was that of carbon sinks. Canada sought, and at Bonn in 2001 eventually obtained, agreement for crediting carbon sinks for removing carbon dioxide from the environment. This was based on the fact that plants and trees "breathe in" and store CO_2 from the atmosphere, and thus forests and agricultural soils that absorb and store CO_2 are known as "carbon sinks" under the Kyoto Protocol. Canada wanted credit for enhancing carbon sinks. Credit for carbon sinks became a key negotiating issue on several levels: scientific, moral, political, and economic. The European Union and many NGOs sought to exclude carbon sinks on the grounds that these were not real GHG emission reductions and that, indeed, they were based, in the view of some, on dubious or not "credible science." Canada and other countries with large forests such as Australia, Japan, Russia, and Brazil supported the inclusion of carbon sinks. In the end the Bonn 2001 agreement included them, subject to certain conditions.

International and related Canadian developments resulted in a package of initiatives designed to reduce and lower GHG emissions but ones that would rely more on the development of alternative technologies and would be calibrated to keep Canadian energy competitive in US markets. This included new technologies to raise fuel efficiency in automobiles, new bio-based energy supplies, and carbon capture and storage technologies (Meadowcroft and Langhelle 2011).

The Kyoto Protocol on the UN Framework Convention on Climate Change was agreed to by 178 nations in December 1997 and came into force in 2005. Canada ratified the agreement in 2002, and the Chrétien Liberals undertook to reduce Canada's GHG emissions by 6 percent below its recorded1990 levels. Given weak and even non-existent actions, the gap between the commitment and actual carbon emitted grew to 26 percent by 2005 (Rivers and Jaccard 2009).

The international climate change debate was increasingly centred on whether countries are planning for, or moving toward, a low carbon economy (Helm 2012). The notion of what *low carbon* means and whether it can be achieved is contentious and bound up in both the climate sciences and socio-economic studies. It involves questions about the end goal and how fast such a movement or progression might occur. The speed of the transition to "low(er) carbon" is driven by the power of various established producer interests, and their alliances with governments, and the counter-pressure of energy producer and manufacturing interests and environmental and science lobbies and their alliances with governments.

The international context of climate change policy shifted and broadened further. The 2007 Stern Report in the United Kingdom broadened the debate through its quite extensive, though controversial, assessment of the economics of climate change (Stern 2007). Stern advocated early and significant action, but critics argued that his choice of a vanishingly small discount rate distorted the conclusion; Nordhaus (2007) showed that more conventional discount rates would justify policy-makers taking a more cautious approach.

Canada has had a challenging time finding its appropriate place in this policy space on climate change. Canada originally signed and then ratified the Kyoto Protocol but, as noted above, undertook few measures to realize its commitments to it. Those international parties most concerned about binding targets led the negotiation in 2009 of the Copenhagen Accord, a non-binding agreement whereby various countries pledged to cut GHG emissions. Pledges ranged up to a maximum reduction of 30 percent from 1990 levels (by the EU). Canada pledged to cut emissions by 17 percent from 2005 levels by 2020. The Harper government formally withdrew from the Kyoto Protocol in 2011, arguing that without the United States and China, the agreement is ill-founded. Harper has since asserted that Canada must follow the US lead, but not get out ahead of US policy.

The G8 Summit in Heiligendamm (Germany), in June 2007, managed to ease the US government into a broader acceptance of the science of climate change though not of any hard regulatory approaches. The internal US debate remains polarized by a powerful and well-organized lobby of anti-climate change science and climate change deniers financed by some business-funded entities. In the first Obama Administration, climate change lay dormant because the Administration had other priorities centred on the banking and fiscal crisis. Following President Obama's re-election in 2012, climate change seems to have moved up the Obama priority list. Pressure is coming, in part, through court cases that forced the Environmental Protection Agency (EPA) to act because of a CO_2 endangerment finding by US courts. Since Obama could get no support from the Republican-controlled House of Representatives, he announced in June 2014 a series of measures he would take using his executive powers regarding power plant carbon emission polluters (Revkin 2014).

In 2014 the United States and China agreed on a common approach: Beijing agreed to cap its GHG emissions by 2030, with an intention to try to peak early and to increase its non-fossil fuel use to 20 percent by 2030; and the United States agreed to lower GHG emissions by 26–28 percent from 2005 levels by 2025. This is significant as it is the first time China accepted any limits, and it committed the United States to longer and more sustained action. This agreement shifted attention in Canada and internationally to the planned UN Climate Change Conference, scheduled for December 2015 in Paris.

Broadly similar dynamics characterized the Lima Conference of the parties in early December 2014, with a focus on paving the way for the definitive agreement in Paris in 2015. It had been preceded by the release of the IPCC's 2014 report with IPCC scientists issuing their starkest warning yet calling for fossil fuels to be phased out by 2100 (McGrath 2014). The report stressed that "continued emission of greenhouse gases will cause further warming and long-lasting changes to all components of the climate system, increasing the likelihood of severe, pervasive and irreversible impacts for people and ecosystems" (quoted in Gillis, 2014). The above-mentioned China-US agreement was seen by both countries as a plan that would pressure all nations to make their own emission plans and reductions (Landler 2014). These developments contributed to agreement at the Lima Conference, where for the first time developing and developed nations signed up to publishing national plans for curbing CO_2 emissions but with environmental groups arguing that these plans were still voluntary. Resolution of those differences will await the final definitive Paris 2015 negotiations (Connor 2014).

Other published research is also important in this changing context regarding climate change, including the implications of limiting and keeping global warming to the level of 2° Celsius throughout the twenty-first century. McGlade and Ekins (2015, 1) report that "our results suggest that, globally, a third of oil reserves, half of gas reserves, and 80 percent of current coal reserves should remain unused from 2010 to 2050." Crucially for Canada, oil sands reserves form a large part of the posited "should remain unused" category; so also should oil and gas reserves in the Arctic. Moreover, carbon capture will not be of much use, according to Harrabin (2015).

In Canada, the developments on global climate change and the related US environmental actions by President Obama put pressure on the Harper government heading into the long-scheduled 2015 federal election with promises to improve its dismal record on climate change (see chapter 5 for more discussion).

INTERNATIONAL SOCIO-ECONOMIC STI AGENCIES AND PROCESSES

In addition to those already mentioned above, a range of other international S&T-based agencies have emerged across the decades, with Canada as both a member and participant. These range across several "science in policy" fields such as environment, health, food, transportation, energy, and policing and security, some anchored in the United Nations but others with separate lineage and development paths. In the context of this chapter, we only discuss four examples, but overall they constitute a complex networked array of relationships, some well known and general with others less known and more specialized. They are all underpinned with diverse kinds of S&T in the core institutions that are fed by scientists and technical staff from national member agencies, including Canadian agencies.

A key agency is the International Council for Science (see ICSU 2013). This non-governmental organization based in Paris, with a global membership of 120 national scientific bodies representing 140 countries and with 31 international scientific unions, is mandated to "strengthen international science for the benefit of society" (ibid., 1). It seeks to do this by identifying and addressing major issues of importance to science and society and by facilitating interaction among scientists across all disciplines and from all countries. One recent position statement important for Canada is entitled *Universality of Science in the Polar Regions* (ICSU 2010). While neither a particularly well-known nor well-funded international agency, the ICSU is not atypical of such broad-based institutions that work in the international background in the day-to-day interactions of scientists that develop the evidence that actually or potentially grounds policy and programming.

Another way of seeing Canada's links to international S&T bodies is through the work of the National Research Council of Canada (NRC) and the Canadian Trade Commissioner Service (CTCS). The NRC has been a member of the ICSU since 1931 but also "adheres" to thirty other international science organizations via partnership agreements with the appropriate Canadian learned society and/or agencies (NRC 2013). These include bodies as varied as those for pure and applied chemistry and nutritional sciences and the psychological sciences. In a quite different way in the trade and economic field, the CTCS has numerous "science, technology and innovation international relations" (CTCS 2013). Twelve framework agreements guide trading relationships with countries such as China, India, Japan, and those of the European Union, to mention only a few, some with funding support and some without.

Operating at a higher level and more in relation to international innovation policy is the Paris-based Organization for Economic Cooperation and Development. The OECD Directorate for Science, Technology, and Industry publishes information and league tables on the main S&T performance indicators of its more than thirty members states (OECD 2013). In this regard, it has functioned as a constant reminder to Canada of its continuing weak level of business R&D and its sluggish innovation record overall. This is in spite of relatively good performance in other areas, such as student performance in reading, math, and science and research support for universities, where Canada placed in the top tier of the league tables.

The OECD, as a club of high-income, democratic countries, delivers a complex interplay of public managerial reform reports and discourse (Pal

2012). Its research and reporting on science, including social science, emerges in its work on science in such policy fields as biotechnology, genomics, telecommunications, the Internet, and public health, as well as in the reports on individual countries and their S&T and innovation records, gaps, and challenges. This complex and frequent rhythm of reporting means that governments have to respond publicly but usually only briefly both to new good and bad performance reports, keeping in mind that other national bodies, governments, think tanks, and universities are a part of the public reporting rhythms and agendas in any given period.

International agencies in particular policy fields with major STI content, including the social sciences, play a significant but at times more diffuse role in S&T and innovation. The United Nations Environmental Programme is a useful example. Formed in 1972, the UNEP currently describes its mission as providing "leadership to encourage partnership in caring for the environment by inspiring, informing and enabling nations and peoples to improve their quality of life without compromising that of future generations" (UNEP 2013). While certainly a significant science-based player, there has always been tension between this kind of historic facilitative role and the desire of some to create a "World Environmental Organization," or WEO, with a more comprehensive regulatory role analogous to what the World Trade Organization does for global trade. A WEO has been both advocated and opposed by various environmental interests, scientists, and scholars (Biermann and Bauer 2005).

In the global environmental field, with a history of multiple separate protocols or agreements centred on a particular pollutant and hazard, science and scientists have been central to the initial hazard identification process and to later stages of negotiation for containment actions, many of which are dependent on the mobilization and application of new technologies (Doern 1993). As of 2012, there were over 1,100 multilateral, 1,500 bilateral, and 250 "other" environmental agreements, each with diverse kinds of S&T underpinnings and features (Mitchell 2013). The key in all of this is that the signatories to environmental agreements generally are countries with interests in a given pollutant either because they (the signatories) are a source of the pollution or are adversely impacted by it.

One final international science-based agency to take note of is the World Health Organization. Formed in 1946, by 2015 the WHO had 193 member countries. It is governed through a World Health Assembly (WHO 2013). Its mandate has changed and become more complex, as it now encompasses not only the control of infectious diseases and the promotion of health, but also, since the mid-1990s, it has become an organization focused on improved capacities for tracing and responding to international health emergencies as a security challenge (Weir 2010). Its mandate also involves general efforts to define and conceptualize health as "well-being" (Holland 2007).

The WHO functions with both an internal science capacity and an array of expert panels and advisory committees. These capacities have been brought to bear on global health problems such as AIDS, prescription drugs and diet, nutrition and the prevention of chronic diseases (WHO 2003). Over time these capacities have been increasingly cast as evidence-based policy advice and guidelines and the dissemination of knowledge. Canada's policy discourse and its scientists in various health agencies have been heavily involved in WHO activities and in responding to its policy ideas and guidance.

The role of the World Health Organization in the detection and management of emerging infectious diseases has had a major effect on Canadian public health policy. A contemporary example that highlights the relationship between the WHO and Canada came in December 2013 with the G8 Dementia Summit, which targeted to double international research funding to increase progress in understanding and treating the disease. The summit discussion keyed off health research that showed diagnosed dementia has increased by 22 percent in three years and that forty-four million people worldwide now had the disease, a figure projected to increase to 76 million by 2030 (Cooper 2013).

Canada's experience with SARS (severe acute respiratory syndrome) in 2003 was also a pivotal event. Although alerted by the WHO, Canada was unable to effectively control the spread of the disease, with corresponding highly visible and serious impacts on Toronto. The crisis showed that Canada's capacities for dealing with such emergencies and related aspects of public health needed significant change. In 2004 a Health Canada report by an advisory committee recommended the establishment of a new Public Health Agency of Canada (PHAC), which was immediately established by the federal Liberal government in 2004. The report also identified weaknesses in public health infrastructure, the need to strengthen the role of laboratories, and the need for improved research on infectious diseases in Canada (Health Canada 2004). A decade later, PHAC's 2013–14 report on plans and priorities gave the highest organizational priority to the need for "strengthened public health capacity and science leadership" (PHAC 2013a, 5), albeit in a period of recession and budget cuts.

CONCLUSIONS

This chapter has provided an account of the changing nature of Canada-US and international STI policy and institutions and Canada's role in shaping and responding to them. The seven representative benchmark examples examined in this chapter show growing international influences and imperatives, greater complexity, intermittent success and failure, and intense political conflict as social, economic, science, and political contours have to be interpreted and decided upon in an ever more complex international S&T and innovation policy and institutional system.

We have surveyed some core features of Canada-US S&T and innovation policy relations across the decades being covered. The United States, as a global superpower and as Canada's closest and most important neighbour, but with a complex and different political system, has had diverse impacts on Canada's STI policy overall. These bilateral and other multilateral international dynamics form a key part of the structure and analysis of our eight domains discussed in Part II of this book.

The international social system of science is an essential underpinning to international S&T and innovation policy and contains important historical and contemporary conceptual and empirical debates about the innovation economy and society nexus that are central to our analysis. It functions as a social system of scientific communication and exchange. But we have also seen how its pivotal system of peer review has been challenged to some extent both because of S&T fraud and the complexity resulting from the changing structure of science and social science disciplines, interactions with grey literature, the sheer volume of variously published research articles and studies, and the emergence of the Internet and social media.

Of more recent importance internationally has been the rapid emergence of Big Data, centred on the Internet and social media. We have traced the early definitional challenges of Big Data and the resulting processes – where it is cheered, feared, and almost universally debated. The resulting new types of data science and policy evidence are more correlational than causal. Issues about privacy, the limits of privacy, and the temptations to exploit pools of information have been previewed, as have the policy and regulatory focus on data collection versus data use.

The discussion of the science-based aspects of international trade agreements points to a major development in national and global economic policy. It has generated an explicit, sometimes mandated, and often entrenched S&T policy discourse. The core story of the international science aspects of the climate change debate is a single illustrative case study but it is also a very big, complex, defining global issue across at least thirty years of change and inertia that has yet to be resolved.

Our illustrative discussion of a few selected international agencies and their STI policy mandates, influences, and challenges shows different degrees of relevance and staying power. Nevertheless, all exhibit continuous and diverse forms of STI content, including science *in* policy, policy *for* science, technology impacts, and innovation defined in various ways. Various kinds of economic and social innovation are increasingly a part of their mandates and aspirations. In this context, we now turn in chapter 4 to an exploration of our first federal government S&T domain, namely, the macro S&T and innovation policy domain.

PART TWO

Empirical Analysis of S&T and Innovation Policy Domains

4

The Macro S&T and Innovation Policy Domain

The macro S&T and innovation policy domain largely defines the scale and scope of the federal government's system-wide and system-defining policies, rules, and budgetary decisions regarding this field of policy, governance, and democracy. We presented in chapter 2 some of the building blocks and continuities in the construction of this domain in each of the prime ministerial eras since the late 1960s, including the way S&T policy is reflected in Speeches from the Throne, the quintessential overall agenda-setting event for the federal government and for Parliament. The emergence of offices or ministers of science have also been traced in an initial way. Some features of this domain predate the past forty to fifty years, including the constitutional reality of Canadian federalism and the division of legislative powers between the federal government and the provinces, territorial governments, and Indigenous people's governments. Canada had policies about S&T well before the formal S&T policy era began in the 1960s, particularly in the context of the Second World War and the early post-war era, when bodies such as the National Research Council (NRC) were reframed and played national policy development and research roles (Doern and Levesque 2002).

We use the shorthand term "S&T policy" to characterize this domain in most sections of this chapter, but the emergence of innovation policy and discourse began to be paired with S&T in the 1990s – hence our STI designation, as already previewed and as later sections of this chapter show in more detail.

This chapter focuses attention on change and inertia in the macro S&T and innovation policy domain by examining five policy and governance histories across the past half-century or so: (1) the prime minister, central agencies, and macro S&T policy statements and strategies; (2) budgets, budget processes, and annual tax and spending S&T policy and decisions; (3) regulatory policy, processes, and S&T and related science activities (RSA); (4) policies and guidance regarding science advice and the role of scientists; and (5) periodic public service renewal review processes, with a particular focus on the S&T complement. In the second major section of the chapter we take a closer look at the three elements in the book's analytical framework: (1) policy ideas, discourse, and agendas; (2) economic and social power; and (3) time, temporal realities, and conflicts. Each element is examined for how it helps track change and inertia in the policy domain. Conclusions then follow, including initial notions of the innovation economy and society nexus and the conflicts inherent in this domain.

FIVE DOMAIN POLICY AND GOVERNANCE HISTORIES

Five policy and governance histories across the past fifty years reveal the scale and scope of the domain and its impact on Canada's broader S&T and innovation policy agenda.

1963–2015: The Prime Ministers, Central Agencies, and Macro S&T Policy Statements and Strategies

The first macro domain history centres on the prime minister and the central agencies, in this case the Privy Council Office (PCO) and Prime Minister's Office (PMO), and in later parts of the chapter, the Department of Finance and the Treasury Board. Change and inertia in the macro S&T and innovation policy domain are characterized by considerable but only periodic S&T policy interest in, and interventions by, the political centre of Canadian Cabinet government and democracy. There have been, as shown below, at least thirty-one macro S&T policy statements, strategies, or announced structural changes that have emerged from the centre across the past fifty years. We are interested first in the structure of S&T power in the macro domain, but also in what the S&T statements reveal about S&T and innovation policy content and discourse.

As discussed in chapter 1, policy governance and power on S&T and innovation policy is centred in prime ministerial, Cabinet, and parliamentary government, where Canada's prime ministers have occasionally taken personal interest and pressed for S&T policy as a priority in Speeches from the Throne (SFTs) or, more often, ignored it as a priority area. As shown in chapter 2, the story for the Trudeau era shows an initial low-key S&T policy presence in 1970, 1972, and 1973, and again in four later Throne Speeches, including the one in 1980 and the one in 1984 – and it was not even mentioned in five Throne Speeches, most of them in the mid- to late 1970s. The account for the Mulroney era revealed two Throne Speeches that had no S&T content and three that offered middle-level attention, particularly the ones in 1986 and 1989. The Chrétien-Martin era overall had the strongest overt S&T and innovation policy mentions in all of its Speeches from the Throne. Its discourse was also caught up in innovation policy language and in frequent reference to the Information Highway. In contrast, the Harper era to 2015 had no S&T mentions in three of its Throne Speeches but considerable reference in three others, including mentions of a sustained focus on Arctic research and clean energy technologies.

The structure of S&T power is further revealed by the usually peripheral nature and role of nominal science ministers whose creation, existence, and varied restructurings are under prime ministerial remit. From the outset ministers of science (with varied titles) in a Cabinet of thirty-five to forty ministers have always been junior ministers on the periphery of power. As mentioned in the introduction to this book, there have been thirty-two such science ministers but their tenure in office averages less than one year each. The first specifically S&T purposed Cabinet-level ministry was the Ministry of State for Science and Technology (MOSST), headed by a minister but with a very small staff and no control over S&T budgets, which made it necessarily reliant on the use of exhortation and persuasion on other ministers and on the public service bureaucracy (Doern 1972). Science ministers, of course, do foster relationships, consultation, and engagement with different changing parts of the S&T community, but for the most part they have been minor and very low-profile players in the Cabinet governance and federal power structure (Dufour 2014).

The overriding reason for this limited role is that several other ministers are de facto S&T ministers because their departmental and agency structures and mandates are significantly S&T-based (see chapter 5 for detailed discussion). MOSST disappeared in the Mulroney era, and S&T policy was strengthened somewhat as it found its way into the name and mandate of Industry, Science and Technology Canada. Although not a macro-economic player in the same league as the Department of Finance, the new ISTC was without doubt more meso-level in its reach and influence compared with MOSST (Doern 1996). This period of shaping and reshaping the nominal leadership illustrates that the S&T policy function was both easily moveable and easily retrievable.

In the present Cabinet structure, at a minimum, ministers and departments such as Industry Canada (IC, the successor to ISTC), Environment Canada (EC), Natural Resources Canada (NRCan), the Department of Fisheries and Oceans (DFO), and Agriculture and Agri-Food Canada (AAFC) all have significant S&T policy mandates vis-à-vis their sectors and industries and vis-à-vis any nominal overall science minister (Clancy 2011; Doern and Kinder 2007; Doern and Reed 2000; Dufour and Gingras 1988).

In pivotal ways, prime ministers are also S&T and innovation policy ministers, through actions and priorities that either support, discourage, or sideline S&T on national agendas. Each prime minister covered in this book has made at least one major speech as new S&T policies were announced. However, the views at the centre are also revealed by the nature and content of actual formal S&T and innovation policy statements and strategies (Dufour 1994). Table 4.1 reveals the pace of the emergence of such statements or strategies (twenty-two in total), their overall discourse, and content, and the timing and number of structural changes to overall macro S&T policy advisory councils and agencies (nine in total). We take note of several key features of this part of the macro S&T and innovation policy story.

First and foremost, a reading of these various policy statements shows the consistent finding that Canada's R&D by business has always been weak and sub-par when compared with other countries, and hence S&T statements have always sought to close this gap but without success (Hayes 1973; Britton and Gilmour 1978; Voyer 1999; Phillips and Castle 2010, 2012; Smarden 2015). Second, especially in the past thirty years, Canada's university research has been rated comparatively highly, both in terms of spending and in terms of some measures of quality and quantity of output (Doern and Stoney 2009; STIC 2012). A third feature is that despite the changes in the relevant literature focusing on innovation traced in chapter 1, S&T is still the main descriptor of most such statements, with innovation only in two policy statement titles in recent years. The many S&T statements, of course, discuss innovation in their detailed discussion but often only very generally.

Also important in an overall way is the fact that S&T and innovation policy statements have testified to the importance of government communications strategies and oversight. This, of course, refers to the overall discourse used, but also what is mentioned with emphasis and what is downplayed or not mentioned at all. For example, strategies in the Liberal era sought to show continuity of focus and good results across the

Table 4.1 Macro S&T policy statements and related advisory structures, 1963–2015

1963	Glassco Commission Report
1964	C.J. Mackenzie Report and establishment of Science Secretariat in PCO
1970	Senate Special Committee on Science Policy (Lamontagne Committee)
1971	Ministry of State for Science and Technology (MOSST) established
1973	Federal make-or-buy science policy
1983	MOSST, *Towards 1990: Technology Development for Canada*
1985	MOSST, *A National Science and Technology Policy for Camada*
1987	Formation of National Advisory Board on Science and Technology (NABST), chaired by prime minister, Feb.
1987	National Science and Technology Policy, signed by federal, provincial, and territorial science ministers, Mar.
1987	Council of S&T Ministers (federal-provincial-territorial) established
1987	Discussion paper on Canada's R&D Effort, Council of S&T Ministers to the Annual Conference of First Ministers, 26–7 Nov.
1987	National Forum of Science and Technology Advisory Councils
1987	Formation of Industry, Science and Technology Canada (ISTC)
1992	Wind up of Science Council of Canada
1994	S&T Review announced and consultation paper released, *Building a Federal Science and Technology Strategy*
1994	Auditor General Report to Parliament on S&T
1995	Internal S&T Policy Review by departments
1995	NABST report, *Healthy, Wealthy and Wise: A Framework for an Integrated Federal Science and Technology Strategy*
1996	*Science and Technology for a New Century* paper
1996	Advisory Council on Science and Technology established (to replace NABST)
1997	*Minding Our Future: A Report on Federal Science and Technology*
2000	Council of Science and Technology Advisors (CSTA), *A Framework for Science and Technology Advice*
2002	Industry Canada, *Achieving Excellence: Investing in People, Knowledge and Opportunity* and *Knowledge Matters: Skills and Training for Canadians Canada's Innovation Strategy*
2003	Industry Canada, *Federal Science and Technology: The Pursuit of Excellence* report.
2004	Office of the National Science Advisor (ONSA) established
2007	*Mobilizing Science and Technology to Canada's Advantage*
2008	Closing of ONSA, CSTA, and Canadian Biotechnology Advisory Committee
2010	Science, Technology and Innovation Council (STIC) *State of the Nation* reports
2011	Report of Expert Panel on Federal Support to R&D (Jenkins Report)
2013–14	Federal S&T consultation paper, *Seizing Canada's Moment: Moving Forward in Science, Technology and Innovation*
2014	S&T and innovation strategy: *Seizing Canada's Moment*

Sources: These policies are discussed and draw from Doern 1972; Dufour 1994; Brassard 1996; and Doern and Kinder 2007; see also references in those books.

Chrétien years but did not want to show the major cuts in S&T spending during Program Review. So have the Harper era's two S&T statements, where progress on key policy themes and programs is shown, but nothing about cuts in the recession era or about the real or alleged muzzling of scientists.

S&T policy is also revealed in the varied new advisory councils where the nomenclature is again S&T or just "science," involving successively the Science Council of Canada (SCC), National Advisory Board on Science and Technology (NABST), the Advisory Council on Science and Technology (ACST), and now the Science, Technology and Innovation Council (STIC), with only the last of these containing innovation in its name (Kinder 2010). The advisory councils, of course, were also frequently reporting publically and advising on any number of general or particular S&T and innovation matters with reports not necessarily aimed at the minister of science or the prime minister but instead at broader audiences in the research community. This meant that the reports and advice were still directed at elite audiences rather than at any kind of mass or general audience.

In terms of other science policy institutions at the centre, only two have been put in place, albeit for limited periods. A Science Secretariat was lodged in the PCO in the 1970s (Doern 1972) and the Office of the National Science Advisor (ONSA) operated from 2004 to 2008 (Hong 2008; Goar 2008). The latter office was headed by Dr Arthur Carty, who brought considerable science and policy capacity from his previous job as head of the National Research Council (see chapter 5). Upon Carty's retirement the office was closed and replaced by the arm's-length STIC, which provides biannual public "state of the nation" reports but is also the only advisory body with a mandate that includes the provision of "confidential" advice including on issues or "charges" referred to it by the government (STIC 2014).

It is difficult, however, to determine how much direct access these offices had to prime ministers. They were "at" the centre but not necessarily "in" the centre or in the centre of the centre. Clerks of the Privy Council as managers of the PCO and permanent heads of the public service clearly have considerable access, but they also have varied interests in, and knowledge of, S&T and innovation matters. In the 2006–09 period, Kevin Lynch brought considerable personal knowledge to the PCO and the Clerk's job, including his experience as a senior official in S&T and innovation policy within Industry Canada from 1992 to 2000. Apart from Lynch, no Clerks have exhibited great knowledge, experience, or interest in the S&T and innovation agenda.

1968–2015: Budgets, Budget Processes, and Annual S&T Policy

Closely aligned in this portfolio and the power equation is the linked role of ministers of finance and their role in managing macro-economic fiscal, economic, and budgetary policy. As mentioned in chapter 2, Paul Martin as Jean Chrétien's minister of finance was pivotal as a de facto S&T and innovation minister, both for good and for ill (Kinder 2010; Doern and Stoney 2009).

Budget Speeches are annual in nature, typically in February or March of each year. Like Throne Speeches they are thematically focused, usually on the macro-economy, and they contain numerous specific tax and expenditure measures – new, altered, or cancelled (Doern 2009). These can be large, medium, or small in size, sometimes only gestures, and can be applied for diverse time periods (e.g., small amounts for a proscribed period, often of only a year, or very big amounts for time commitments of seven to ten years and thus likely well beyond the current government's tenure in office). Most are traceable, but some disappear into the fiscal mists.

Budget Speeches and budget formation are preceded by often quite elaborate and extensive pre-budget consultations but also by strictures regarding budget secrecy that are designed to ensure that no player or interested party can profit from prior knowledge of specific new fiscal measures, especially new tax measures. But within the intra-Cabinet and senior bureaucracy, the successful appearance of S&T and innovation fiscal initiatives also depends on which ministers or departments can gain the attention of the minister of finance (and the prime minister) and convince them of the value (politically, economically, or socially) of their preferred S&T and innovation measures.

Budget Speeches are usually much longer and more detailed than Speeches from the Throne, and usually they include a relatively greater number of S&T and related measures in budgets on both the tax and spending sides of the fiscal equation than would surface in Throne Speeches. But these are typically also listed as secondary or tertiary measures in support of the macro-economic orientation of the overall budget (Doern, Maslove, and Prince 2013).

Table 4.2 shows S&T and innovation priorities and agendas as reflected in forty-nine Budget Speeches across the four main prime ministerial eras from 1968 to 2015. As is the case with Throne Speeches, S&T themes and initiatives are rarely shown as highly ranked items in most Budget Speeches. Rather, they tend to be expressed in the body of the Speeches, sometimes

Table 4.2 S&T and innovation policy as expressed priority in budget speeches in the Trudeau, Mulroney, Chrétien-Martin, and Harper eras

TRUDEAU BUDGET SPEECHES (18), 1968–1984

- No direct mentions/items in 13 of 18 Budget Speeches
- 1970 (12 Mar.) Budget: commitment to "improvement of technology in industry" and $40 million to Atomic Energy of Canada Ltd (AECL)
- 1972 (8 May) Budget: provision for scientific research equipment for use in testing new products to be exempt from sales tax
- 1978 (10 Apr.) Budget: R&D support measures in form of tax write-offs and investment tax credit
- 1978 (16 Nov.) Budget: doubling of investment tax credit for R&D from 5% to 10% (20% in Atlantic Canada and Gaspe); R&D tax credit of 25% for small businesses
- 1983 (19 Apr.) Budget: $290 million (m) for research and training facilities: $180 m on high tech procurement; $100 m in areas of new technology; allow R&D companies to transfer the value of tax incentives to outside investors

MULRONEY BUDGET SPEECHES (9), 1985–1993

- No direct mentions in 3 of 9 Budget Speeches
- 1985 (29 May) Budget: R&D support for small companies by refunding all of the tax credit they earn on their first $2 m of qualifying R&D each year; changes in the definition of qualifying R&D; reductions in nuclear power system research
- 1986 (26 Feb.) Budget: R&D theme of "investing in the future" and "investing in knowledge" including $300 m for granting councils over 4 years
- 1990 (20 Feb.) Budget: S&T programs "constrained to 5% annual growth for 2 years"
- 1991 (26 Feb.) Budget: S&T programs "will grow at 3% per year"
- 1992 (25 Feb.) Budget: R&D support to increase by $230 m over 5 years; but government to wind-up SCC, the Economic Council of Canada (ECC), and the Law Reform Commission
- 1993 (26 Mar.) Budget: new rules to streamline tax treatment of corporate R&D, enriching it by $250 m

CHRÉTIEN-MARTIN BUDGET SPEECHES (11), 1994–2005

- No direct mentions in only one Budget Speech
- 1994 (22 Feb.) Budget: Seeks "a Canada that leads in technology rather than leaning on the technology of others" and that "makes innovation a more effective engine of economic growth." Support for satellite technology and a New Technology Partnership Program for universities and government labs; strategy for developing Information Highway, increased funding for NRC; strategy for growth of environmental technology industries
- 1995 (27 Feb.) Budget: restrictions announced on Scientific Research and Experimental Development Tax Credit (SR&ED)
- 1996 (6 Mar.) Budget: departmental spending cuts, including on S&T, of 20% on average but higher in some, including Environment Canada. But technology and innovation funding increases by reallocation $270 m from budget savings; Technology Partnerships Canada established ($150–$250 m); SchoolNet program expanded with Community Access Program (CAP) component for Internet services; development of Information Highway with affordable access and Canadian presence
- 1997 (18 Feb.) Budget: Canada Foundation for Innovation (CFI) funded with $800 m; also Networks of Centres of Excellence (NCE) program

- 1998 (24 Feb.) Budget: $400 m increase for granting councils; $205 m for SchoolNet and CAP and $55 m for Canadian Network for Advanced Research in Industry and Education (CANARIE)
- 1999 (Feb.) Budget: new health-related research and innovation fund; $200 m more for CFI; and $176 m for granting councils
- 2000 (28 Feb.) Budget: innovation economy theme; $900 m over 5 years to create 2,000 Canada Research Chairs (CRC), $900 m more for CFI, $160 m for Genome Canada, $700 m for environmental technology, and $90 m for biotechnology regulation
- 2003 (18 Feb.) Budget: theme of "strengthening research and innovation"; $125 m for granting councils; funding for indirect costs of granting bodies by universities; $10 m more for NRC
- 2004 (23 Mar.) Budget (Martin as PM): theme of "importance of learning" and also knowledge and communities; $90 m for granting councils, $60 m for Genome Canada
- 2005 (23 Feb.) Budget: theme of "investing in ideas and enabling technology"; $375 m for granting councils, $165 m for Genome Canada, $126 m for UBC TRIUMF science facility

HARPER BUDGET SPEECHES (11), 2006–2015

- No direct mentions in only one Budget Speech (its first one)
- 2007 (19 Mar.) Budget: theme of the "knowledge advantage"; $800 m annually for provinces to strengthen quality and competitiveness of universities, $510 m for CFI, $350 m for new Centres of Excellence for Commercialization and Research (CECR), $100 m for Genome Canada
- 2008 (27 Feb.) Budget: $80 m for granting councils, $140 m for Genome Canada, $250 m for auto sector R&D
- 2009 (27 Jan.) Budget: theme of "investing in knowledge infrastructure"; $750 m for CFI, $50 m for University of Waterloo's Institute for Quantum Computing, $87 m for Arctic research facility, $500 m for Canada Health Infoway; extended broadband coverage
- 2010 (4 Mar.) Budget: theme of "economic growth through innovation"; $222 m for TRIUMF, $75 m for Genome Canada, $125 m for NRC clusters, $48 m for R&D in medical isotopes
- 2011 (22 Mar.) Budget: tied to federal Action Plan; $80 m for Industrial Research Assistance Program (IRAP); 10 new research Canada Excellence Research Chairs; $47 m for granting councils; digital economy strategy and funding; Canada Brain Research Fund; Perimeter Institute funds
- 2011 (6 June) Budget: post-election, essentially same budget as 22 Mar. budget
- 2012 (29 Mar.) Budget: theme of "supporting entrepreneurs, innovation, and world-class research"; increased funding for R&D in small business; NRC to refocus on business; plans to streamline the SR&ED; increased support for granting councils and Genome Canada
- 2013 (21 Mar.) Budget: again, world-class research theme; $12 m for community colleges, $165 m for Genome Canada, $225 m for CFI, $121 m for NRC, $325 m over 8 years to Sustainable Development Technology Canada (SDTC), $100 m to enhance Canadian venture capital system
- 2014 (11 Feb.) Budget: creating new Canada First Research Excellence Fund, $1.5 billion (b) over next decade; additional $500 m over 2 years to the Automotive Innovation Fund to support new strategic R&D projects
- 2015 (21 Apr.) Budget: tied "investing in advanced research" to major budget section on "supporting the manufacturing sector"; $1.33 b for CFI, $105 m over 5 years for CANARIE, $46 m per year for granting councils, $243.5 m for Canada's participation in the international Thirty Meter Telescope, $119 m over 2 years to NRC

themed separately but more often just quickly mentioned and listed as a general measure on the tax and spending side. Specific tax measures often take the form of accelerated capital cost allowances, some of which are intended to support particular industries or industrial categories such as small business or business activity in particular regions or sectors.

In terms of S&T and innovation policy mentions in Budget Speeches, the Trudeau era saw the weakest record in that in two-thirds of Budget Speeches S&T was not mentioned. One probably should put this into global context, where S&T and innovation were just emerging as a priority and seldom were the focus of major announced effort. The Mulroney era saw S&T initiatives in six of its nine Budget Speeches. In both the Chrétien-Martin and Harper eras, S&T items were present in virtually all of their Budget Speeches. In only one Budget Speech, the Chrétien budget of 2000, was S&T and innovation a leading and defining theme, but this was also at the peak of the budgetary surpluses during the Chrétien era where there was money to spend. For large parts of the Trudeau, Mulroney, and Harper eras, large fiscal deficits were a central feature for budgets.

In more specific terms, the S&T and innovation initiatives that did emerge in budgets reflected the changing state of play about S&T policy ideas, albeit couched in particular kinds of tax and spending constraints and preferred tax instruments. The hesitant Trudeau agenda focused on the "improvement of technology in industry," supporting equipment used in testing new products and overall R&D, particularly in small businesses. The most significant measure was disappointing: the 1983 S&T budget allowed R&D companies to transfer R&D tax incentives to outside investors and was gamed, leading to major revenue losses with little measureable change in performance. Although largely unsuccessful, this failed exercise eventually led to the Scientific Research and Experimental Development Tax Credit (SR&ED) program (see chapter 7).

Budgets in the Mulroney era began with the replacement of the refundable R&D tax incentive and the introduction of the SR&ED focused on supporting business, albeit couched in the room for manoeuvrability left by or set by recession and fiscal deficits. The budget was also the first effort to cast S&T policy in terms of investing in knowledge, which in this case came in the form of increased support for the granting councils and for university research. Apart from this spark of policy innovation, most of the rest of the Mulroney S&T programming was iterative and minor.

Budgets in the Chrétien-Martin era began with themes that Canada needed to lead in technology rather than leaning on the technology of others and to make innovation a more effective engine of economic development. Technology and innovation funding emerged in early budgets from the "savings" of quite severe program cuts in departmental budgets. Frequent mentions of the Internet or Information Highway and related smallish funding initiatives followed.

After budgetary surpluses emerged in 1997, the innovation focus became central to S&T budgetary strategy until the end of the Chrétien-Martin era. This involved a combination of renewed spending through the granting councils and new forms of partnered funding, both with the universities and beyond. Rather than simply replacing the cuts to postsecondary education, the Chrétien-Martin government introduced more strategic direction to the new funds. First, universities were required to self-identify areas of pre-eminence or emerging strength in order to gain access to funding for research chairs, through the Canada Research Chairs (CRC) program, Industry Research Chairs, and more recently the Canada Excellence Research Chairs (CERC); infrastructure, via the Canada Foundation for Innovation (CFI); and network funding for research, for example, through the Network of Centres of Excellence (NCE) programs and Genome Canada. Each institution was allowed up to six themed areas.

Henceforth, applications would be matched to areas of strength rather than simply spread about in a formulaic way, either based on share of an area or on some notion of industrial and regional benefits. Second, increasingly, the new funds were allocated based on international peer review, rather than simply national review. The combination of these two measures opened up the distribution of funds, in many ways consolidating much of the research investments in the nation's top-ranked institutions, helping raise their global competitiveness, while still enabling other lower-ranked provincial universities to compete and win support for their research strengths. Many of the recent national success stories in research can be traced back to these investments and these decisions.

Budgets in the Harper era spoke of the need to strengthen the quality and competitiveness of universities and the need for much greater commercialization of research. But there was little change from the decision structures and institutions created in the Chrétien-Martin era. This involved support for the councils, research chairs programs, NCES, CFI, and Genome Canada. It also included support for the TRIUMF project, an initiative established in the Pearson Liberal era (Doern 1972). The key shift under Harper was a move to begin initiatives on innovation and commercialization, including support for venture capital funding and a considerable penchant to announce new capital, big-science, and targeted measures such as the Automotive Innovation Fund (AIF), the

University of Waterloo's Institute for Quantum Computing, and frequent mention of an Arctic research facility – culminating in the announcement in 2015 of the establishment of Polar Knowledge Canada (2015). This greater overall focus on commercialization led to what some view as radical changes at the National Research Council, with the elimination of institutes and the introduction of research platforms and partnerships, which were intended to link the upstream scientific capacity with industrial and business needs and opportunities. Given the dominant focus on the recession and on deficit reduction in the Harper government's Action Plan, and its focus on infrastructure and stimulus spending, some important but little-heralded innovations and new technology investments were embedded in other support measures that span specific sectors and regions of the economy.

1980–2015: Regulation, Regulatory Processes, and S&T and Related Science Activities

Central to the macro S&T and innovation policy domain are the dynamics of regulation making and regulatory processes, including rules regarding implementation and enforcement as they relate to both S&T and related science activities and hence to the personal knowledge, technical expertise, and advice of scientists. The Treasury Board Secretariat (TBS) is the central agency responsible for regulatory affairs, but most federal departments and independent agencies and their varied science staffs and capacities, including those of the social sciences, have a role to play. Regulatory policy and governance practices are based on a series of TBS and other documents and guidelines (Doern, Prince, and Schultz 2014). The basic dynamics are also influenced in different direct or indirect ways by developments in the macro regulatory regimes of other countries and by the OECD, which has been exhorting and advising about such regimes over the past two decades (Jakobi 2012; Pal 2012).

A key contrast and difference between the regulatory process and the budgetary process is that the latter has, as set out above, a quite formal and elaborate agenda process leading to the Budget Speech, while there is no such government-wide and public process for regulation making, for developing a formal regulatory agenda, or for delivering a "regulatory agenda speech" (Doern 2007; see further discussion below regarding departmental or agency regulatory plans).

Table 4.3 sets out the key features of the regulatory policy and governance system, each of which has real and potential S&T and RSA dimensions and dynamics. We focus here on the Cabinet Directive on Regulatory

Table 4.3 Key features of regulatory policy governance

- Cabinet Directive on Regulatory Management (CDRM)
- Statutory Instruments Act
- *Canada Gazette* consultation processes
- Constitutional and quasi-constitutional provisions
  - Federal-provincial constitutional division of powers
  - Charter of Rights and Freedoms
  - Canadian and international human rights laws
  - Major UN conventions and protocols
  - International trade agreements
  - Agreement on Internal Trade (AIT)
- Oversight and challenge functions
  - Prime minister, ministers, and Cabinet committees
  - Minister of justice
  - Central agencies (especially Treasury Board Secretariat)
  - Parliament
  - The courts

Management (CDRM), in effect since 2012. Under the CDRM, the federal government commits when regulating that it will (TBS 2013, 2):

1 Protect and advance the public interest in health, safety, and security, the quality of the environment, and the social and economic well-being of Canadians, as expressed by Parliament in legislation.
2 Advance the efficiency and effectiveness of regulation by ascertaining that the benefits of regulation justify the costs, by focusing human and financial resources where they can do the most good, and by demonstrating tangible results.
3 Make decisions based on evidence and on the best available knowledge and science in Canada and worldwide, while recognizing that the application of precaution may be necessary when there is an absence of full scientific certainty and a risk of serious or irreversible harm.
4 Promote a fair and competitive market economy that encourages entrepreneurship, investment, and innovation.
5 Monitor and control the administrative burden (i.e., red tape) of regulations on business and be sensitive to the burden that regulations place on small business.
6 Create accessible, understandable, and responsive regulation through engagement, transparency, accountability, and public scrutiny.
7 Require timeliness, policy coherence, and minimal duplication throughout the regulatory process by consulting, coordinating, and cooperating across the federal government, with other governments and jurisdictions in Canada and abroad, and with business and Canadians.

These seven expressed commitments in the current CDRM, developed and approved by the Harper government in 2012, are much more elaborate than the two commitments in the previous 2007 Harper government's Cabinet Directive on Streamlining Regulation (CDSR; TBS 2007, 1) which were to:

- Protect and advance the public interest in health, safety, and security, the quality of the environment, and social and economic well-being of Canadians as expressed by Parliament in legislation.
- Promote a fair and competitive market economy that encourages entrepreneurship, investment, and innovation.

The seven CDRM commitments suggest somewhat more "managerial" content, but at their core, they still set out policies and values about regulation and, hence, norms about rule making and compliance. Evidence, knowledge, science, and precaution in the face of lack of absolute scientific certainty are mentioned in that order in Item 3 of the CDRM but so also are potentially conflicting notions of regulations being understandable and timely (in Items 6 and 7).

The CDRM also contains provisions whereby departments and agencies must prepare annual regulatory plans and priorities and practise a life-cycle approach to regulation. This includes managing the life cycle of the measure, including provisions requiring re-evaluations of all regulations be conducted every five years, a policy that built on earlier spending program evaluation requirements. But broader notions of the life-cycle approach involve, as discussed further below, approaches that extend regulation beyond the pre-market approval stage for products to post-market monitoring. The CDRM encourages harmonization and mutual recognition initiatives (both federal-provincial and international) and advances the principle of regulating only if there is evidence that regulation is necessary and beneficial, confirmed through the greater use of quantitative cost-benefit analysis. Each of these has implications for the presence or absence of S&T in regulation. The CDRM overall provides considerable room for innovation society values, but they remain hedged in by typically more frequently ranked innovation economy norms and imperatives.

The CDRM also elevated into the macro regulatory framework policy the "one for one" rule, whereby departments and agencies must control the number of regulations by "repealing at least one existing regulation every time a new one imposes an administrative burden (i.e., red tape) on business is introduced" (TBS 2013, 9). There are, however, "carve outs" or exceptions to this policy, including regulations related to tax or tax administration and in situations of emergencies and crises.

The macro regulatory system also includes an eleven-step process for developing a Governor-in-Council regulation that begins with planning, then analysis, and ends, after Governor-in-Council approval (or rejection), with review by the Parliamentary Standing Joint Committee for the Scrutiny of Regulations (TBS 2011). These steps in total constitute the front end of the development phase for new regulations (delegated laws).

Regulatory proposals can form part of the basic system of Memoranda to Cabinet (MCs) and thus be examined and debated within different Cabinet committees. As a pillar of the Cabinet's regulatory oversight system, the Treasury Board's review focuses, mainly through its Regulatory Affairs Sector (RAS), on the new specific individual regulatory proposals; the aim is to ensure that regulatory proposals conform to the CDSR, the Statutory Instruments Act, and the *Canada Gazette* process. The Treasury Board also ensures that required Regulatory Impact Analysis Statements (RIAS) are carried out by departments and agencies, as well as the Treasury Board's early stage Triage Statements intended to differentiate regulations that have medium and high impact from those that have low to medium impact (TBS 2012).

These approval and challenge roles of the Treasury Board Secretariat are complemented by advisory, educational, and facilitating roles, mainly through its Centre of Regulatory Expertise (CORE), a small six-person unit designed to help in implementing the CDRM. Its annual planning process, based on discussions with departments, seeks to identify and improve analytical capacity, with attention also being paid to the different challenges faced by small versus larger departments. The Regulatory Affairs Sector unit itself is also fairly small in terms of staff (just over thirty professional analysts).

The role of this TBS central agency challenge function is not to second-guess every regulatory proposal or the science and evidence base underpinning it, given that numerous regulations (new or amendments, the latter being as high as 80 percent of the total) proceed through the system each year. On regulatory proposals with expected high impacts, it offers a somewhat limited scrutiny, with an emphasis on early advice, watchdog observations, commentary, and guidance, including on the quality of regulatory analysis.

One of the correlated problematical issues relates to exactly how many new regulations are approved each year. The federal government does not publish such annual data. The only occasions when it has revealed such information is when the OECD is doing a country regulatory study on Canada. The last of these overall country reports was in 2002, when Canada was advancing nearly two thousand new regulatory approvals annually (OECD 2002; Doern 2007). These data do not include rules within new legislation nor do they necessarily do a good job of differentiating major rules from minor amendments. It is probably worth mentioning that federal rules are only the tip of the iceberg, as provinces and municipalities implement many of the regulations and policies that influence S&T and innovation policy (Atkinson et al. 2013).

The more limited or constrained notion of a challenge function (with S&T implications) is the product of two realities of modern Cabinet government and delegated bureaucratic administration. The first is that regulatory departments and agencies overwhelmingly have the main substantive regulatory S&T and RSA expertise and familiarity with their regulatory clientele and, moreover, Cabinet government functions on large amounts of delegation in regulatory development and the crafting of proposals. The second, and equally important, reinforcing reality is that most regulatory proposals depend on science-based and evidence-based capacity and capacities for risk assessment and risk management within and across departments (Doern and Kinder 2007; Leiss 2000; Doern and Reed 2000; Hood et al. 2001). Central challenge function entities (such as TBS and the Cabinet) have some needed expertise and important perspectives of their own, but in the total scheme of things and in the context of a large Cabinet of thirty-five to forty members and extended amounts of legal and policy delegation, there is little doubt of where the preponderance of S&T expertise lies. When it is not clear where the power resides, special inquiries and studies, such as on epidemics or hazards such as severe acute respiratory syndrome (SARS), bovine spongiform encephalopathy (BSE, or mad cow disease), and listeriosis, have been developed and led to regulatory change and/or new regulatory agencies being formed (Doern and Reed 2000; Doern 2010).

Finally, there is the expressed goal of adopting life-cycle approaches to regulatory oversight, involving a movement *from just pre-market assessment to greater post-market monitoring* of products. Because it deals with regulatory implementation in the form of product assessment, it is much more in the hands of departmental and arm's-length regulators than under the control of regulatory policy and governance ministers and officials at the centre. Moreover, life-cycle assessment is a set of monitoring activities that requires networks of co-participants in business, non-governmental organizations (NGOs), and among diverse science and health professions, patients, caregivers, and users, and thus a broader innovation economy and society nexus of players. We referred in chapter 2 to Health Canada's life-cycle approach to regulating in 2006–07 under the Harper Conservatives, but it has a longer lineage in environmental regulatory policy with regard to environmental assessment of projects and policies and in the regulation of nuclear reactors and the long-term storage of nuclear wastes. Federal environmental assessment policy focuses on assessing proposed projects that involve federal laws or funding, often imposing requirements for life-cycle regimes to follow projects through time, including final stages such as the closing of a mining operation or a manufacturing plant.

A report by the National Round Table on the Environment and the Economy (NRTEE 2012) stressed the great value of life-cycle approaches to fostering sustainable economic development in Canada but pointed to many practical obstacles along the way, including incomplete conceptual understanding, complexity, and serious gaps in science and front-line science-based regulatory and policy capacity. Concepts of sustainable development (SD) and the precautionary principle are implied in life-cycle notions and aspirations, although these norms are not always realized (Bregha 2011; CEAA 2010).

1990–2015: Policies and Guidance Regarding Science Advice and the Role of Scientists

Although the overall macro issue of science advice and the role of scientists has been present since the dawn of Canada's S&T policy history, it has taken on a higher profile in two periods: the late 1990s and early 2000s under the Chrétien Liberals and from 2006 to the present in the Harper Conservative era. We examine it here as a macro domain issue but so-called science-policy interfaces (SPI) occur throughout the federal government horizontally and vertically and vis-à-vis external interests and participants (Saner 2007, 2014).

Concerns about policy for science advice emerged quite sharply and visibly in the late 1990s in the wake of public controversies such as the safety of blood supplies, declining fish stocks, aspects of nuclear reactor safety, and biotechnology (Enros 2013; Public Policy Forum 1998; Doern 1999). But they also flowed from the cuts in federal spending on science in the 1990s, from focused debates on the general state of science in government, and from the concerns about newer partnership-centred ways of linking government science with university and industrial science. These concerns are not unique to Canada; they are global and international, with most countries and jurisdictions having to deal with similar problems and challenges. International controversies such as BSE, foot and mouth disease, stem cell research, biotechnology, nuclear power, disaster management, and general debates about the use of the precautionary principle have made science advice a global issue. These concerns and debates have raised concerns about ethical issues in science advice and policy-making as a whole and have highlighted all the things that can go wrong. At one level, the concern is about learning from our experiences and sharing best practices so that mistakes are not repeated or at least are minimized.

The Chrétien government responded to these challenges and concerns in several ways, including the formation of the Council of Science and Technology Advisors and the publication in May 1999 of the report entitled *Science Advice for Government Effectiveness*, or *SAGE* report (CSTA 1999). The report advocated the adoption of the six *SAGE* principles set out in Table 4.4. The detailed subsidiary guidelines under each principle are not shown here.

It is worth pointing out here that a key reality of science advice in policy- and decision-making is that, in general, the further up the hierarchy of the federal government one goes the fewer the number of scientists there are (Doern 1999). Even basic scientific and technological literacy might not reside near the top in any consistent fashion. This is viewed by many as an institutional weakness, especially in the context of a knowledge-based and innovation economy. However, in other respects, this situation is simply a reflection of the fact that the upper levels of government must and should embrace a broader level of values and points of view as more diverse trade-offs are confronted in the formation of public policy. Winston Churchill, who made great use of science and scientific advisers in the Second Wold War, was often quoted as saying scientists should be "on tap and not on top"; see Porter and Phillips (2007) for a detailed discussion of the role of science in liberal democracy.

Another key distinction to note here is the difference between transmitting scientific advice and knowledge *in writing versus in person or verbally*. At some stages, a paper (or electronic) trail of written knowledge and advice is the norm and includes memos, reports, and Cabinet documents. Having said this, it is also true that as decisions move to the top of the federal government, in-person and verbal summaries of scientific advice, indeed, verbal "summaries of summaries of summaries,"

Table 4.4 Federal framework on S&T advice in federal policy and decision making

Principle I: Early Issue Identification. The government needs to anticipate, as early as possible, those issues for which science advice will be required, in order to facilitate timely and informed decision making.
Principle II: Inclusiveness. Advice should be drawn from a variety of scientific sources and from experts in relevant disciplines, in order to capture the full diversity of scientific schools of thought and opinion.
Principle III: Sound Science and Science Advice. The government should employ measures to ensure the quality, integrity, and objectivity of the science and science advice it uses, and ensure that science advice is considered in decision making.
Principle IV: Uncertainty and Risk. Science in public policy always contains uncertainty that must be assessed, communicated, and managed. Government should develop a risk management framework that includes guidance on how and when precautionary approaches should be applied.
Principle V: Transparency and Openness. The government is expected to employ decision-making processes that are open, as well as transparent, to stakeholders and the public.
Principle VI: Review. Subsequent review of science-based decisions is required to determine whether recent advances in scientific knowledge have an impact on the science advice used to reach the decision.

become the norm. Although Cabinet and detailed background documents are carefully prepared and read, there is also the world of "slide-decks," "power points," "tweets," and person-to-person meetings and briefings in hallways or in taxi cabs as officials rush from one meeting to another. Science advice in policy and decision making must fit into these overarching realities. Some assert the complexity of science is ill-suited to these venues.

A further important reason for thinking about science advice at key stages of policy and decision making is that it raises the broader concern about the duties and responsibilities of a *good public servant in his or her role as policy analyst and adviser.* Duties and responsibilities and ethical issues relate to how the official advises or deals with the minister, the government in power, Parliament, opposition parties, the media, stakeholders, colleagues, Canadian citizens, and "the public," including through the now pervasive social media.

Although the SAGE principles are useful and garnered some support in the 1999–2002 period, they immediately faced other practical challenges and puzzles when thinking about science advice and nominal policy stages. For example, regarding *problem identification and goal clarification* there are lots of candidates for potential policy action or inaction. The first issue is one of seeing whether a given perceived problem is a policy problem and whether it can be defined and understood in order to begin to deal with it through policy. Where a given problem fits with departmental priorities or government priorities is crucial. Expressing the goals of government is complicated. First, what goals should be expressed substantively in terms of actually solving problems? Second, what ideas need to be expressed to gain support for the initiative? Third, should there be a single goal or multiple goals?

Consider the dynamics of *stakeholder consultation and/or negotiation.* The key questions in consultation include the following: Who should be consulted and why? What exactly is the purpose of the consultation? What input is sought? How early in the process should consultation occur? Should another government be treated as just another stakeholder? Transparency and openness argues for processes that are open and transparent to stakeholders and the public. But a key question is: How much time should be taken for such consultation? Democratic processes require *timely* decisions, but timely in whose interests or from whose perspective? Some science-based controversies may genuinely need a lot of time and a very considered process. Science advice in consultation processes crucially has to deal with human resource demands regarding which scientists or science advisers are skilled and competent in functioning in these consultative processes (O'Doherty and Einsiedel 2012).

Last but not least in this sample list of challenges is the task of *policy communication.* As policies are developed and emerge as agreed policies, there are continuous needs to deal with communication at several levels: communication of general information to the public and those being regulated; science input in Memoranda to Cabinet and Treasury Board Submissions; opinion polling; the withholding of information and the preservation of confidentiality; dealing 24/7 with the media and social media; communication with members of Parliament (MPs) and caucus; and demarcating the policy from past policies with appropriate new labelling and or just political spin. For a science-based department there are also crucial issues regarding the communication of risk (linked closely to and in practice an aspect of both risk assessment and risk management). Policy communication ultimately involves communicating entire policy agendas and key departmental themes.

Although the *SAGE* guidelines were accepted, they did not stay for long in an explicit, noticeable way (Enros 2013, chapter 11). Other issues and agendas climbed to centre stage, including Program Review, and budget cuts, and then budgetary surpluses in the Liberal era. When the Harper Conservative government came to power in 2006 it brought a quite different view of S&T, scientists, and science advice that had to be fit into the new reality of continuous campaigning and attack-politics directed from the centre by the prime minister and the PMO.

The Harper government's views and practices about science advice and the role of scientists per se was not central or explicit in its first S&T policy strategy, published in 2007 and entitled *Mobilizing Science and Technology to Canada's Advantage* (Industry Canada 2007). The strategy offers general support for the overall work of Canada's scientific and technological experts and research centres that "provide solutions to many of the most important issues to Canadians, giving us the knowledge and the means to preserve the quality of our environment, protect endangered species, improve our health, enhance public safety and security and manage our natural and energy resources" (ibid., 7). There were some misgivings articulated when Harper in early 2006 moved the Office of the National Science Advisor to Industry Canada, out of the PCO where it reported to the prime minister. Antagonism flared and remained potent when the ONSA office was closed permanently upon the retirement of Arthur Carty in 2008. Battle lines were eventually drawn and trench warfare has been waged between the government and the science community in the public service ever since (Hong 2008; Goar 2008).

Although the fate of the Office of the National Science Advisor was visible, far more significantly, in 2007, the Harper government quietly introduced new internal communications rules to guide public servants, including scientists talking about science and public science matters. As Douglas points out, the new rules required "that all public servants, including scientists" got "approval before speaking in public or to a journalist on any topic. These rules were rolled back for routine queries about the weather, but other than that, scientists were to check with media relations or communications officers before any public contact" (Douglas 2013, 1). These controls were especially detailed on sensitive S&T and related policy issues such as climate change and, later, oil sands pipeline developments. As general practice, all external communications by public servants, including scientists, required pre-approval and were only authorized based on pre-determined talking points. As these policies and practices took hold, it was reported that there were instances where limits were placed on government scientists' ability to publish their research in peer-reviewed journals. These communications policies are perceived by many as an attempt by the Harper government to "muzzle science and scientists." Critics argue that a new era was underway, citing the closure and dismantling of a number of regional Department of

Fisheries and Oceans (DFO) libraries as an example of the anti-science orientation of the government (Salutin 2014). Each and every cut in science staff and budgets that emerged in the Action Plan and subsequent budgets was pounced upon and used to reaffirm the concern that the Harper deficit reduction program was being used as a smokescreen to quietly dismantle the public science infrastructure (Gatehouse 2013; O'Hara and Dufour 2014).

This so-called muzzling and harassing process has been a core theme in Canada's recent debate on climate change policy. Critics argue cuts to science infrastructure, especially in fisheries, devolution of environmental review to provinces for developments that lie solely in provincial jurisdictions, and the strict communications control on public scientists are all attempts to bolster and support oil sands development and policies intended to make Canada an energy superpower. To make matters worse, the measures impacting on many of these policies were enacted through two abnormally large omnibus budget bills in 2012 following the re-election of the Harper Conservatives with a majority following the 2011 election. Compounding this is the constant attack-politics mode of behaviour in Canada. Science and more broadly all policy debate has been immeasurably weakened. There are vanishingly few venues or examples of open dialogue; debate has largely degenerated into disconnected volleys of arguments based on made-up or assumed facts, values masquerading as arguments, name calling, character assassination, and innuendo. Although the government representatives are at the centre of this problem, they are not the only ones behaving badly. Some science groups have argued in public demonstrations that this signals the "death of evidence" (Toner and McKee 2014). This dispute has drawn considerable attention and comment in the United States and among journalists internationally (Findlay and Dufour 2013; Woods, 2013).

The "muzzling" theme has been linked to recent cuts in science by a number of groups, including the Professional Institute of the Public Service of Canada (PIPSC), the union that represents sixty thousand scientists and professionals employed at the federal and in some provincial and territorial governments. In their report entitled *Vanishing Science* (PIPSC 2014, 1), drawing on polling data by Environics Research, the PIPSC concluded that "over 9 of 10 federal government scientists (91%) believe cuts to federal science budgets – most of which take effect over the next few years – will have a detrimental effect on the federal government's ability to serve the public (over half – 51% already believe the impact is very detrimental)."

As with all public controversies, this one is not as clear-cut as it might appear. In the first instance, although there have been real cuts in some departments and some subject areas, the government science complement overall has continued to rise, at least until 2013. Moreover, the number of scientists in universities, colleges, and institutes has risen sharply over the past decade, so overall there are now more publicly funded scientists doing research in Canada than at any previous time in our history. Nevertheless, some key areas of science capacity in government have undoubtedly been weakened. As noted in our introductory chapter, however, Canada's public sector science complement is but a tiny portion of a much larger and faster growing world science community, so the capacity may still exist, albeit not within the federal bureaucracy. Second, the debate about the role of scientists in the federal system is only a part of a larger debate about the appropriate role for public servants in general in a world of governance rather than government.

Donald Savoie (2003a, 5–6) has asserted that the traditional government system involved something like a bargain, where "under the arrangement, public servants exchanged overt partisanship, some political rights, and a public profile in return for permanent careers ... anonymity, selection by merit ... Politicians exchanged the ability to appoint and dismiss public servants ... for professional competence and non-partisan obedience to the government of the day." In short, as Aucoin and Jarvis (2005) argue, public servants do not possess authority on their own; they are agents of their ministers. In that context, Alex Smith (2006, 3) submits that "public servants do not have a public voice, or identity, distinct from their minister; they are anonymous." This nostrum firmly guides practices for most policy advisers in executive government, such as economists, lawyers, accountants, and engineers, but Canadian government scientists have by tradition been given more liberty to speak publicly in order to contribute to the larger public interest; in most countries this tradition has been under pressure for quite a while. In Canada, rightly or wrongly, the Harper government has decided to invoke the implicit bargain for all its employees, including scientists.

Regardless of who is right about the public role for public scientists, the quality of science policy discourse has deteriorated due to the recent changes and conflict. Some simply want to return to the old ways where public scientists have liberty to speak about their work. No governments, but especially the federal government, seem keen on this option. The federal government has quite explicitly decided to proactively seek advice from outside government – expert advice through references to the Canadian Council of the Academies and policy advice from arm's-length advisory groups such as the STIC. Paul Dufour, a long-time (now former) federal science policy adviser, suggests that this approach may not solve much, as "formal science advice provided by the Science, Technology and Innovation Council operates in stealth mode, with no public accountability" (Dufour 2014, 2). Others hope scientists in universities and NGOs, in Canada and abroad, will take up the slack – and in some ways they have. Moreover, some have called for the establishment of a new science watchdog, via an independent Parliamentary Science Officer (PSO; Owens 2013; Findlay and Dufour 2013).

1963–2015: Public Service Renewal and S&T

Our fifth history in the macro S&T and innovation policy domain centres briefly on issues and exercises regarding federal public service renewal. The federal public service as a politically neutral but expert merit-based part of the democratic governance system is composed in the main of non-scientists. Scientists are a minority component of the public service as a whole but a strong majority in some key departments and agencies, such as Natural Resources Canada, Environment Canada, the Department of Fisheries and Oceans, Health Canada, and Agriculture and Agri-Food Canada. The notion of S&T staff in the federal public service also becomes broader if one defines such staff to include other social science disciplines and qualified professional personnel.

Thus, the question arises about how S&T and S&T staff are thought of (if at all) in major periodic public service renewal strategies and processes. Such renewal reviews and processes have included the following:

- 1963 Glassco Commission (Royal Commission on Government Organization, RCGO)
- 1979 D'Avignon Special Committee on Personnel Management and the Merit Principle
- 1997–98 La Relève (re-awakening) government-wide public service renewal process
- 2001 Task Force on Modernizing Human Resource Management in the Public Service
- 1992–2012 Advisory Committee on the Public Service Reports and Reporting Process
- 2013–15 Blueprint 2020 process underway (including Government Science 2020)

Each renewal process was different and took place in changing historical circumstances and contexts; nevertheless, there is a kind of book-end symmetry to the renewal processes in relation to S&T and scientists in the public service. Only the Glassco Commission in 1963 and the current 2013–14 Blue Print 2020 process has any significant S&T content, discussion, or expressed concerns.

The Glassco Commission's report on scientific R&D was very critical of the then existing science policy machinery. It offered many recommendations on science in particular departments (RCGO 1963; Doern 1972, 5–7). Fast forward fifty years and one finds only the second example of significant explicit S&T content in the Blue Print 2020 exercise underway in 2013–15 (Privy Council 2014a). This S&T content centres on "Government Science 2020: Re-thinking Public Science in a Networked Age" (2013), a discussion paper authored by Jeffrey Kinder, of Natural Resources Canada, but with considerable consultation and input from other science-based departments and personnel. The Kinder document examines a number of themes and developments around changes and challenges such as a "whole of government approach," "open innovation in government science," the smart use of technology, including social media, and the "rise of social science" (Kinder 2013, 2014). Recent writing by Amanda Clarke (2014, 1) reports that in the federal government recently "civil servants passionate about improving public management have formed a vibrant online and offline community, fuelling an unprecedented grassroots public sector reform movement that has emerged outside the formal structures of government." Not surprisingly, new information and social media technologies are a part of the story of government S&T and in the public service overall. At one level, these stand in contrast to the central controlling style of executive governance (both in the Harper era and earlier).

In between these book-end public service renewal exercises, other studies have given short shrift to S&T matters. The La Relève initiative had scarcely any focus on S&T in its overall work, with only a brief mention in a discussion of needs regarding the "strengthening of professional communities" – policy, communications, scientific professionals, human resources, infomatics, and information management (Stilborn 1998, 9).

In one sense, it is not surprising that S&T and scientists are not central to public service renewal initiatives and processes. Such exercises have dealt with a broad array of issues, including the merit system, management styles, integrity and reputation, leadership, modernization, recruitment and retention, gender, and varied notions of streamlining. Most of the considered published reviews of federal public service renewal processes also do not include S&T issues and are variously critical of their work regarding what they failed to address (Zussman 2010; Johnson and Molloy 2009; Juillet and Rasmussen 2008; Savoie 1999, 2003). Among these issues that are buried in such review processes are major changes in politics. Reviews led by or centred on the civil service, of course, refer to the core principles of a neutral independent merit-based public service, but they do not dare speak openly or candidly about political imperatives such as ever more centralized prime ministerial control or more extensive partisan communications strategies, let alone the related excesses of unrelenting attack-politics.

In a recent report to the prime minister, the Prime Minister's Advisory Committee on the Public Service (PCO 2013) spoke reassuringly of how the public service was showing "signs of progress" in that it was more "innovative, collaborative, streamlined, high performance and adaptable, diverse" and that its future priorities needed to be a "focus on performance and productivity embracing enterprise approaches, and working together for better outcomes" (5). Innovation and risk are mentioned as discourses but without any mention of science and scientists nor of their appropriate role in public dialogue, as framed in the debate about muzzling scientists.

THE THREE DOMAIN ELEMENTS

With the aid of and in the context of evidence supplied in our five policy and governance histories, we now offer a summary account of the three elements in this book's analytical framework as factors for change and inertia in the evolution of the macro S&T and innovation policy domain.

Policy Ideas, Discourse, and Agendas

With regard to ideas and discourse, S&T is the common title and terminology in almost all the macro policy statements and in the changing structure of the macro S&T policy advisory agencies. The same is true in the overall discourse revealed in Budget Speeches. Innovation emerges in later statements and Budget Speeches, especially in the Chrétien-Martin era, where it is linked broadly to ideas about the knowledge economy and to universities. The Harper era, and to some extent the Mulroney era, focused more on commercialization and competitiveness ideas and goals though with awareness of innovation ideas looming larger in the Harper era.

Table 4.5 Policy and governance histories in the macro S&T and innovation policy domain: Three analytical elements

Policy and governance history	*Ideas, discourse, and agendas*	*Economic and social power*	*Time, temporal realities, and conflicts*
1963–2015: Prime ministers, central agencies, and macro S&T policy statements and strategies	• S&T titles in all but 2 of the S&T statements • Innovation as idea discussed in later statements but rarely in terms of its nonlinear nature • Industrial R&D gap and weakness for Canada stressed in all statements across 50 years • statements in Mulroney and Harper eras had more focus on commercialization ideas but some innovation ideas too • Chrétien-Martin era stressed broad themes about innovation and knowledge economy with focus on university and partnered research	• Prime ministerial power on S&T for either positive or negative reasons but also only periodic Science ministers are junior and peripheral, don't stay for long • Middle-level power and influence of ministers of several science-based departments/agencies, social and economic; Industry Canada becomes leading S&T minister and department • S&T advisory councils report quite widely on socioeconomic S&T and innovation issues, but still to elite audiences; Science Secretariat and then later the ONSA are lodged first in PCO but then abolished.	• S&T statements seek to capture and reflect changing S&T discourse in context mainly of changing economy • Industrial R&D gap and weakness stressed in medium and short term, but essentially not solved in performance terms over the full 50-year period
1968–2015: Budgets, budget processes, and annual S&T policy	• Budget as central policy and agenda process, but S&T items quite specific and rarely given top focus across 49 Budget Speeches • S&T and innovation ideas and agenda given top focus only in 2000 Liberal budget • Trudeau era weakest in S&T mentions in budgets; Mulroney budgets stronger, but both in deficit contexts • Knowledge economy and society emerge in Chrétien-Martin era with emphasis on universities and granting councils, and also partnered research; budgetary surpluses helped a lot • Competitiveness and commercialization ideas emerge more strongly in Harper era and in deficit-recession context	• Finance minister as S&T minister, especially Martin in Chrétien era Budgets as primarily an economic power occasion and process • Business R&D lobby and emerging small business lobby, in particular • Impact and context of deficit/recession and later surplus realities and degrees of flexibility; Internet and Information Highway as socio-economic lobby • Universities lobby as growing socio-economic lobby at federal level • Community college lobby and presence in Harper era • Severe cuts in S&T spending for departments in Liberal Program Review and in Harper era recession	• Budgets as annual and medium-term focus processes and also vis-à-vis the electoral cycle • Witnessing, guessing at managing new technologies

Table 4.5 Policy and governance histories in the macro S&T and innovation policy domain: Three analytical elements (*Continued*)

Policy and governance history	*Ideas, discourse, and agendas*	*Economic and social power*	*Time, temporal realities, and conflicts*
1980–2015: Regulation, regulatory processes, and S&T and related science activities	• Current CDRM and earlier macro regulation policy give first priority to protecting and advancing public interest, but also efficiency and economic effectiveness and growth • Science-based ideas about regulation making and enforcement, but also about evidence-based and use of precaution • Fair and competitive markets and hence regulatory impact socio-economic science and evidence • Risk-benefit and life-cycle approaches in health and environment • Harmonization and mutual recognition of rules internationally and in Canadian federalism	• Treasury Board as leading central agency but also longstanding *Canada Gazette* and constitutional imperatives of law and practice • But core S&T knowledge and expertise resides across socio-economic and risk-benefit departments and agencies • Economic/business power in pre-market regulation, but broader socio-economic power in post-market regulation • Role of RSA and thus frontline science and also dispersed networked knowledge professions • Business power lobbying regarding red-tape reduction and presence of "one for one" regulatory policies and processes	• No central regulatory agenda as in budgetary process • Regulatory plans are • required, but very uneven as to establishment and time frames • Time scales vary for RIAS processes
1990–2015: Policies and guidance regarding science advice and the role of scientists	• Emerge as ideas and discourse on creating, abolishing, or restructuring macro S&T advisory bodies (from the SCC to the current STIC • Concerns about nature of S&T advice arise with crises/failures such as blood supplies, fish stocks, nuclear waste, biotech, BSE, and SARS • 1999–2002 *SAGE* and its proposed federal S&T advice principles and guidelines • Muzzling of science issues in Harper era focused on central communications policies and practices about what federal scientists could say • Issue linked also to S&T budget and staff cuts and to practices of continuous government-led attack-politics, including use of social media	• Usually implies breadth and expansion of values and points of view • Societal interests raising concerns and evident in a serious failure • But rise in social influence and power often quite episodic • Issue of S&T as written vs. verbal advice in complex hierarchies and networks • Role of burgeoning social media • Power in the design of consultation and engagement processes • Forging of continuous attack-politics and controlling strategies to limit criticisms and critics	• Length of advice giving and consultation processes in age of "streamlining"

Policy and governance history	*Ideas, discourse, and agendas*	*Economic and social power*	*Time, temporal realities, and conflicts*
1963–2015: Public service renewal and S&T	• Preserving and enhancing a politically neutral and merit-based federal public service • La Relève and other renewal processes dealing with ideas and concerns about management, integrity, recruitment, and retention • S&T and scientists as public servants not usually central except in early 1960s Glassco Commission and in current Blue Print 2020 processes • Power of PMs as "breaking the bargain," but not frontally addressed in public service renewal processes/reviews	• Influence of PCO led public service but which reports to the PM and is influenced by PM • Role and internal lobbying of the PIPSC • Rapport and/or malaise of the public service in some overall way • Social media technology role and growth within the public service at lower and mid-levels	• Roughly once a decade for such renewal processes Uneven responses to changes, threats, and challenges already well underway • Playing catch-up with core needed competences, including S&T as part of public service professions • Long-term implications of the muzzling of scientists for policy and for democracy

Note: see the chapter text for expansion of abbreviations in this table.

When seen in regulatory policy terms, ideas about science-based protection and advancing the public interest become more explicit but also usually are quickly paired with economic efficiency, growth, and fair and competitive markets. Ideas about risk-benefit as paired goals emerge, mostly in support for ideas such as life-cycle approaches and precaution. While S&T and innovation ideas are a part of this regulatory structure, S&T and related science activities are not usually precisely discussed or linked. Indeed, broader ideas about evidence emerge including that supplied by the social sciences in and outside of RIAS analysis.

Policies and guidance regarding science advice and the role of scientists per se is shown to be quite general in the case of changing science policy councils and bodies. But it is also very episodic when one sees how the 1999–2002 *SAGE* S&T advice principles and guidelines emerged in response to policy- and science-related crises and failures but then exhibited limited sustained support for dealing with the complexities and puzzles of science advice. The muzzling of science under the Harper government also emerged under its communications policy dictums and its strident attack politics, strategies, and practices.

The public service renewal history is mainly one of difficulties in even getting S&T policy and scientists into the consciousness of the public service renewal leadership and advocates. There is little doubt that S&T and scientists are central to a politically neutral and merit-based public service but only in two such exercises, the 1963 Glassco Commission and the current Blue Print 2020 processes, have S&T concerns been expressed or been able to muscle in on the agenda.

On agendas as a whole, the story of the macro S&T and innovation policy domain shows the steady presence of S&T and innovation policy and ideas, but for the most part it is not a high-level priority, regardless of whether it involves prime ministers, ministers of finance, or the Treasury Board in regulatory governance terms. S&T and innovation policy operates in the middle and lower realms of the priority-setting system.

Economic and Social Power

These five histories tend to show the predominance of economic political power over broader forms of social power in the macro S&T and innovation policy domain. S&T policy is cast more often as aimed at elites and communicated to elites. The role of ministers of science per se is weak in the overall structure, and power has been more frequently sited within Industry Canada and its predecessor departments.

Social issues and social interests are more likely to find some presence and traction in the macro power structure in regulatory processes involving both regulation making and regulatory compliance. It is possible to see that when innovation ideas gain traction, as in the Chrétien-Martin era, social issues and values gain a greater presence particularly in some discussions about policy for universities and knowledge economy contours. Support for the Internet or Information Highway and early aspects of new social media in the main always found some favour as a technological and socio-economic development that was of interest to a wide swath of Canadians as users, citizens, and voters, in many ways melding the economic and social dimensions.

Time, Temporal Realities, and Conflicts

Time, temporal realities, and conflicts as an element traversing the five policy histories require detective work to interpret their changing presence. Overall S&T policy statements are periodic (22 in 50 years) as are changes in S&T central advisory structures (9 in 50 years). In many ways the tentative nature of the changes reflects the difficulty of addressing the coming medium-term or even longer-term challenges as they are seen or guessed at.

A strong temporal reality is that one sustaining policy goal – closing Canada's R&D and/or innovation gap – is stressed in virtually all S&T policy statements and in many Budget Speeches over the fifty-year period. But it remains unsolved. With respect to budgets and S&T, Budget Speeches are in essence focused on annual choices couched in some medium-term goals but rarely on longer-term objectives. They can be and are easily influenced by the electoral calculus both of the pre-election kind and the post-election results kind, including minority versus majority government status.

The policy story on science advice and the role of scientists was itself very episodic in nature and hence hard to get a grip on as a policy matter per se. Policy crises and noticeable failures, one at a time, or in clusters, as in the late 1990s, are themselves sudden and unexpected triggers for change. Moreover, as noted, the policies and practices of streamlining regulatory and related S&T-informed regulatory processes raise important issues of how long proper regulatory processes ought to or actually do take. Temporal rhythms were central to the entire set of public service renewal exercises including the need to catch up with challenges and capacities (S&T and otherwise) of the modern public service as it serves a changing democratic system.

CONCLUSIONS

Our analysis of the macro S&T and innovation policy domain has shown the basic contours of how it is structured and how it makes the federal government's system-wide and system-defining policies, rules, and budgetary decisions regarding S&T and innovation policy. Such policy, governance, and democratic dynamics function, ultimately, in the basic arenas and rhythms of Cabinet government, the growing controlling nature of prime ministerial power, and the power of other central players including the minister and department of finance and the Treasury Board and Privy Council Office. The chapter has shown that S&T and innovation policy overall has been important at a middle level of agendas and decision making but rarely at the top.

At a minimum, at least five policy and governance histories make up the domain's evolution, and each must be understood at some basic level, not only to define the domain but also to see how its influences, positive and negative, need to be revisited in relevant ways in the other S&T and innovation domain chapters that follow. The core prime ministerial and central agency story shows the importance of periodic interest in S&T and then its shifting to other policy and political priorities, especially fiscal and economic ones, but also others where S&T is linked only tenuously, or ignored altogether. The periodic S&T policy statements mark efforts to define, redefine, stay ahead, or play catch-up in an area of policy that is for the most part the concern of elites rather than ordinary Canadians as citizens and voters.

The analysis has shown how the central budgetary and regulatory processes function quite differently in S&T contexts and on an overall basis. The discussion of policy and guidance on S&T advice and the role of scientists has shown an irregular but routine periodicity. S&T advisory councils have been created and abandoned, often in the wake of S&T-related major controversies, crises, and failures. The current development of policies and practices regarding the muzzling of science is linked to prime ministerial–centred communications strategies and nonstop attack-politics strategies that are demeaning to both democratic politics and effective science "in" policy-making. In the periodic public service renewal processes examined, S&T policy and the role of scientists seldom arises or receives attention, which is a challenge: scientists may be a minority of the total public service career system, but in several key science-based departments they are a significant majority.

The chapter has tracked the ways in which the three elements of our analytical framework show both change and inertia. The elements are (1) policy ideas, discourse, and agendas, (2) economic and social power, and (3) time, temporal dynamics, and conflicts, and each offers an important analytical probe to help us understand the macro S&T and innovation policy domain story. They are also an important lead-in to chapter 5, where we examine the government S&T departments and agencies domain.

The chapter has brought out some initial empirical evidence regarding our central innovation economy and society nexus. Overall, the innovation economy is the larger theme in the S&T strategy statements even though S&T is the overall nomenclature of choice. One can find social themes in some Budget Speeches, especially when new knowledge is emphasized and linked to education and universities and in the context of early Information Highway Internet initiatives that were tied to support for communities and schools. Under the Cabinet Directive on Regulatory Management, social issues, if not the innovation society, have gained prominence as stated values, including those regarding health, safety, and the well-being of Canadians.

5

The Government S&T Departments and Agencies Domain

The federal government's S&T departments and agencies domain is both broad and complex. There are about ten departments headed by Cabinet ministers and upwards of two hundred laboratories, agencies, and subunits, including divisions, institutes, or directorates of these departments or their arm's-length agencies that perform varied S&T activities of both an R&D and related science activities (RSA) nature. If S&T is defined to include the social sciences, then many more federal departments and agencies would be a part of this domain.

Historically, the domain was dominated by entities such as the Geological Survey of Canada that mapped Canada's resources and land mass (Vodden 1992; Zaslow 1975) and those that collect data, undertake research, and interpret results, such as Statistics Canada. The overall strong S&T links to Canada's geological and resource economy warrant emphasis here particularly regarding the oil and gas industry, including the oil sands historically, but also regarding mining (Gray 2005; McKenzie-Brown, in press; Chastko 2004; Breen 1993). As chapter 9 shows, S&T is also embedded in Canada's agricultural and farming industry, including via the historical and current role of federal experimental farms. This strong historical linkage is important in countering the later and recent simplistic claims that the resource economy is "the old economy" and is thus contrasted rhetorically but not factually with the "new" economy or the "innovation" economy.

In this domain the typical science-based departments include Health Canada, Environment Canada, Natural Resources Canada, Fisheries and Oceans Canada, Agriculture and Agri-Food Canada, National Defence, and Industry Canada. Key agencies include the National Research Council, the Canadian Space Agency, the Canadian Food Inspection Agency, the Public Health Agency of Canada, and Atomic Energy Canada Ltd.

Federal STI policies are directed in various ways at these departments and agencies, but such policies do not enter their consciousness and operations on their own (Kinder 2010; Doern and Kinder 2007; Doern and Reed 2001). They are immediately and necessarily conjoined with and interpreted through the imperatives of specific departmental mandates, laws, and regulations as well as the established cultures and rhythms of their traditions, which are reflected in their stories of success and memories of decisions and policies that went wrong. Moreover, they are influenced and reshaped in relation to the inherent number and content of the statutes that they are responsible for, on their own or in concert with other departments (de la Mothe 2000). Finally, they are influenced by different volumes and types of projects, products, and services that are a part of their regulatory tasks (both pre-market and post-market) and their larger societal roles.

Across the fifty-year account of this domain, there have been numerous STI policies and discourses aimed at these departments and agencies. These have included the initial federal "make or buy" science policy of the early 1970s, where the "make" imperative was centred on the need for science directed and provided by the government and centred on the public interest. The "buy" aspect of the then newly expressed policy thrust was to use procurement of S&T from the private sector as a form of extended support for industrial research (Doern 1972). Later, in the 1980s and 1990s, the "buy" coinage shifted to the innovation policy discourse and related newer forms of organizing science then emerging, including those linked to international trade policy and globalization (de la Mothe 2003). In 2008 the Independent Panel of Experts, chaired by Arnold Naimark, reported on how and why more intersectoral partnerships with business and universities should be developed for non-regulatory federal laboratories (IPE 2008).

In this chapter, we explore key features of change and inertia in this domain by looking at S&T and innovation policy and governance histories centred on the National Research Council of Canada (NRC), Health Canada (with a particular focus on regulating drugs and related health products), and Environment Canada.

THREE DOMAIN POLICY AND GOVERNANCE HISTORIES

Although there are upwards of a dozen departments and more than two hundred agencies or labs that engage in this domain, the National Research Council, as Canada's largest single research entity, Health Canada as the single largest investor in health research, and Environment Canada as the most far-reaching agency, in combination, offer a good way to see inside and understand the dynamics of the government S&T departments and agencies domain. The different bases for the narrative account of each of the three entities help portray the domain as a whole.

Thus, the National Research Council as the longest established agency is looked at through three periods highlighted by different agency leaders and various efforts to develop and support S&T and innovation for nation-building purposes amid cyclical pressures to be more focused on the economy and business. The Health Canada S&T and innovation history has a focus on the 1960s to the present day, but is presented here mainly through the pivotal Food Products and Health Branch and its regulation of drugs and related products. This story is complicated by an ever-broadening policy scope amid global changes in Big Pharma. The Environment Canada story, which has been referred to in part in previous chapters, does not focus on any of its branches or research units per se but looks instead at a broad history of its organization and management of S&T in relation to a statutory and policy mandate that has its own departmental core but also where, in various ways, it depends on other departments and agencies.

1945–2015: National Research Council

Change and inertia in the National Research Council can be explored in three broad periods: its pre–Second World War birth and post-war prominence, ending in the early 1970s; the Mulroney era in the late 1980s to early 1990s and the effort to make the NRC more industry-focused, immediately followed by consolidation in the Chrétien era, in 1995–2001, and corresponding push-back from the industry focus; and finally, the effort in the Harper era to restructure the NRC to be mainly an industrial R&D and market-pull commercialization entity.

Previous histories have presented the National Research Council as evolving through its formative years from 1917 to 1939, its "golden years" from 1939 to 1969, and its "fall from grace years," 1970 to early 1990s (Phillipson 1991; Jarrell and Gingras 1991; Eggleston 1978). But these histories tend to focus on the NRC's role as a research institution, ignoring other core key institutional roles, including the following: policy advise and de facto policy-maker, research granting body, provider of incentives for industry, technology transfer and information institution, and manager of intellectual property and research ethics.

In 1917 the Research Council Act established an advisory council to the newly created Privy Council Committee on Scientific and Industrial

Research headed by the minister of trade and commerce. In the period immediately following the First World War, the National Research Council advised that Canada's scientific manpower and infrastructures were minimal and that Canada was ill-equipped to carry out scientific research (Thistle 1965).

The NRC's policy advisory role changed markedly during and after the Second World War. As NRC laboratories were established (see below) and the Council's prestige grew, it became without doubt Canada's main science policy adviser, with that role extending into the 1960s. This preeminent policy role was a least partly due to the leadership and prestige, both of its key scientists such as E.W.R. Steacie, president of the National Research Council from 1952 to 1962 (King 1989), and due to the close and supportive relationship between the NRC and succeeding Liberal governments, particularly through powerful ministers such as C.D. Howe.

In the 1960s, first under the Diefenbaker Conservatives and then under the Pearson Liberals, dissatisfaction with the prevailing science policy and the post–Second World War consensus increased. The National Research Council quickly lost the overall lead role in science policy, but remained a very important contributor to strategic policy. The NRC's own survival and adaptation depended then, as now, upon its own strategic policy capacity both to plan and run its own affairs and to tread its way through, and influence, the broader S&T and innovation policy process and, crucially, the larger macro- and micro-economic policy process.

The National Research Council acted as a granting council until 1978 and the establishment of the Natural Sciences and Engineering Research Council (NSERC). In response to its loss of granting duties, the NRC established a system of industrial incentives and advisory services to smaller firms via its Industrial Research Assistance Program (IRAP). However, the NRC in its soul and corporate culture remained convinced that building the scientific community, as well as nation building and development was still its primary role. Hence universities and academic scientists were generally privileged beneficiaries of money granted by the NRC. But this particular emphasis on building and nurturing the academic scientific community was criticized, first in the 1960s and off and on ever since – the NRC was cast critically as being a university without students.

As its roles changed, the NRC laboratories and their performance of research became central to the Council's role and influence. NRC lab and related staff grew from three hundred in 1939 to nearly three thousand by the end of the war in 1945. Its wartime scientific research efforts included subjects such as radar technology, wartime food, war gases, experimental aircraft, and atomic research (Phillipson 1991; Thistle

1965). After the war, the NRC's laboratories moved away from military research, which was transferred to the newly created Defence Research Board in 1947. Like the modern Canadian university, the NRC undertook both pure and applied research in its institutes, but there is no doubt that much of its agenda was internally generated and tended to favour upstream basic research topics (Jarrell and Gingras 1991).

To alter that gravitational pull, the National Research Council tried to change the way it was governed, opening up more space for members of the business community at the highest levels of the decision-making process via advisory boards (NRC 1990–91). These boards, however, produced widely varying kinds of support and resistence and thus the NRC itself remained powerful.

A related structural change occurred in 1990–91, when all research divisions became institutes, many targeted on specific sectoral interests, such as the Plant Biotechnology Institute (PBI) in Saskatoon. The less focused research divisions were reorganized to create five new service institutes "focused on areas of strategic importance: molecular sciences, national measurement standards, environmental chemistry, microstructural sciences and information technology" (NRC 1990–91, 9). More will be said later about the nature of the NRC internal structure. Suffice it to say at this stage that the transformation from discipline-based divisions to multidisciplinary institutes focused on clusters of science and new technologies was often symbolic rather than substantive. The NRC as research performer was changing but in precisely what direction in practice has always remained an open question. This is because the NRC is always forced to balance top-down S&T policy-driven directives with the bottom-up, science-driven initiative of front-line research scientists. Some institutes reconciled the inherent conflict better than others, making for a somewhat uneven internal landscape as it relates to the institutes' missions and their execution.

Table 5.1 elaborates on the three key phases in the more recent development of the National Research Council of Canada: (1) 1990–95 and the search for a competitive edge; (2) 1996–2001 and the pursuit of a new vision; and (3) 2013–18 and the strategic realignment of the NRC into research and commercialization platforms. In the first instance, these three periods reveal the importance of leadership as an instrument of policy change, first the appointment by Mulroney of Pierre Perron as NRC president in 1989, then of Arthur Carty by Chrétien in 1994, and finally of John McDougall by Harper in 2010. The changes governments wanted were partly embodied in the already known characteristics of the individuals being appointed. But in another sense, governments often cannot know exactly what their appointees will in fact do or whether they can control their actions. At times they may not even approve of all that they do (Doern and Levesque 2002).

Perron believed that, although the National Research Council had made some structural changes in the 1980s, its culture had not changed enough. The research divisions were still very independent and exhibited an insufficient awareness of the NRC as a corporate entity and of the changes in its environment that were increasingly apparent at the head office. Perron had been a member of an NRC associate committee in the 1980s and thus had formed critical views of what he saw as its leisurely management. These views formed his hard performance-oriented management instincts and helped shape his strategy that the labs would henceforth be required to make real choices. As already mentioned, Perron helped to increase the business focus in the advisory committees and shaped the institutes as a way to further the changes he felt were needed. He also challenged the public service unions, especially the Professional Institute of the Public Service of Canada (PIPSC).

When Arthur J. Carty became president of the National Research Council of Canada in 1994 he brought a new set of priorities and a style of leadership and management that were intended to mark a break from the Perron era. Although some of the business-oriented directions were maintained, they were now framed in the context of the broader rubric of local, regional, and national innovation systems. This fit well with the growing number and size of the NRC labs situated across Canada, many in communities that had a concentration of firms in the same area as the R&D focus of the local labs.

Carty made it clear to colleagues that he wanted to move away from what was seen as the "command and control" era of the early 1990s toward a more participatory style of science and research management. Carty explicitly targeted entrepreneurship and began to encourage the creation of NRC spin-off companies. Indeed, the style that Carty brought to the National Research Council could be seen as being a product of his twenty-seven years of experience at the University of Waterloo, the university that was created to be and was fast becoming Canada's most entrepreneurial university (Doern and Levesque 2002).

The Carty era saw expansion of international initiatives from the NRC's president, especially in the East Asian economies, which by the mid-1990s were the global growth leaders. Carty saw the need to develop such links as important in themselves, and he made a point of personally being a

Table 5.1 Five-year plans or strategies of the National Research Council of Canada: A comparison

	1990–1995	*1996–2001*	*2013–2018*
	The Competitive Edge[a]	NRC *Vision to 2001*[b]	*Strategy 2013–2018*[c]
Structure, themes, and discourse	The National Challenge • The need for change • The competitiveness challenge Corporate Values and Objectives • To double by 2000 the total financial investment in R&D that NRC stimulates and jointly sponsors with others • World-class research • Research that is relevant • National competitiveness • Action strategies preparing for the 21st century	The Challenge • The national need: innovation • A powerful national resource NRC's Vision • Linking S&T to an innovative economy • Renewed commitment to excellence and relevance • Collaboration and focus on key technologies • National leadership in the system of innovation • Entrepreneurial and innovative organization • Measuring progress • NRC's core competences	The Strategy • Cast in context of a failing Canadian situation • Vision for NRC "to be the most effective research and technology organization in the world, stimulating sustainable domestic prosperity" • To quickly assemble cross-functional teams in response to urgent industry challenges • Clear, measurable goals and outcomes relevant to our clients • Using a market pull vs. technological push mentality in everything • 3 R&D divisions headed by vice presidents covering 12 R&D portfolios overall
Key traits	• Hard-nosed focus on competitiveness paradigm • Greater focus on internal change • More detailed plan with action focus on budgetary stability, revenue raising, and structural change • Necessary reference to both research and relevant research • More explicit reference to need for responsiveness to client needs	• Adoption of systems of innovation paradigm • Greater focus on external opportunity • Greater use of entrepreneurship ethos • Stronger overt mention of national leadership • Less detailed plan and more focus on vision themes, almost no reference to budgets • More specific examples of NRC achievements given	• 4 broad "business lines": strategic R&D, technical and advisory services, the IRAP, and scientific infrastructure • NRC will not "pursue research that does not respond to market pull or client need; will not invest in programs whose benefits will not remain in Canada"

Source: Adapted and updated from Doern and Levesque (2002), 63.
Notes: [a] NRC (1990) [b] NRC (1996) [c] NRC (2013b).

part of some of the mid-1990s Team Canada trade missions. These international initiatives elicited mixed views within the National Research Council. Some agreed with the need for a greater international profile of this kind, but others saw too many of the initiatives as being unplanned. Missions often involved a sudden request by Carty for seed money, funding, or people, which would come out of institute budgets or out of the NRC's revolving fund, which held the retained earnings from revenue-generating work done by the institutes. In contrast to the official view of NRC management, these initiatives were sometimes seen by others as not particularly outcome-oriented projects (Doern and Levesque 2002).

Arthur Carty also brought more attention to the communication activities of the National Research Council and to the practical problems of how to garner support for the NRC from its relevant publics. Carrying on from Perron, Carty gave considerable emphasis to the local presence of the NRC in the areas where NRC institutes are based. This was both a conceptual interest in that Carty had experience from his University of Waterloo days on local-regional systems of innovation and an applied tactical strategy because the NRC was seen as largely focused on Central Canada interests, being led from Ottawa and largely located in Ottawa and centres in southern Ontario and Quebec. Complicating this was that in the past the National Research Council had been reluctant to market itself politically across Canada and in new ways.

As shown in Table 5.1, the plan in the Perron era had a harder-nosed focus on the competitiveness paradigm and discourse favoured by the Mulroney Conservatives in the early 1990s. It had much greater focus on internal change within the National Research Council. The Carty era to 2001 picked up on the language of development and innovation systems, following the cues of the Liberal governments' preferred innovation policy discourse, as well as the continuously evolving body of academic knowledge about the way in which S&T-based innovation works to support economic growth and development.

By 2000 the National Research Council had sixteen institutes that presented a daunting problem of S&T policy and governance complexity. The NRC was confronted with a number of issues, including managing in the context of diverse regional-local economies with which they aspired to be, or were actually, networked; working with institutes that emerged out of earlier science discipline divisions versus those that emerged more de novo in recent years around leading technology clusters; dealing with newer and older industries each with different innovation challenges; and an array of other unique or simply interesting features about the disparate institutes. These summary concerns are cast here as NRC features, but they could equally be a summary of overall federal government S&T and innovation policy challenges.

The evolution of the National Research Council in the Harper era from 2006 to 2015 reveals a move away from arm's-length labs toward a more intensely business-focused innovation entity. Indeed, the Martin era had considered this but not made any progress (Di Salle, van Beek, and Baskerville 2006). The Harper focus was reinforced by the appointment in 2010 of John McDougall as NRC president. As a professional engineer, McDougall had previously been president of the Alberta Research Council for twelve years and had worked for most of his career in the oil and gas sector. He brought a strong technology commercialization focus. In 2011 he initiated a major shift, moving the NRC away from upstream basic research and strongly toward industry-relevant research that offered a clear pathway to the market.

The resulting NRC *Strategy 2013–2018* is overtly cast in the context of a "failing Canadian situation" where "studies, statistics, and reports all point to the expectation that our relative economic standing will fall over the coming decades" (NRC 2013b, 2). The new "vision" for the National Research Council of Canada is "to be the most effective research and technology organization in the world, stimulating sustainable domestic prosperity" (4). The NRC is now cast as "Canada's National Research and Technology Organization," partly emulating the German Fraunhofer model; its new business model "provides the holistic approach, industry focus and expertise required to address the Canadian situation" (ibid.). Science and innovation were no longer part of the new brand.

The *Strategy 2013–2018* (ibid.) says that the NRC has "retooled its approach to ensure that:

- We can understand the needs of industry and drivers for industry change
- We have the structures and culture required to quickly assemble cross-functional teams in response to urgent industry challenges
- We work towards clear, measureable goals and outcomes that are relevant to our clients
- We plan and execute with client needs in mind, using a market pull vs. technological push mentality in everything we do
- We hire, train and retain the absolute best people and expertise available."

The five-year strategy announces that the National Research Council has four broad "business lines," namely, strategic research and development, technical and advisory services, the Industrial Research Assistance Program (IRAP), and scientific infrastructure. In its discussion of strategic research, the NRC states what it will *not* do, namely, that it will not "pursue research that does not respond to market pull or client need; [or] invest in programs whose benefits will not remain in Canada" (ibid., 6). Structurally, the NRC now has three R&D divisions headed by vice presidents. These cover twelve "portfolios" and provide access to thirty-seven research infrastructure facilities. The division and portfolio structure is grouped as follows (8–9):

Emerging Technologies
- Information and communications technologies
- Measurement science and standards
- National science infrastructure
- Security and disruptive technologies

Life Sciences
- Aquatic and crop resource development
- Human health therapeutics
- Medical devices

Engineering
- Aerospace
- Automotive and surface transportation
- Construction
- Energy, mining, and environment
- Ocean, coastal, and river engineering.

The NRC plan specifically mentions its links with the earlier Harper 2007 S&T strategy but does not explicitly mention the study by Naimark's Independent Expert Panel (2008) panel on non-regulatory federal laboratories, even though the NRC is one of the targeted non-regulatory agencies. Nor does the plan mention the Harper-appointed Expert Panel on Federal Support to Research and Development, chaired by Tom Jenkins, and its study (the Jenkins Report, EPFSR&D 2011) that emphasized a strong business focus but made two specific recommendations for the NRC. The panel wanted, in effect, to change the NRC into its preferred Industrial Research and Innovation Council to deliver the federal government's business innovation programs and to become the federal "common application portal to help businesses find the right programs for their needs (a 'concierge' service)" (ibid., 6).

The title change was not adopted. And, although the portal idea was not in the plan itself, the concierge discourse is now highlighted in the NRC main website. The Jenkins panel also sought to transform the "institutes of the National Research Council into a series of large-scale collaborative centres involving business, universities and the provinces" (ibid., 7), without seeming to recognize any partial contradictions in these two recommendations. This recommendation has come to fruition. The creation of the division and portfolios signalled the end of the NRC institutes. Some of their work was recast into an array of "flagship programs" that leveraged provincial, university, and private capital in pursuit of five-to-ten-year commercial objectives. By 2015 four flagships were operating: Algal Carbon Conversion, Canadian Wheat Improvement, Industrial Biomaterials, and Printable Electronics.

The reactions to the announced changes to the National Research Council in 2013 to reform its historically opaque managerial approaches were diverse, with business interests being broadly positive. The science minister at the time, Gary Goodyear "cast a wide net, inviting every business in Canada, regardless of size, to henceforth consider the NRC as a kind of benevolent, wise helper" (Den Tandt 2013, 1). Other reactions simply interpreted these changes as an attack on the NRC and basic research, and as yet another example of the Harper government's war on science (Ivison 2013). Still others drew attention to "existential confusion" as the changes were tested out among NRC science employees and among current or possible business or university clients (O'Malley 2013). Complicating this was the shift in managerial power, as the leadership for the divisions and portfolios resided in Ottawa, and the regional and local leadership that had been embedded in the institutes diminished as the regional and institute directors retired or were reassigned. Adaptive gaming of the system is also occurring such as through the greater intermingling of NRC budgets for infrastructure and for operating budgets and personnel.

The 2015 Harper government budget plan (Minister of Finance 2015, 102–10) treated the NRC quite well by increasing its funding by $119 million over two years in support of its announced NRC reform strategy. Some of NRC (and other granting council) funding is being reallocated to enable federal support and involvement in the international Thirty Meter Telescope, but total federal support is planned at $243.5 million over ten years. The latter support was linked to Canadian and NRC strength in astronomy and astrophysics.

Overall in these periods of change, the National Research Council of Canada has been in different ways and in different degrees mandated, required, and induced to become more of an R&D agency for business and for commercialization. The management and structure has unambiguously changed, from an entity of research divisions and then institutes, now to divisions with research portfolios. But there always was and still is some resistance by front-line researchers as they adhere to the traditional long-term mission that attracted them to the NRC: to ensure that Canada has a strong S&T foundation to build the nation.

1980–2015: Health Canada and Drug and Health Product Regulation

Health Canada has, to say the least, a complex and wide-ranging mandate both as a department and in terms of the agencies in its larger ministerial portfolio (Health Canada 2014a, 2014b, 2014c). The mandate overall is derived from sixteen statutes and a much larger number of regulations. The ministerial portfolio includes the following: the department itself; the Public Health Agency of Canada (PHAC), which emerged after the SARS (severe acute respiratory syndrome) crisis in 2003 (Health Canada 2004); and the Canadian Institutes of Health Research (CIHR, discussed in chapter 6) as part of the granting, universities, and levered-money domain. In this brief look at Health Canada as a department we focus on its overall regulatory approaches and the related S&T base, with a specific examination of the Food Products and Health Branch and the changing dynamics of drug regulation in its Therapeutic Products Directorate (TPD; Health Canada 2014d). This in turn involves ethical issues, especially since the thalidomide tragedy in the 1960s. The Therapeutic Products Directorate has many linkages, through its product reviews, in drawing on science, evidence, and expertise found in its newer sister directorates, such as those for biologics and genetic therapies, veterinary drugs, food, and natural health products (Health Canada 2014f; Murphy 2006). In addition, biotechnology and related genomic products and processes have been particularly complex and problematical as were conceptions of what a "product" is when it is a human being (Doern and Prince 2012; Doern and Phillips 2012).

At present the overall mission, vision, and goal of Health Canada "is for Canada to be among the countries with the healthiest people in the world" (Health Canada 2014a, 1). To achieve this Health Canada (ibid., 1) emphasizes that it:

- Relies on high-quality scientific research as the basis for our work
- Conducts ongoing consultations with Canadians to determine how to best meet their long-term health care needs
- Communicates information about disease prevention to protect Canadians from avoidable risks
- Encourages Canadians to take an active role in their health, such as increasing their level of physical activity and eating well.

The focus in the Food Products and Health Branch has been on its Regulatory Modernization program begun in 2006 (Health Canada 2014d). This program is influenced and ultimately governed by the following: the Government-Wide Forward Regulatory Plans process, which requires publishing plans for two to three years ahead;, the Cabinet Directive on Regulatory Management (CDRM, as described in chapter 4); the Red Tape Reduction Action Plan, which is part of Canada's Economic Action Plan for fiscal and economic recovery, initiated in 2008; and the Action Plan's efficiency-oriented Service Standards for High-Volume Regulatory Authorizations, which is subject to a 2013 Treasury Board Directive. We map the Regulatory Modernization program reforms first and then look back at Health Canada's science- and evidence-based regulatory and institutional history regarding the regulation of drugs and related health products.

Health Canada's initial modernization paper, *Blueprint for Renewal* (2006, 2007), anchors key reforms being carried out since then. The *Blueprint* agenda was linked to changes underway in the United States at the US Food and Drug Agency (FDA), which added new guidelines for post-market review that required new forms of evidence to support the assessments (Evans 2009; Fox 2008). The Health Canada *Blueprint* covers both health products and food, but our summary here focuses on the drug regime aspects of the report. A recent status report emphasizes that a number of *Blueprint* initiatives have been achieved or are underway after several consultation processes (Health Canada 2014a), including initiatives and changes relating to registration of and disclosure of information about clinical trials, cost recovery, progressive licensing, and the review of public input concerning regulated products and related policies.

The *Blueprint* pointed out that since 1953, the Food and Drugs Act and its regulations have "largely intended to be a consumer protection statute" (Health Canada 2006, 6). Accordingly, Health Canada's new approach to regulation identified the following five challenges (6–7):

1 An outdated regulatory toolkit that is increasingly limited and inflexible in responding to today's health products and food environment
2 The regulatory system's current incapacity to consider a given product through its entire life cycle, from discovery through to examining the "real world" benefits and risks of a health product or a food on the market
3 The impact of social and economic changes such as accelerating scientific and technological advances, the rise of transborder health and environmental threats, and more informed and engaged citizenry
4 A regulatory system that currently works in isolation from the activities and policies at the stage of research and development, and those of the broader health care system
5 A regulatory system with insufficient resources for long-term efficiency and sustainability.

Among the recommendations for change in the *Blueprint for Renewal* (2006, 7–24) are suggestions for moving to:

- A product life-cycle approach, including suggestions such as the introduction of pharmaco-vigilance plans as a requirement for pre-market submission review
- Regulatory interventions proportional to risk, a process that would require the revamping of the product categorization system
- A proactive and enabling regulatory system, so as to not only keep pace but to be ahead of the trend where possible (partly through greater regulatory foresight programs and activities regarding new and changing technologies)
- A system that makes the best use of all types of evidence, through a process of complementing the current pre-market assessments with more extensive post-market monitoring
- An emphasis on specific populations, regarding patient drug responses, and including a capacity to deal with new tailor-made products for diseases that affect smaller patient or genetically specific populations
- An integrated system.

Each of these separately and in combination involves complex changes in Health Canada's S&T and evidence-based capacities overall, and at specific stages requires different kinds of science- and evidence-based decision making and discretion. Health Canada's Science Policy Directorate (SPD) is a key part of the department's S&T capacity (Health Canada 2014b). The Science Policy Directorate states that it is tasked with ensuring that "science is reflected fully and appropriately in departmental programs and planning" and that "science is considered in evidence-based departmental decision making" (ibid., 1). But after recent expenditure reviews, this directorate had its staff cut massively so that it now has fewer than fifteen persons.

The full drug review under the Progressive Licensing Framework (PLF) as set out by Health Canada (2009b, 2) consists of the following six stages:

1 Pre-clinical studies
2 Clinical trial authorization
3 Special access program
4 Submission review
    - Regulatory product submission
    - Safety, efficacy, and quality review
        - Drug identification number application
        - New drug submission
    - Market authorization decision
5 Public access
6 Post-market
    - Surveillance, inspection, and investigation
        - Adverse reaction, post-market changes
        - Drug identification number change
        - Changes to a new drug.

The drug development and review process ultimately starts with the drug firms and applicants, where new candidate drug products emerge through a form of internal self-regulation and scientific review (Doern 2000). For the rest of the pre-market stage, the core relationships tend to be between drug firms and other research and medical applicants (and their S&T staff) and Health Canada science assessors and reviewers. Consequently, these relationships are intensely front-line, bilateral, and scientifically and technologically detailed in nature.

At the heart of the broadened system is an extended post-market process under the Progressive Licensing Framework (Health Canada 2009b). Health Canada set out the logic and key features of the PLF:

- The central concept of Progressive Licensing is that, over time, there is a progression in knowledge about a drug. The emphasis of the new framework is to identify opportunities within this progression over the full *life cycle* of a drug, rather than placing the focus primarily upon pre-market assessment, safety, and efficacy. This represents a fundamental shift from the idea that the pre-market testing of a drug assures its safety and efficacy. The new proposed model is that a drug should be evaluated throughout its life cycle for its benefit-risk profile.
- The framework is meant to support *evidence-based* decision making (throughout the life cycle); this is later explained in relation to how a scientific standard of evidence exists among regulators for clinical efficacy and safety and that this standard "requires positive validated outcomes from adequate, randomized, controlled, and confirmatory studies" (8). However, the broadened notion of evidence-based analysis arises from the fact that "benefit-risk is founded upon scientific evidence of safety and efficacy but also encompasses a larger scope of evidence regarding contributing circumstances, including effectiveness" (8).
- *Good planning* involves planning at "every viable step in the regulatory process, which would allow for a proactive approach to managing both expected and unexpected issues" (2).
- *Accountability* is associated with all aspects of the framework, including the underlying accountability of both Health Canada and drug manufacturers. These include "the ability to make enforceable conditions upon issuing an initial market licence, so that, for example, certain field reporting commitments or further studies are required to be completed" (3).

The aspirational ambition of the new extended system is understandable when set out in the above manner; nevertheless, it must be put in the context of earlier arguments, ideas, and pressures in the Canadian and international context and in relation to the earlier pre-market safety assessment processes embedded in the regulatory system for drugs and for other health products.

Earlier institutional change came with the formation of Health Canada's Therapeutic Products Programme (TPP) in 1997 (Doern 2000). The Therapeutic Products Programme was an amalgamation of the former Drugs Directorate and the Medical Devices Bureau of Health Canada to better deal with three product groupings: pharmaceuticals; medical devices; and biologics, radio-pharmaceuticals, and any product interactions. The 1997 changes supported traditional safety mandate issues but also introduced more explicitly a range of new goals, such as timely access to products and a more innovative regulatory system. During the 1990s, three studies – the Gagnon and Hearn reviews of programs regulating drugs and medical devices and the Krever Commission of Inquiry on the Blood System of Canada (1997) – began to nudge the TPP from its traditional role as health- and safety-focused, science-based regulator to a science-based risk-benefit manager (Doern 2000).

The report of the External Advisory Committee on Smart Regulation (EACSR 2004) was also influential, as it used the drug review process as one of its five case studies. The Advisory Committee agreed that safety should be the system's paramount concern but argued that in spite of recent improvements in the drug review and approval process, as of 2004, it still took too long. The report, entitled *Smart Regulation: A Regulatory Strategy for Canada*, highlighted the need to improve the regulatory regime to remove "disincentives for the development of new pharmaceuticals in Canada and lost opportunities for innovation" (79). It argued that "Canada cannot support a review agency as large as the US Food and Drug Administration (FDA), nor can it afford to carry out drug reviews as extensive as those of its American counterpart. It must be strategic in its use of its limited resources" (82). *Smart Regulation* advised Health Canada and its regulatory partners to look more strategically at using international regulatory cooperation, particularly with the United States but as well with others. It called for Health Canada to consider specializing in certain areas of product regulation and accept the decisions of other national regulators (especially those of the Americans) in realms where they had comparative advantage in expertise and overall capacity.

There was also growing pressure from the drug industry about the need for greater regulatory speed and efficiency, particularly given the increasing number of products queuing up for review and approval. Globally, the situation was (and is) characterized by Big Pharma firms whose patents on highly profitable large market blockbuster drugs are now ending and which, therefore, are looking for new drug inventions to fill their pipelines. Big Pharma has two strategies: developing new drugs internally through their own research and testing or through buying up smaller, more-innovative but capital-starved companies, especially biotechnology start-up firms (*Economist* 2007a, 2007b).

Canada's pharma sector is largely characterized by multinational research drug firms located in Montreal doing discovery research, and generic drug firms doing production-based research mainly in the Toronto area. Drug treatment breakthroughs are increasingly fewer in number, and the costs of developing them are soaring (Jack 2009, 2010), which has added pressures on regulators to speed up clinical trials on patients. As Andrew Jack observes, "The more quickly that promising new medicines are launched, the greater the revenues before patents expire. By the same token, the more swiftly that drugs with problems are identified, and abandoned before expensive late-stage testing begins, the lower the wasted development costs" (2009, 7).

In terms of product volumes, the logic of drug development suggests that there will be even higher volumes of pre-clinical studies and clinical trial authorization processes than actual eventual new drug applications, all of which have to be assessed by firms, applicants, and Health Canada. Some analysts are critical of the innovation-driven and access-driven logic of the Health Canada *Blueprint* reforms and of the *Smart Regulation* logic, arguing that it is too much a business-driven way of thinking at the expense of the public health and safety approach. By conforming to the market approach, Heath Canada is ultimately weakening its core safety-focused regulatory mandate forged after the 1960s thalidomide catastrophe (Lemmens and Bouchard 2007).

At the post-market stage, the players, expertise, and knowledge involved widen enormously: a complex network of tens of thousands of patients, health professionals, and health institutions become engaged in a daily, even hourly, part of the post-market review process, thereby generating complex information and reporting networks. Moreover, the system is fundamentally multilevel, including federal, provincial, local, and international agency involvement (Murphy 2006). Relationships with the FDA in the United States are also extremely close and frequent at the technical and regulatory level at both the pre- and post-market stages.

As indicated above, the post-market phase includes processes for reporting on adverse reactions. Much of the adverse reaction reporting has been a part of the concept of *pharmaco-vigilance*. Under the Health Canada reforms, pharmaco-vigilance is being sought even earlier in the process through planning to detect adverse reactions earlier in patient care, possibly even in the very early non-clinical testing stage. Many of the planned provisions for post-market oversight are likely to produce more information and data demands by applicants at the pre-market stage as well.

There is little doubt that Health Canada as a department overall, and particularly in the drug regulation area, has become a very different science-based department than it was in the 1960s. The shift from pre-market safety evaluation is only part of the challenge. The imperative to respond to higher numbers of new products and the need to adopt post-market regulatory oversight is daunting. The forms of science and evidence have changed markedly, making the evidence and decision processes less accountable and transparent and more open to claims of undue power and influence.

1971–2015: Environment Canada and Its S&T Strategies

Environment Canada is a major S&T departmental entity that has been transformed by its changed and expanded environmental policy and governance mandate. Our account of its policy and governance history and its S&T and innovation features centres on three broad periods where there were both continuities and changes: the early gestation in the 1970s to late 1980s period culminating with the Mulroney Green Plan; the Chrétien-Martin era from 1990 to 2005, when sustainable development (SD) and related innovation policy pressures and discourse emerged; and the Harper era since 2006, which both continued some of these policies but also via its ten-year environment science plan sought to capture the changing nature of risk analysis and its greater focused interest on natural resources, as highlighted in earlier chapters.

In each of the eras we are discussing in this book, we are interested, as was the case with our National Research Council and Health Canada histories, in how the management of S&T evolved structurally and in relation to managerial reorganization and discourse. In an overall sense, in each era Environment Canada's S&T and innovation policy sought to prepare for the science needed in the coming years, but it found itself playing a continual game of catch-up where its reach exceeded its grasp. In contrast to the Health Canada S&T history, the analysis does not delve into any particular agency or lab within Environment Canada.

When created in 1971, Environment Canada had a mandate driven by the Department of the Environment Act, the Fisheries Act, the Canada Water Act, and the Clean Air Act, and thus with S&T in relation to overall realms of water, air, and land (Whittington 1980). The department was from the start both a regulatory agency and an advisory body, buoyed by initial strong public support for resource conservation that provided backing for its own initially quite broad areas of authority, including managing the environmental aspects of statutes administered by other federal departments, early on, including the Motor Vehicles Act and the Pest Control Products Act.

From the outset, Environment Canada has had to accommodate science- and evidence-based decision making with socio-economic imperatives. Although scientific knowledge, expertise, and evidence are crucial, the department has always had to complement them with front-line evidence from citizens, NGOs, and professional knowledge holders (Enros 2013; Doern, Auld, and Stoney 2015; Doern and Conway 1994).

In its early years, the science and evidence base of Environment Canada came from the scientific and technical staff that joined the department as other government departments became a part of the new portfolio. Whittington (1974, 212) observed that "although the functions of the new department officially include the carrying on of established programmes related to the 'management' of natural resources such as marketing research and assistance to the fishing and forest industries, the overall focus of the department has shifted significantly towards the protection and enhancement of the quality of the environment." The establishment of the Environmental Protection Service in 1971 was intended to provide several points of contact between the department and the public, including an education role.

The $3 billion 1990 Green Plan was the culminating event of the Mulroney era in that it marked the apex of its priority and a financial heyday for Environment Canada and to some extent its ability to support new emerging S&T demands. The impetus for the Green Plan was helped considerably by a series of environmental crises and accidents internationally, including the Bhopal chemical disaster in India in December 1984 and the Chernobyl nuclear reactor meltdown in the then–Soviet Union in April 1986 (Doern and Conway 1994).

The Green Plan emphasized the new global realities of environmental problems and took advantage of the strong support for environmental issues in public opinion polling both before and after its announcement. The Green Plan advocated the concept of sustainable development and contained a comprehensive series of goals and more than a hundred specific initiatives (Environment Canada 1990; Hoberg and Harrison, 1994). The projects included the regulation of up to forty-four priority toxic substances within five years and plans to reduce the generation of waste in Canada by 50 percent by the year 2000. The Green Plan helped give Canada a lead role in the 1992 Rio Earth Summit.

By the early 1990s, Environment Canada's direct legislative base had been extended to include the Canada Wildlife Act, the Canadian Environmental Protection Act, the International Rivers Improvement Act, and the Migratory Birds Convention Act. Its advisory role expanded to include the Arctic Waters Pollution Prevention Act, the Canada Shipping Act, and the Transportation of Dangerous Goods Act, to name only a few. As the department expanded and absorbed other statutory mandates it again had to find strategic ways to acquire and share science, analysis, and data with other departments with whom it had shared mandate roles on a statute-by-statute basis.

Key insights on this period but also extending into the Chrétien-Martin era from 1994 to 2005 are offered by Philip Enros (2013), a former senior Environment Canada official. His focus overall is on "policy for science" in the department and in particular on the extent to which Environment Canada could or could not develop what Enros characterizes as "general strategy" for science, a level of crucially needed thinking and action on science in an environmental department. In one sense this involves for Enros potential stages such as a policy statement on science, setting R&D priorities, and a research agenda. Enros relates these empirically to Environment Canada's strategic approaches or the lack thereof, northern and international S&T, ensuring capacity, providing infrastructure, and managing personnel. He also examines Environment Canada's efforts regarding collaboration with other federal departments and agencies, with universities, and in relation to developing innovation in the environmental industry.

Enros concludes that "policy work on overarching strategies for science did not meet with great success" (2013, 228). Environment Canada's ability to make the case for its scientific capacity was a struggle in part due to the "subordinate position of science in the department, its fragmentation into Services, and the inherent difficulty in determining the level of capacity that was required" but even more due to the federal government's "struggles with deficits and its inclination to build up S&T capacity in the university and industrial sectors" (228). In essence, universities and industries were Environment Canada's direct budgetary competitors for S&T funding and staff.

Overall, Enros observes that policy work on Environment Canada's science "was episodic" and was carried out in "response to specific departmental management needs, the struggles of senior managers to get the most from tight or declining budgets, and the ambitions or frustrations of the Department's scientific personnel" (2013, 227). He does not discuss the role of Cabinet ministers, but the fact that environment ministers changed every one and a half years on average during this period certainly was not conducive to developing long-term views and plans on any kind of environmental science strategy (Doern, Auld, and Stoney 2015).

Analysis of Environment Canada in the Chrétien era throws light on three trends: a "stretching of green science," the reduction of "patient science," and the expansion of "science-on-demand" (Doern 2000b). The analysis pointed to five drivers of change: (1) less money and fewer internal staff resources because of heavy Program Review S&T cuts; (2) the need for scientific effort to match immediate needs within the department and across the government (leading to the science-on-demand dynamic); (3) less institutional freedom to "roam" and explore longer-term research based on the instincts of scientists (leading to a decline in patient science); (4) a rise in "global" science, which put departmental scientists in touch on a daily basis with counterparts in other countries; and (5) a shift away from repeated measuring and monitoring to the greater use of statistical sampling approaches (ibid., 293–5). The combination of these five pressures was interwoven with the need to respond effectively to the sustainable development "triple bottom line" discourse regarding the balanced analytical consideration of environmental, economic, and social factors in decision making.

In the mid-2000s Environment Canada's view of its S&T goals and structure changed to meet the changing times nationally and globally. In 2005 in the Martin Liberal era a departmental science review panel recommendation led to the establishment of the Science and Technology Branch and brought more of Environment Canada's science under a single assistant deputy minister. The Harper era began with the development and announcement in 2007 of *Environment Canada's Science Plan* as a "a vision for managing and conducting Environment Canada's science over the next 10 years" (Environment Canada 2007, vii). The Harper government sought to take a longer view of its S&T in environmental matters. The department's plan is "to deliver the high quality knowledge, information and data that enable the Minister, the Government, the Department, and other decision makers to enhance the *safety* of Canadians, protect the *quality of the natural environment*, and advance Canada's *long-term competitiveness*" (ibid., emphasis added). This trio of purposes and values, especially the first and third stated purposes, had not been the preferred discourse in earlier periods.

Environment Canada's three strategic directions that were identified for the department's science were expressed as (1) developing an integrated monitoring and prediction capability, including "moving to finer time and spatial scales for predictions and analysis"; (2) understanding cumulative risks, including "multiple stressors interacting over time" and also the "risks to, and opportunities for, Canada's long-term competitiveness resulting from a changing environment"; and (3) managing risks, optimizing opportunities, and building resilience, including "understanding the costs and benefits of the various risks" (ibid., viii–ix).

The *Science Plan* reiterated that science represented about 70 percent of the budget and two-thirds of Environment Canada's employees. But it noted that two-thirds of the science was not R&D but rather covers "a variety of *non-research science activities* including weather forecasting, risk assessments, regulatory activities, data collection and environmental monitoring, emergency preparedness, and *S&T knowledge brokering*" (ibid., 1, emphasis added). "Non-research" science essentially meant related science activities as set out in other federal S&T reports and thus included the science knowledge in the brains and experience of mainly front-line science and knowledge professionals typically interacting on a daily basis with their counterparts in other national and intergovernmental bodies but also universities, businesses, and interest groups (Kinder 2010; Doern and Kinder 2007).

The *Science Plan* further highlighted that Environment Canada's science had to reflect the government's 2007 Federal S&T Framework and its three core principles of *alignment* to reflect and support the priorities of Canadians, *linkages* built on effective collaborative relationships, and *excellence* by producing the "highest quality, leading edge, credible and unbiased environmental science" (ibid., 4). The *Plan* stated that "above all, we must demonstrate transparency and openness in how we conduct our scientific activities" (4).

Structurally, Environment Canada in the Harper era became a more complex managerial entity. It now had eight assistant deputy ministers, each heading a unit, five of which are more policy and managerially oriented and three of which are the modern locales of its historic three initial science-focused services. The main regulatory environmental protection roles are headed by an assistant deputy minister for "environmental stewardship." The assistant deputy minister for S&T has the main atmospheric, water, and wildlife realms plus the overall S&T policy coordination role. A third assistant deputy minister heads up the Meteorological Service of Canada (Environment Canada 2013).

In the Harper era the notions of science and broader kinds of evidence were given more play and focus in two further senses. First, the Harper government embraced, aspirationally at least, the life-cycle approach that it intended to adopt in all health and environmental regulation and indeed in overall regulatory policy (as discussed in chapter 4). These implied, as seen in the Health Canada story, both pre-market and post-market science that draws on knowledge and evidence-based capacity of a very extended and networked nature. The second sense in which the "evidence" debate emerged in the Harper era centred on claims protesting the "death of evidence," in short, the notion that environmental policy and perhaps policies of many other related kinds were increasingly centred not on systematic knowledge and research but simply on opinions and on made-up facts of the moment, often tied to continuous attack-politics and the muzzling of scientists by communications staff at the centre of prime ministerial government (Gatehouse 2013; O'Hara and Dufour 2014; Toner and McKee 2014).

THE THREE ELEMENTS

With the aid of Table 5.2, we now trace the three elements of our analytical framework in relation to the evidence provided by the above three policy and governance histories in the government S&T departments and agencies domain.

Policy Ideas, Discourse, and Agendas

Not surprisingly, the three policy and governance histories reveal both diverging and cross-over ideas, discourse, and agendas. As the longest established entity, the National Research Council of Canada had an initial long period of supporting nation building through science. By the late 1960s, however, episodes of advocacy and pressure helped to make the NRC more commercially focused. This led to some internal change but also considerable resistance from NRC's scientists in its divisions and institutes. New pressures emerged in the 1990s and beyond. Ideas about local and regional systems of innovation emerged in the Chrétien era. Most recently, the current NRC leadership has pushed hard on a business and commercialization focus, recasting the NRC as an industry-facing research and technology organization offering "concierge" services, designed to respond to "market pull" rather than "technology push."

Health Canada's overall regulatory approach to health products and drugs begins in the 1960s with a focus on pre-market consumer safety in response to the thalidomide crisis globally and nationally. Although those safety ideas and values are still a key part of the underlying culture in Health Canada, in the late 1990s it increasingly saw itself also as a science-based, risk-benefit regulatory manager. Not until the 2006 *Blueprint for Renewal*, however, did its discourse and agenda recognize

Table 5.2 Policy and governance histories in the S&T departments and agencies domain: Three analytical elements

Policy and governance history	*Policy ideas, discourse, and agendas*	*Economic and social power*	*Time, temporal realities, and conflicts*
1945–2015: National Research Council	• Building Canada's national science capacity • Science policy • Second World War defence research • Granting council role and peer review; • Small firms support via IRAP program • NRC as a university without students • Calls for greater business focus • Spin-off companies • East Asia and greater international profile • Local/regional systems of innovation • Holistic approach in industry focus • Market pull, not technology push • Scientific infrastructure • "Concierge" service portal role	• Early combined economic and nation-building power anchored by key • Liberal ministers in Second World War and post-war era • NRC as first science policy maker and power centre • Periodic pressure by business and business-oriented • Conservative governments for NRC to be more business-focused • But variously described labs/institutes/divisions/portfolios played diverse socio-economic research and innovation roles • 1990s–2000s emphasis on contributions to local-regional innovation systems	• Science capacity building from Second World War to late 1960s took decades to achieve or show progress • Calls and pressures to make NRC more business and commercially focused was cyclical but across each recent decade there was some progress but also institutional push-back • Managerial push to change research divisions to institutes to portfolios but also underlying continuity for some of the these units across the decades despite the name change • Temporal issues and complexities in determining when something is market pull vs. technology push
1980–2015: Health Canada and drug and health product regulation	• 1960s thalidomide crisis and pre-market safety primacy • Science-based risk-benefit manager • Regulatory modernization and Blueprint for Renewal since 2006 • Canada as among healthiest people in the world; • Product life cycle, pre-market and post-market • Boundary blurring regarding types of products such as drugs vs. medical devices • Transborder threats • Pharmaco-vigilance;	• A broad socio-economic department with health seen more as a sector than an industry • Consumer safety and protection borne out in thalidomide crisis in 1960s and related consumer movement • Greater economic and industry focus as access to drug and health products was pushed and as high volumes occurred • Pre-market and then post-market regulatory push and also Smart Regulation approach came from social and economic interests and also global and US FDA reforms	• Long-term staying power of thalidomide pre-market safety idea and structure • Took up to two decades to acknowledge weaknesses in regulatory system via Blueprint for Renewal • Post-market product monitoring approach is inherently longer term (decades and inter-generationally) • More pharmaco-vigilance at pre-market stage means longer pre-testing and assessment processes • Pressure for faster approvals cast as a service delivery standard

	• Regulatory interventions proportional to risk • Specific populations focus as well • Progressive licensing • Evidence-based policy and regulation • Smart Regulation agenda • Excessive business-driven reform agenda	• Department's multiple directorates subordinate in hierarchical terms, but also have power based on expert knowledge and science of a more detailed kind than those at higher managerial level and also even more so, of Cabinet ministers	
1971–2015: Environment Canada and its S&T strategies	• Overall early reference to key natural realms of water, air, and land • Natural resource conservation • S&T ideas always in context of 20+ mandate statutes, and role regarding other department's statutes with environmental content • Science- and evidence-based from the outset, including public education role • Environmental quality • SD as triple bottom-line balance of environmental, social, economic values in decision making • Safety, quality, and long term competitiveness • Cumulative risks and multiple stressors interacting over time • Life-cycle ideas of environmental monitoring • Knowledge brokering • R&D and RSA as science based, especially RSA • Strategic science policy for environment • Death of evidence and muzzling of science critique	• Formation in 1971 due to initial socio-environmental interest group and voter pressure, also surge at time of 1990 Green Plan, due in part to international environmental disasters • Strength of nature and conservation interests • SD as international concept • Periodic influence of citizen science • S&T budget cuts in the department in 1990s Program Review and in 2008–15 recession • Overall subordinate position of science in the department • Responsible resource regulation as pro-business power in Harper era • More dispersed environmental and SD interests as statutes widened and deepened mandate • Death of evidence and muzzling of science actions, pressures, and external criticism	• Different time frames inherent in sustainable development approaches and their S&T underpinnings • Overt life-cycle regulatory and science-based approaches, involving decades of monitoring capacity • Long-term aspirations of 1990 Green Plan • Difficulty of developing medium-term strategic policies for environmental science within Environment Canada • Harper era 2007 environmental science plan as a 10-year plan -Short-term turnover of environment ministers (1.5 years on average) • Meaning of long-term competitiveness goal as expressed in 2007 plan

Note: see the chapter text for expansion of abbreviations in this table.

the need for more complex ideas that picked up on both the new concepts of smart regulation and the compelling need to add a product lifecycle dimension that extended oversight massively into post-market use and monitoring. Ideas such as pharmaco-vigilance, progressive licensing, and proportionality in rule making and enforcement are now part of the department and its food and health products ethos and mandate. In the context of global and national business pressure and the reality of much higher volumes and diversity of products, commercial innovation has become a key part of both its agendas and discourse. A new part of the discourse is the notion of a much more complex evidence-based system involving, especially in the post-market realms, numerous networks and types of medical and health knowledge professions.

For Environment Canada, evidence-based notions were inherent from the outset, in part because of the complex nature of its early statutes and their focus on broad systems of air, water, and land. Strong conservation ideas were also present and indeed predate the establishment of the department per se. Environmental S&T and innovation ideas shifted when concepts of sustainable development emerged in the 1980s. These required approaches embodied by the triple bottom-line of environmental, social, and economic sustainability. In the Harper era, environmental priorities have shifted to encompass "responsible" resource development, environmental safety, and long-term economic competitiveness. Some critics assert the dystopian side of this change is that it has muzzled science and overtly attacked some environmental groups, heralding for some "the death of science."

Economic and Social Power

The economic and social configurations and dynamics across the three policy histories reveal instances of economic or social power ebbing and flowing, leading to changing institutional mandates and new departmental and agency managerial cultures. The NRC story shows a period of combined economic and nation building anchored in a period of postwar reconstruction centred around then-dominant Liberal ministers and science leaders in the Liberal era. They insured that the National Research Council of Canada was the power centre regarding science policy well before the larger science policy era began. Business pressure was present in some of this period, but the business lobby tended to be explicitly engaged only under later Conservative governments, where we can see them pushing for a business and commercialization focus.

In the cases of Health Canada and Environment Canada, socio-economic interests and lobbying both outside and inside the state were observed. For Health Canada this came in the form of the thalidomide crisis and the related mobilization of early stages of the consumer movement. For Environment Canada it was the birth of the environmental lobby in the late 1960s that led to the formation of Environment Canada and its early science-based capacities. For both departments, business and industry interests and different notions of consumerism changed the power configuration in terms of both regulatory mandates and their statutory authority over products or projects. The shift from a pre-market focus to a combined pre- and post-market focus in Health Canada product regulation processes came in the past twenty years, but it was complicated by pressure from business, especially Big Pharma, and consumers, both of which pushed for more timely access for the latest health products.

For Environment Canada, its expanding mandate came into direct conflict with those seeking to operate in the economy. Over the entire period of our history, there was growing business concern regarding the nature and conduct of environment policy and its underlying science, evidence, and management under the ethos of sustainable development. Throughout the period, firms remained interested in how smart regulation and reforms and better concepts of risk-benefit might improve the speed and efficiency of regulatory assessments of industrial developments. The relative role and influence of S&T was thought to be in decline and under attack, partly due to two recession-era budget cuts and partly because, as the S&T and innovation knowledge economy became ever more complex, it was more difficult to discern what kinds of successes or failures may be occurring.

One remarkable feature of these stories was the relative influence, stability, and staying power of the National Research Council, Health Canada, and Environment Canada, especially their labs and their institutes, divisions, portfolios, and directorates. These units played core roles for Health Canada and Environment Canada, in particular because of their statutory and regulatory mandates, which helped to protect front-line R&D, related science activities, and monitoring capacities. Although they were a part of hierarchical structures that senior management and ministers could "reform," they also possessed more direct operational science- and evidence-based knowledge and expertise than those further up the hierarchy.

Time, Temporal Realities, and Conflicts

Time, temporal realities, and conflicts are more difficult to trace in the three policy histories. Currently in the NRC, the notion of supporting research and analysis only or mainly when proposals exhibit market pull versus technology push are replete with temporal judgments or assumptions on a case-by-case basis. Temporal features were also inherent in the periodic pressure to transform the NRC into a more business-focused and commercial entity, often followed by considerable push-back from front-line scientists. In some ways, this is a generational battle that exists in many similar institutions around the world.

For Health Canada, temporal issues and factors were exhibited by the staying power of the primacy of safety in the post-thalidomide era. A different sense of time can be seen in the 2006 *Blueprint for Renewal* process that advocated a much more elongated time scale of progressive licensing, including post-market monitoring, in response to drug recalls and re-licensing. The time dimension has also been at the centre of demands for speedier access to new health and drug products, many of which are now available earlier through the global trading world.

For Environment Canada, there is both pressure for speedier project approvals and for life-cycle or post-market risk management, with the result that active engagement and monitoring can span decades. The department has always had to deal with different time scales: pollution clean-up and remediation operate on a completely different schedule than preventative evaluations and reviews, even those conducted under the logic of sustainable development.

CONCLUSIONS

This chapter has mapped the government S&T departments and agencies domain through a closer look at the above policy histories of the National Research Council of Canada, Health Canada, and Environment Canada across several decades. It is worth reiterating that this analysis represents the tip of a much larger effort that includes a number of other science-based departments and agencies, such as Natural Resources Canada, Agriculture and Agri-food Canada, and Fisheries and Oceans Canada.

Although our focus overall is on S&T and innovation policy in this domain, it has been shown that it is impossible to tell the S&T and innovation policy and governance story without a clear and constant sense of the substantive statutory, regulatory, and policy mandates of these departments and agencies. Although federal S&T and innovation policy overall is directed at all such entities in the federal government, it is nonetheless seen by them mainly in terms of how it plays out, makes sense, must be finessed, or even ignored in relation to their mandates. They are each driven by science and evidence but only in the context of the practical mandated ideas, agendas, and power structures they must work through and are shaped by. Of the three entities we have explored, only the NRC has "research" in its title. But each of the three has had to devise its own S&T and innovation policies and de facto practices as a public interest S&T entity.

The array of S&T and innovation ideas, discourse, and agendas varies across the organizations, but each in their own way has exhibited both periodic narrowing and broadening of its mandate as it attempts to respond to official government-wide STI policy. Business power has arguably grown stronger across the entire domain, but there have also been changes in the nature of consumer and health interests in the face of both safety concerns and desires for better and faster access to new drugs and health products. The time and temporal dimensions of the domain have been explicitly recognized largely in the shift from a pre-market to a combined pre- and post-market life-cycle approach. We also see this in the National Research Council of Canada, which has had to adapt its workflow to the urgency of market pull (rather than the more internally scripted pace of technology push), while Environment Canada has faced an atmosphere of rising concern (some would say a sense of urgency) of the need to come to grips with the cumulative risks and multiple stressors in the environment.

With regard to our overall conceptual and empirical interest in the innovation economy and society nexus, this domain demonstrates interesting dynamics and tensions. Government exhibits a more obvious presence of innovation society dynamics as a partial counterweight to the norms and pressures of an innovation economy. The Health Canada policy history, where at present the stated mandate relates to making Canadians the healthiest people in the world, remains historically anchored in lessons from the 1960s thalidomide crisis. This is reflected in the current broad-based aspirations to create a full-scale pre- and post-market progressive licensing system for health and food products and its needed involvement by health and medical professions and patient and consumer interests nationally and provincially. But the analysis also shows an increasing recognition of health as an industry, including, of course, the pharmaceutical industry, which has generated the need for speedier approvals for drugs and products. The Environment Canada policy history showed the socio-ecological breadth of its mandate and the presence of social science in its functioning in addition to the natural and environmental sciences. It also showed the difficult managerial dynamics in developing viable S&T strategies in the midst of budget cuts and intense competition for funding across the Environment Canada divisions and independent agencies. There was also push-back in economic terms, seen in the Harper era's combination of safety, quality, and "long-run competitiveness" goals and aspirations. The NRC policy history showed that major reform efforts were each aimed in different eras at realizing better business and economic innovation and performance regarding industrial R&D both through its changing divisions and institutes, through programs such as the Industrial Research Assistance Program, and through support for innovation clusters. But there was always related health and safety content and push-back, including in the area of research into the life sciences. The role of the National Research Council in nation building also remains salient, including in the ways that it seeks to show how its research and its researchers are distributed across Canada.

In each of the three domain policy histories, there is ample evidence of the three organizations functioning as both hierarchies and networks, but their constituent divisions, institutes, portfolios, and directorates have had in some ways the independence and space to pursue separate agendas driven by front-line S&T needs. Later chapters will add to this evidence and these realities by examining related entities functioning within the industrial S&T, intellectual property, agriculture, food, biosciences, and biotechnology, genomics, and Internet domains.

6

The Granting, University, and Levered-Money Domain

The granting, university, and levered-money domain is a key domain in the federal S&T and innovation policy and governance system. The granting part of the domain refers initially to the three current, independent federal granting councils, the Natural Sciences and Engineering Research Council of Canada (NSERC), the Canadian Institutes of Health Research (CIHR), and the Social Sciences and Humanities Research Council of Canada (SSHRC) that provide funds to university academics and graduate students. The funding is based on competitions via research proposals that are peer or merit reviewed, ranked, and awarded or rejected. In addition to dispensing funds to researchers (and therefore to universities), these bodies are also increasingly quasi-regulatory bodies in that they set the rules regarding research ethics and define what is good or desired research through the style of peer and merit review and the structure of competitions themselves (e.g., selection committees, targeted fields, required mixes of disciplines, and leveraging requirements).

These institutions define and create two types of funding competitions: smaller, curiosity-driven projects centred on individual or small teams of investigators, and large- scale, thematic projects, networks, or programs that involve a diverse team of investigators working on a designated priority area. Both types of competitions, but especially the large-scale competitions, increasingly require for approval some leveraging from university, industry, government, or foundations of infrastructure, money, expertise, or other in-kind contributions (Doern and Stoney 2009). Overall, in governance and political terms, the domain involves discourses about the meaning of research independence; the appropriate role of executive government, business, and society in research; and the changing role of scientists themselves, who in this new model become their own advocates and active funding lobbyists (MacKinnon 2014).

The domain story overall begins with the creation of the foundational granting bodies in 1977 and follows key changes in the system. In the late 1980s the Networks of Centres of Excellence (NCE) programs started. In 1997 the Canada Foundation for Innovation (CFI) was created to fund research infrastructure for universities but also hospitals and related medical research facilities and equipment. In 1999 Genome Canada, the largest funder of the biosciences, began to operate, spending almost $1 billion over its first twelve years. In 2000 the Canadian Institutes of Health Research replaced the former Medical Research Council (MRC), and the Canada Research Chairs (CRC) program started. What began as a logically structured set of single-purpose research funding organizations is now a portfolio of aggressive research agencies that variously compete and complement each other's efforts to build Canada's science capacity.

Federal government funding is the main focus in this domain chapter. Provinces are responding, albeit from a modest base. As recently as 2004, provincial expenditure on R&D was just under $1.4 billion (Statistics Canada 2014). By 2014 it is estimated that provincial funding for R&D jumped by about 50 percent to more than $2 billion, compared with an overall rise of only about 15 percent from all sources. Federal expenditures over the corresponding period rose about 20 percent, which led to an overall decline in their share of a correspondingly weak national R&D performance. Although provinces posted a large relative change, provincial expenditure on R&D represented only about 6 percent of national gross expenditure on R&D, as share of GDP (GERD), with more than 80 percent coming from provincial governments in Ontario, Quebec, and Alberta. Provinces are even more insignificant as performers of R&D. Provincial departments and agencies in-house R&D grew about 17 percent over the 2004–14 period, but in aggregate their direct efforts contributed only about 1 percent of total national effort. Some might argue that provinces exert influence through the higher education sector, which in aggregate funds about 17 percent of national outlays and undertakes just under 40 percent of the total research effort in Canada. Although universities are unambiguously creations of provincial governments, all universities and many provinces would assert that their academic independence severely limits the potential for provinces to have much real influence on research efforts. For the most part, universities treat their home provinces much as any other funder – if they provide funding they might undertake the desired work. Hence, while provinces are becoming more important in a number of discrete ways, the relatively small and divided efforts pale in comparison with the federal effort.

The provinces, however, are important because they are the sole regulators and key funders of universities (Tupper 2009). The provinces also have their own granting agencies though typically these are smaller and combine disciplines and fields in one agency. This would imply that much of the politics and governance in this sector would be directed through formally structured, federal-provincial and interprovincial intergovernmental fora; however, the reality is that universities project an independent voice, directly lobbying in Ottawa through their national lobby group, the Association of Universities and Colleges of Canada, now Universities Canada (AUCC 2014; Morgan 2009). Complicating this story is that each province's set of universities (and community colleges, technical schools, or polytechnics) has its own traditions and trajectories of development, playing different local, provincial, regional, national, and international roles socially, educationally, and economically (Fisher et al. 2014: MacKinnon 2014).

A further key driver for the new roles of universities is that they are being compared and ranked on capacity for research and teaching excellence both nationally and internationally with those of other trading or neighbouring countries, including the United States, the United Kingdom, and the countries of Europe (*Economist* 2015a). They are, in effect, competing as more commercialized, market-focused entities in a knowledge-based innovation economy and society, both in relation to their research and their production of graduates (often labelled by policy-makers as highly qualified personnel, or HQP). Universities are also increasingly seen as a key part of Canada's human capital system (Berman 2012; Chan and Fisher 2008; Doern 2008).

When thinking contextually about universities in this chapter, it is useful to see universities in diverse ways, including as corporate entities that are creatures of the provinces; partners with business, government, and communities; agglomerations of teams of scholars rather than simply hosts for individual researchers; producers of highly qualified personnel; sites of basic and applied science, engineering, health and social science researchers, both in disciplinary silos and in interdisciplinary units and teams; and knowledge transfer and teaching entities (Doern and Stoney 2009).

As we examine the domain as a whole, we need to be alert to the changing nature of the sciences and disciplinary and multidisciplinary boundaries. The definitions and boundaries are quite fluid, as the fields are being buffeted by developments within science; new technologies used in research; varied political pressures inside universities and outside of them; the changing balance between "bottom-up" investigator-led research and "top-down" thematic or strategically directed research; the changing structure and nature of peer-review and related merit-review systems; competition for funding among and between granting bodies and programs; pressures to increase the visibility of grant decisions, research outcomes, and success stories locally, regionally, and nationally; changing relationships between the granting agencies and other federal government departments, business, and commercial interests; and changing notions of agency independence and interdependence in the conduct of research (Doern 2006).

In the light of strong interdisciplinary and innovation pressures in all three granting councils, it is unclear whether Canada needs three separate councils or whether instead, as in some Canadian provinces and in other countries, Canada should move toward a more explicit, single, coordinated, governance structure (Dufour 2013b). There are questions about whether the three councils in their own realms may be too conservative and risk averse for the current environment.

We examine the key developments of this domain through four policy and governance histories, namely, the emergence and management of the three foundational granting bodies, the Medical Research Council, the Natural Sciences and Engineering Research Council, and the Social Sciences and Humanities Research Council; the creation of the Networks of Centres of Excellence; the refashioning of the MRC into the Canadian Institutes of Health Research; and the establishment of the Canada Foundation for Innovation as Canada's major infrastructure granting and investment body. The foundational granting bodies are a key contextual part of these later histories, including the NCE system, which they jointly administer. They also are involved with other granting programs, such as the Canada Research Chairs program, the recently established Canada First Excellence Research Fund (CFREF) and Genome Canada (discussed in chapter 10). The final part of the chapter summarizes these four histories in relation to the three elements being deployed in this book to trace both change and inertia.

FOUR DOMAIN POLICY AND GOVERNANCE HISTORIES

The granting space is arguably the most important pillar for Canadian S&T and innovation policy and performance. Canada invests on average more of its gross expenditure on higher education R&D (HERD) through universities than most OECD countries do, so getting that system right is critical to the ability of the nation to compete in rapidly advancing technologies and markets.

1960–1980: The Foundational Granting Councils

The origins of the Medical Research Council in 1960, and Natural Sciences and Engineering Research Council and the Social Sciences and Humanities Research Council in 1978 are important. Natural sciences, engineering, and medical research granting gestated in the National Research Council (see chapter 5), while SSHRC emerged out of the previous Canada Council for the arts and humanities, which had both taxpayer funding and funding from charitable foundations. As the three initial independent granting councils emerged as separate entitles, they took some of the key norms from their parent council – peer review and competitive granting in particular. Their internal structure and modus operandi centred on peer-review selection committees, which assessed research grant proposals that were initially overwhelmingly bottom-up, curiosity-driven initiatives presented by individual researchers or small teams of researchers.

Peer review (as discussed in chapter 1) had previously emerged as a central value and process in relation to the publication of research in academic journals. Peer review in the granting process was centred on the applications for research funding rather than on publication, although an applicant's prior publishing record was a core variable in the granting review process. We draw attention later in this history to the emergence of *merit review* in the granting bodies and also globally – the prestigious US National Science Foundation (NSF) merit-review system asserts peer review is the fifth phase in its nine-phase overall merit-review process (US National Science Foundation 2014). Less obvious but fundamentally important, as these institutions emerged as separate granting bodies they were essentially defined by the scientific discipline or, more accurately, sets of disciplines that were assigned to each council. These demarcations helped shape the evolution of the universities and hospitals, which competed for and used the grants.

From 1960 to 1989 these were the core features of the foundational councils. After 1989 interdisciplinary and networked research were introduced and became the norm for these agencies. Both models were foundational to the Networks of Centres of Excellence in 1989, the Canada Foundation for Innovation in 1997, Genome Canada in 1999, and the Canadian Institutes of Health Research in 2000. Moreover, as we see below, these pressures merged to some extent into the operations of the three foundational councils, where bottom-up curiosity-based research continued, but new network and themed research programs instituted top-down, thematic, networked research in support of the government's strategic priorities.

The Medical Research Council of Canada as a separate independent granting council was the first chronologically of the three being profiled here; it was established formally in 1960 and refashioned in 2000 as the Canadian Institutes of Health Research (Murphy 2007; MRC 2000). Health research had emerged out of the National Research Council where it had been given considerable autonomy, although it functioned largely in the still basic research ethos of the NRC. The Medical Research Council was established by Order in Council in 1960 as a granting council although its terms of reference also included the possibility of establishing its own laboratories, a power it never acted on (Aucoin 1971, 1972). The MRC reported to and became an adviser to the minister of national health and welfare. This locale became important as medicare took hold in the1960s as a federal initiative in an otherwise provincial constitutional realm. The operational centre for health research was in fast-growing Canadian nursing and medical schools and their affiliated hospitals, referred to as "health school complexes" (Aucoin 1971, 158). Pressure initially from these research centres was to undertake basic research, but the realities of applied health care began to take hold, although only with reluctant support initially from MRC. As a result, MRC President Malcolm Brown reported in 1970, "We are on the whole weak in the applied and developmental end of the spectrum as compared to the fundamental research end" (Brown 1970, 1).

The Medical Research Council of Canada had a good international reputation for health science but "the perceived rigidity of its hierarchical decision-making structures and its siloed sciences were considered incapable of adjusting to the challenges and demands of the knowledge-based economy and society" (Murphy 2007, 241). Medical researcher Donald Forsdyke (2000, 135) concluded that more funding support is required for those researchers at "interfaces between disciplines." The MRC responded to some of these pressures and realities, creating small programs in the 1970s referred to as "Research Centres of Excellence," but they had very mixed and low profile impacts (Murphy 2007). Meanwhile, traditional biomedical sciences were being affected by biotechnology and information and communications technology (ICT). The science of health systems, which recognized the role of both personal and public health choices, brought forth greater integrated social science research, including the subfield of researcher and research ethics (Murphy 2007; MRC 2000).

In the 1980s and 1990s, the Medical Research Council came under increasing pressure from the private sector as the health sector was recast as a "health business," including the pharmaceutical industry and the emergence of new drugs, medical devices, and diagnostic tools (Doern and Prince 2012). But, unlike the other two foundational councils, the MRC had strong links with, and support from, major charities, such as the Canadian Cancer Society and the Heart and Stroke Foundation of Canada. They, too, saw the need for change in the granting system to support their particular disease and research concerns. These pressures for change culminated in the creation of the Canadian Institutes of Health Research in 2000; we discuss it below.

When formed in 1978, the Natural Sciences and Engineering Research Council was mandated to meet the needs of a university system that had been rapidly growing. There was a genuine concern about possible shortages of highly qualified personnel, in short, persons with doctoral and graduate degrees and training in the sciences and engineering. Because of this concern, NSERC created three key programs to complement the former NRC's Program of Grants and Contributions in Aid of Research: Undergraduate Student Research Awards (USRA), University Research Fellowships (URF), and Industrial Research Fellowships (IRF). Also of interest in the context of this chapter as a whole is that NSERC's first five-year plan published in 1979 had two further major stated features, namely, "to develop a program to update equipment and infrastructure, and to expand focus on targeted research" (NSERC 1979, 3).

In the late 1980s NSERC was seeking a greater role in supporting the knowledge-based economy (KBE), which was not unreasonable given its policy and budgetary link to Industry Canada. As Lopreite (2006, 106) notes, "NSERC emphasized the need for more integration between Canadian universities and the private sector." NSERC's priorities then and into the early 1990s related partly to three new core technologies that had the potential to transform all industrial sectors: biotechnology, microelectronics, and new materials. These technologies were gaining increased emphasis in the federal government's competitiveness strategies of the early 1990s, when the discourse of competitiveness rather than innovation was used. As Lopreite argues, "The development of these new technologies affected a mix of research across the R&D spectrum but also across all the traditional disciplines of the natural sciences and engineering. In many ways, they were helping to break down some of the boundaries that traditionally separated or distinguished core scientific disciplines" (ibid., 108).

As the millennium approached, the main objective of NSERC's programs was to promote and support both research and the formation of highly qualified personnel for the Canadian system of science and technology. Its main sub-objectives were the following: (1) to support a diversified base of high quality research in the natural sciences and engineering; (2) to assist in the development of highly qualified personnel; (3) to promote and support targeted research in selected fields of national importance; and (4) to encourage closer links between the university community and other sectors of the economy (NSERC 1999). Its mission was "to provide the largest possible number of Canadians with leading-edge knowledge and skills to help Canada flourish in the 21st Century" (ibid., 2).

By 2002, the innovation agenda had been brought to the front and centre of overall federal S&T and innovation policy. Lopreite (2006, 109) sums up this period, asserting:

> NSERC conducted its own consultation strategy in the spring of 2002 with more than 300 stakeholders, including students, university and college professors and administrators, industry leaders and federal and provincial public agencies. Following these and other consultations, NSERC announced its five-year *Investing in People – An Action Plan.* These plans built on past programs but also were clearly geared to dealing with more specific challenges that emerged from the consultations (and from the larger federal innovation agenda). Smaller initiatives were promised to deal with: 1) increasing the pipeline of young people interested in science and engineering, through science promotion activities and increasing the number of undergraduate student research awards; 2) ensuring that Canada develops a skilled and talented labour force to satisfy the anticipated demand for HQP (in part by increasing the value of NSERC postgraduate scholarships but also by studies to determine why research fellows often do not return to Canada); 3) ensuring that Canadian research is world-class and internationally competitive (through measures such as improving its work in the intellectual property management program); and 4) minimizing the time a researcher (or student) spends in applying for funding and peer review funding proposals.

NSERC (2013a) describes its overall research granting review process as its "peer review" process. It stresses that "international representation within the peer review process is crucial to fostering world-class excellence and in ensuring that supported research is gauged against the highest international standards" (NSERC 2013b).

NSERC (2014, 1) now has a mix of goals: "advancing knowledge, seizing opportunities; building prosperity through research; inspiring the next generation; showing the value of R&D investments; and increasing the visibility of research." In describing its current $1 billion budget, NSERC says that 38.4 percent is for "discovery" (what it previously called basic research), 31.3 percent for "innovation," 25.7 percent for "people," and 4.6 percent for administration. The fact that two of NSERC's goals are communication-oriented (showing value and increasing visibility) is important in the current context (Phoenix Strategic Perspectives Inc 2013).

The Social Sciences and Humanities Research Council was also established in 1978 (Miller 2006). Previously, funding for the social sciences and the humanities had been a part of the mandate of the Canada Council, whose focus had been on funding the arts using a mix of taxpayer and endowment funds. SSHRC, in contrast, is entirely taxpayer funded and its statutory purpose is to promote and assist research and scholarship in the social sciences and the humanities.

At present, in the Harper era, SSHRC (2013, 7) asserts that the following three key principles inform its strategic priorities for 2013–16 in "strengthening Canada's cultures of innovation": (1) knowledge and understanding about human thought and behaviour are critical to helping Canada and the world address current and future challenges; (2) SSHRC must be responsive to the changing research enterprise and the changing context for research; and (3) SSHRC must work in close collaboration with research institutions and other organizations that have a stake in the future of Canada's higher education and innovation system. SSHRC's "program architecture renewal" involves (ibid., 8):

- Emphasizing impacts through new merit review criteria that consider a broader range of research contributions, as well the likelihood of influence on critical social, cultural, economic, technological, and environmental issues
- Encouraging broader participation in merit review by experts from many sectors
- Fostering partnerships across institutions and sectors that encourage a collaborative approach to designing and conducting the research
- Implementing a flexible, responsive framework for funding research in priority areas
- Providing support for emerging scholars

- Revising reporting requirements for completed grants that focus on outputs, results, and impacts
- Offering flexible programming and a more flexible competition cycle.

SSHRC (2014a, 1) speaks of its overall *merit review* process rather than its peer-review process, citing an "acclaimed review" of it in 2008 by an international blue ribbon panel. That panel report, however, actually refers to SSHRC's "peer review" process. It is evident that at some point SSHRC simply decided that "merit review" was the more accurate descriptor of its system that assesses not just research proposals in social science terms but also their impacts in different kinds of thematic realms. SSHRC also contextualizes its reviews in the very broad program groupings defined as the talent, insight, and connection programs. Overall in 2012–13 SSHRC (2014b) offered funding in eleven areas of mainly interdisciplinary research projects ranging from environmental science and technologies to social development, education, and learning.

Russell LaPointe's (2006, 128) analysis of SSHRC notes that "throughout the existence of SSHRC there has been criticism by parts of the academic community about thematic research. But other parts also lobby for key themes or get government officials to plant the seed as to why a given topic or issue deserves attention. The criticism has been ... that the Council's decision to move to thematic research meant that the Council was bending to the desires of the federal government." SSHRC initially funded such thematic areas as the Human Context of Science and Technology, Women in the Labour Force, the New Economy, and a range of grants focused on community-university research (CURA) linkages.

In 2004–05 Social Sciences and Humanities Research Council began an elaborate process to change itself from a granting council to a "knowledge council." SSHRC (2004, 10) stated, "What we are aiming for is a new council – one that remains in charge of delivering grants awarded through peer review, but one that also directly supports and facilitates the sharing, synthesis and impact of research knowledge. In short, we are aiming for a knowledge agency. We need to work out concretely what it means to the human sciences to contribute to a knowledge society. Everyone has to take stock, both those who produce knowledge and those who rely on it to do their work effectively."

LaPointe (2006) points to a number of pressures that were propelling the SSHRC initiative: demographic changes in the research community and the movement toward multidisciplinary research, new accountability norms, and rising demand for research in and from the social sciences and humanities. Complicating this was the fact that the social sciences and humanities community was large, but the budget to support it was smaller than the other two councils, and the demand from the community for grants was rising sharply. Moreover, the Council estimated that over the next decade about one-third of full-time professors would retire, and there was concern that new researchers were not being hired into tenure-track positions or being funded adequately to fill this gap.

1989–2015: Networks of Centres of Excellence

The NCE programs were established in 1989 as a joint initiative of the three granting councils, in partnership with Industry Canada and the Department of National Health and Welfare (now Health Canada). Now more than twenty-five years old, the NCE programs had their mandate extended under the Harper government to include Centres of Excellence for Commercialization and Research (CEHR), business-led Networks of Centres of Excellence (BL-NCE), and the Industrial Research and Development Internship (IRDI) program (NCE 2014a).

The 2013 annual report stresses the radical nature of the NCE program from the outset, arguing that it was uniquely centred on "virtual teams of academics working with government, industry and communities to solve problems that mattered most to Canadians" (NCE 2013, 1). In 2012–13 NCE partner institutions contributed more than $245 million, "leveraged by expenditures from NCE grants of about $150 million" (ibid., 4). NCE draws attention to the 2,841 peer-reviewed articles published in 2012–13 by NCE-supported research and to the 398 new companies launched since its inception, as well as its role in training more than 42,000 highly qualified personnel (NCE 2014a).

The NCE's twenty-fifth anniversary report (NCE 2014b) opens with a joint statement by the presidents of the three granting councils that argues that NCE's work shows that "collaboration and convergence are defining features of 21st century research. New fields of are emerging from the cross-fertilization of discoveries, ideas, tools and knowledge, changing the landscape of research and opening new opportunities to address the complex challenges of our modern world" (ibid., 3). The report gives examples of six particular network achievements and success stories, including the Quebec Consortium for Drug Discovery, ArcticNet, and the Promoting Relationships and Eliminating Violence Network (6–17).

Although such ways of setting out and claiming success by examples and related data are increasingly a part of granting agency publications and websites, formal evaluations are also instructive. These are provided on contract by professional consulting firms. Such evaluations have occurred in 2002, 2007, and 2013. The review of NCE by Performance Management Network Inc (2013) looked at the 2007–08 to 2011–12 period using a variety of sources and data, including interviews, reports for the twelve funded networks, and surveys of twenty-one network researchers and fifty-six partners. The Performance Management Network (ii –iv) concluded with several key findings regarding relevance, effectiveness, training highly qualified personnel, and efficiency of administration:

- The network approach offers an effective mechanism for multidisciplinary and multisectoral research; brings together a critical mass of people and resources to achieve results that cannot be achieved separately; offers an opportunity for parties to share risks; [and] increases the visibility and credibility of the research.
- The NCE program's investments ... have enhanced research, development and innovation in areas targeted by the funded networks ... the [partner] organizations are truly engaged in these networks as evidenced by their significant cash and in-kind contributions; ... and that the research undertaken by the networks appears to meet the research needs of partner organizations.
- Networks have adopted a range of mechanisms to help ensure that research results are shared and can be used ... about two-thirds of partner organizations increased their knowledge base as a result of their participation ...
- The NCE program has been effective in providing extensive opportunities for the training of highly qualified personnel by involving thousands of PhD and master's students.

Although broadly positive, the reviews contain the normal and needed cautionary points about inadequate data, for example, in the extent to which evidence about knowledge transfer and impact can be inferred through the number of patents issued.

The only major academic analysis of the NCE program is Janet Atkinson-Grosjean's book (2006), which covers the period from its origins in 1997 to 2005, by which time numerous Networks of Centres of Excellence had been established, and matured. She begins her analysis by arguing that "when the Canadian government created the ... NCE experiment in 1989, it introduced the most dramatic change in Canadian science policy since the National Research Council was established in 1916" (xiii). The policy establishment's "radical goal (later modified) was to turn university researchers away from basic science and towards commercial application ... research should not only be 'managed' – a novel concept – but managed on private sector rather than academic principles" (xiv). The Atkinson-Grosjean analysis examines how these macro-level changes in policy and ideology are "playing out at the micro-level of the university or hospital laboratory" and especially "how has the research culture changed after two decades of strategic science policy?" (14).

Atkinson-Grosjean's (2006) approach is both "top-down" in that it explores how the new science policies were forged and administered and "bottom-up" through detailed case study analysis of the Canadian Genetic Diseases Network. The CGDN was funded for fourteen years as a Network of Centres of Excellence but in 1997, when the NCE program announced all networks would be capped at fourteen years, it incorporated as CGDN Inc with "a corporate organizational form and an aggressively commercial focus" (7). Atkinson-Grosjean observes that "the tension between the public and private faces of NCEs became increasingly apparent over the course of the study (of the CGDN). Alternating currents of confidentiality and openness ebbed and flowed around the project. Scientists spoke to me freely, while gatekeepers erected barricades. The contradictions and barriers to access illustrated the normative and ethical barriers that are constantly negotiated in these networks" (ibid., 7).

Atkins-Grosjean deployed a "two by two" conceptual matrix for scientific research, plotting both the public and private domains, on the one hand, and basic and applied science, in an effort to find "where the public interest is located" and to focus "on the fundamental tension between 'open science' and proprietary knowledge" (2006, 11). The analysis paid attention to various gatekeepers both within the Canadian Genetic Diseases Network and between the network and the NCE program administrators and the partner universities, for whom the NCE network seems to "float" above its normal structures of accountability.

Atkinson-Grosjean's (2006) analysis is valuable in many ways and raises valid concerns about what public university research could or should be about. She is critical of the Networks of Centres of Excellence and the science policies behind it but, given the focus on a single case study, there are inevitably analytical and empirical boundary grey zones in the two-by-two matrix that cannot be resolved with such a small sample. But this is a dilemma for all studies of S&T and innovation policy, including the present book, which examines several domains with illustrative policy and governance histories.

There are some obvious issues that go begging for analytical attention. The NCE program is jointly administered by the three granting councils, which themselves, as seen above, have fostered their own network-like granting activities within their own changing program structures. One can't help but wonder whether the NCE program is kept relatively small (with about $150 million budget) so that there is more room for the granting councils per se (the CIHR and NSERC are each appropriated about $1 billion annually).

It is also useful to mention here the common interest among the three granting councils and the NCE program in the $265 million annual Canada Research Chairs program. Established in 2000 it is "a permanent program to establish 2000 research professorships in eligible degree-granting institutions across the country … to attract and retain some of the world's most accomplished and promising minds" (CRC 2014a, 1). It is a highly visible federal program based on peer review centred on a core "college of reviewers." The number of filled chair positions early in 2014 was 1,743, including 278 or 15.9 percent from outside of Canada (expatriates or international recruits), with 26.6 percent female scholars and 73.4 percent male (CRC 2014b, 1). In terms of areas of research 45 percent of the chairs are in NSERC areas of research, 35 percent CIHR, and 20 percent SSHRC (CRC 2014c).

A 2010 evaluation of the program in its tenth year was quite positive in terms of its attraction and retention goals and in other goals such as publicity and comparative advantage in strategic areas of research. The review, however, raised issues about some resentment within universities from other faculty who were not being as well recognized for their contributions or who may not have competitive funding provided (Science-Metrix 2010). Obviously the three granting councils can see gains from the program, but it also always leaves unanswered what each may have done, or how the system may have benefited, if an equivalent amount of money had instead gone into existing granting programs in the three core agencies.

2000–2015: Canadian Institutes of Health Research

The Canadian Institutes of Health Research was formed in 2000 by merging the former Medical Research Council of Canada and the National Health and Development Research Program (NHDRP) from Health Canada. In the process it created thirteen "virtual reality" research institutes devoted to interdisciplinary research on applied health and wellness problems and challenges (Murphy 2007).

The specific genesis in the formation of the CIHR was the work of a task force of leaders in the Canadian health research community that met in 1998 and discussed ways to better link researchers from all disciplines and ways in which resources could be focused on Canada's major health challenges. The dilemmas of the Medical Research Council were pivotal in this review. The task force was aware of the Networks of Centres of Excellence program and the compelling example of the US National Institutes of Health (NIH) as alternate models.

The guiding principles for the CIHR, as stated in *A New Approach to Health Research for the 21st Century* (CIHR 1999, 9), are to:

- Adopt research priorities that are linked with Canadian health priorities and complement the provincial investment in research, education, and health
- Encompass and support the full spectrum of health research – from basic science to clinical research to population health – recognizing the important role of investigator- initiated research
- Ensure Canadian researchers succeed in the worldwide research community through the application of peer review as fundamental to the evaluation of research excellence and internationally competitive levels of funding
- Encourage individual institutes within the network to conduct unique programs – from capacity-building to third-party partnerships – in pursuit of the improved health and well-being of Canadians
- Collaborate with all organizations that have demonstrated a capacity to support or conduct health research. CIHR supports and recognizes the major contributions to health research by voluntary health organizations, universities, provincial granting bodies, and individual research centres
- Recognize and support the central role that universities and associated health science centres play in education, in training, and in creating interdisciplinary research opportunities.

In structural terms, the Canadian Institutes of Health Research, like the predecessor Medical Research Council, is an agency of the federal government reporting to Parliament through the minister of health. The

CIHR incorporated the operations of the MRC itself but its core new structural feature is its series of institutes. The initial set of thirteen institutes emerged in 2000 following significant consultation; the set includes institutes on traditional or familiar areas of research such as cancer research, but also several that met the test of broader interdisciplinary realms such as the Institute of Aging and the Institute of Aboriginal People's Health. The thirteen institutes, still in existence today, include the following (CIHR 2014b):

- Aboriginal People's Health
- Aging
- Cancer Research
- Circulatory and Respiratory Health
- Gender and Health
- Genetics
- Health Services and Policy Research
- Human Development, Child and Youth Health
- Infection and Immunology
- Musculoskeletal Health and Arthritis
- Neurosciences, Mental Health and Addiction
- Nutrition, Metabolism and Diabetes
- Population and Public Health.

Federal policy stressed that "the Institutes would not be centralized 'bricks and mortar' facilities. Instead these virtual organizations would support and link researchers who may be located in universities, hospitals and other research centres across Canada" (TBS 2005, 4). A key feature is that the institutes "would support researchers who approach health challenges from different disciplinary perspectives" (ibid.). The vision statement differentiated the CIHR from the existing system of "dispersed research efforts, disciplinary separation, separate from delivery, and multiple agendas"; the new model anchoring the CIHR was "integrated: across geography/institutions; across scientific disciplines; into the health system; and with the national health agenda" (ibid., 6).

The stated objective of the Canadian Institutes of Health Research at present is "to excel, according to internationally accepted standards of scientific excellence, in the creation of new knowledge and its translation into improved health for Canadians, more effective health services and products and a strengthened Canadian health care system" (CIHR 2014a, 1). Each of its thirteen "virtual" institutes "include fundamental bio-medical research, clinical research, and research respecting health systems, health services, the health of populations, societal and cultural dimensions of health and environmental influences on health" (ibid.).

The CIHR's *2014–15 Report on Plans and Priorities* stresses that the CIHR's work involves the following (CIHR 2014b, 5):

- Funding both investigator-initiated research and research on targeted priority areas
- Building research capacity in underdeveloped areas and training the next generation of health researchers
- Focusing on knowledge translation that facilitates the application of the results of research and their transformation into new policies, practices, procedures, products, and services.

The CIHR functions through forty-eight peer review committees in the core Open Operating Grant Program (OOGP; CIHR 2014c). This is a massively larger number of committees than operated within the former Medical Research Council. The CIHR budget of approximately $1 billion is divided into 75 percent for investigator-driven health research and 25 percent for priority-driven health research (CIHR 2014b, 10). The CIHR has its own Science Council composed of its president and the science directors of its thirteen institutes. The slate of thirteen institutes (established in 2000) has not been reviewed, though individual institutes have been assessed. An institutes "model review" was underway in 2014–15 involving both an external working group of CIHR stakeholders (excluding researchers) and an internal working group composed of the scientific directors of the thirteen institutes (CIHR 2014d).

Joan Murphy's (2007, 51) overall commentary on the CIHR institutes is worth quoting at length:

> The Institutes concept provides researchers with an opportunity to align their research strengths to priority areas in the Institutes. Institutes improve the support for well-developed research communities and may be able to meet many of their goals through the structure of grants for basic and applied research under the CIHR's peer review system. In this way, the Institutes assist in shaping the research environment, while promoting investigator-driven research. Through the organization of the Institutes structure, Institutes with well-established and well-organized research communities can also more easily identify and promote key strategic initiatives. The flexibility to

> support strategic initiatives allows the Institutes to address the priority needs of their researchers and stakeholder communities. Preferably, strategic initiatives should be supported by multidisciplinary research teams. Others may support efforts to build a research base in emerging fields.
>
> Other Institutes, simply given their focus area, have broad responsibility to encourage the development of accomplished and experienced investigators in research fields that are underdeveloped and have a small research base.
>
> Each Institute's multi-year strategic plan details a proposed mixture of strategic grants. Once approved by the Governing Council, the Institutes are expected to receive a direct allocation to support its strategic initiatives. Institutes can also receive funding for Institutes' development initiatives.
>
> The Institute's budget will also indirectly accrue funding from the investigator-initiated researchers that are assigned to it. All Institutes will be able to count on a floor level for research funding, to support a base amount of research. Moreover, Institutes will have the budgetary flexibility to invest more money in research that is relevant to their approved strategic plan – above and beyond their allocations from the CIHR pool.

The larger part of the CIHR budget provides grants based on investigator-driven ideas and proposals through open competitions judged through peer-review processes. In a sense this competitions-focused approach is consistent with the discipline-based research funded by the former Medical Research Council.

By 2006 the total CIHR budget had more than doubled since 2000; the open competitions as a proportion of the total budget decreased from 87 to 66 percent, while the strategic initiatives component increased significantly, moving from 8 to 27 percent of the budget (CIHR 2006). Operational costs increased four-fold largely because managing the virtual reality institutes is more complex in transactional terms. It must be stressed that there is no clear-cut way to differentiate the proportions of grant spending that go to the institutes versus investigator-driven research per se. This is because there are significant overlaps making it difficult to assign them to one category versus the other.

By 2006, among the three granting councils, the Canadian Institutes of Health Research had garnered by far the largest percentage increases in spending since 2000, in part due to the greater inherent political saliency of health issues among Canadians compared with the natural sciences and engineering and the social sciences and humanities. Health Canada's blueprint set out five areas of focus (TBS 2005, 2):

- Developing national research platforms and initiatives
- Supporting strategic research through CIHR's institutes
- Strengthening Canada's health research communities
- Knowledge translation
- Commercialization.

The commercialization focus reflected the fact that the CIHR's formative years coincided with the emergence of the federal innovation strategy in 2002 and the greater expectation that research would support a health "industry" and not just a health sector. The commercialization goals have been a particular challenge given that the large majority of the health research community sees itself as being in a public interest-centred research endeavour overall rather than in a commercial one. One application of this is the CIHR backing for university-industry partnerships in support of research on biotechnology and research on clinical trials (Murphy 2007).

Explicit rules and practices regarding research ethics are central to all the granting councils, but especially the CIHR (Levasseur 2009). In 1998 a joint AUCC-federal effort led to the Tri-Council Policy Statement: Ethical Conduct for Research Involving Humans (TCPS), which provides policy guidelines that are conditions of eligibility for federal grants from the three granting bodies. Though care was taken not to cast these as regulations per se, universities signed up to them and held grant-holding faculty and students to the goals of the TCPS. Ethics rules have created some cross-granting council concerns and differences because some in the natural sciences and social sciences believe that the rules are too heavily dominated by the biomedical model which is skewed in favour of the needs found in research on human subjects (SSHRC 2006; van den Hoonard and Hamilton 2016).

Overall, the dominant policy narrative about health as a defining feature of national identity has kept the Canadian Institutes of Health Research in a good position, even in the face of changes in government. In the 2 May 2006 federal budget the CIHR received a 2.4 percent increase in its budget plus a special award of $21.5 million over five years to fund research into pandemic preparedness (CIHR 2006, 1–2). Overall, in the Harper era the CIHR has maintained is leading position among the three granting councils in terms of budgetary success. However, this could change when the above-mentioned institute "model review" is completed. Some of the existing thirteen institutes are likely to have more staying power in both government and health research circles (e.g., cancer), while others may have to give way to other new institutes, especially if the total number of institutes has to stay in the current range. If and when some institutes have their funding reduced or ended, it is not clear they will vanish. As with the Canadian Genetic Diseases Network discussed earlier, they might secure replacement or new funding from business, provincial governments, and charities or foundations. Provincial network involvement emerged on CIHR funding on the CIHR Strategy for Patient Oriented Research. But otherwise the CIHR was mainly a federal dynamic rather than one led by the provinces per se.

Finally, while the CIHR (2014d) refers to its overall review process as a peer-review system, its manual for grant applications also refers to a "new merit review rating scale" applying to all funding opportunities. Merit review "requires that both the scientific merit and potential impact of the projects be assessed using separate scores … the potential impact score reflects the relevance/importance of the project to the knowledge users and the likelihood that the project will have a substantive impact on health outcomes, practice, programs and/or policy in the study context" (ibid., 12). Of interest here is that the US National Institutes of Health, on which Canada's CIHR is partially modelled, describes its statute-mandated, peer-review system as "dual peer review," with a second level of peer review by an institute's advisory council or board which focuses on applicant impact scores (US NIH 2014). As CIHR explores merit review, more aspects of the NIH system may find their way into CIHR debates.

The CIHR has also been revamping its review system processes from the previously mentioned system of institutes-based committees of reviewers that meet in person. The new model adopted by CIHR is electronic peer review of a large college of virtual peer reviewers. The CIHR maintains the college using tools such as Elsevier's PURE database to verify reviewer expertise via publications and citations, and it has also undertaken a study to determine the "optimal" number of reviewers. The effects of these changes are being monitored by the CIHR – and, of course, by the community of researchers directly affected by the changes.

For Canada's granting councils, international developments are always important. For example, Canada via the Natural Sciences and Engineering Research Council has joined the recently established Global Research Council (GRC) which describes itself as "a virtual organization composed of the heads of science and engineering funding agencies from around the world, dedicated to promoting the sharing of data and best practices for high-quality collaboration among funding agencies world-wide" (GRC 2014, 1). At its 2012 Global Summit on Merit Review hosted by the US National Science Foundation (NSF) and attended by funding agency presidents from fifty countries or regions, the Global Research Council published a *Statement of Principles for Scientific Merit Review* asserting that "the terms Merit Review and Peer Review are used interchangeably" (GRC 2012, 1). The issue that framed the granting agencies and has occupied them for their entire existence continues, only in international fora.

1997–2015: Canada Foundation for Innovation

In 2012–13 the Canada Foundation for Innovation completed its fifteenth year as Canada's first and main funder and supporter of research infrastructure in universities, colleges, research hospitals, and non-profit research institutes. It was established in 1997 by the Chrétien government with strong influence from Paul Martin as minister of finance and supported by strong lobbying by some key university presidents. It emerged at least partly because the federal fiscal coffers were overflowing after several years of deficits. It was among other early federal efforts to fund infrastructure more generally in the Canadian economy, mostly recently at the municipal level.

In S&T policy terms, the Canada Foundation for Innovation was a quite different funding focal point than the granting councils whose focus was on the operational costs of research. The CFI is federally funded but has greater structural independence, for example, it is not audited by the Auditor General of Canada (Lopreite and Murphy 2009). It works in conjunction with the granting councils. In contrast to the US National Science Foundation, which funds infrastructure and operations, in the Canadian context NSERC or Genome Canada operating funds would be awarded separately from CFI infrastructure funds, requiring researchers to create separate but interrelated funding applications.

In both the *CFI Strategic Roadmap 2012–17* (CFI 2012b) and its *Annual Report 2012–2013* (CFI 2012a, 8–10), the CFI set out as its four main goals to:

- Attract and retain the world's top research talent
- Enable world-class research and technology development that leads to social, economic, environmental, and health benefits for Canada

- Support private sector innovation and commercialization
- Enhance accountability to responsibly steward public funds.

The Canada Foundation for Innovation reported that over its first fifteen years it committed more than $5.9 billion in support of 7,879 projects at 138 research institutions in 66 municipalities across Canada, which leveraged from institutions and their partners a further $12 billion for research infrastructure (CFI 2012a, 3). On average over the fifteen years being reported, the CFI invested $393 million per year and levered an additional $800 million per year from other sources.

With regard to the medium-term future, the Canada Foundation for Innovation pointed to its road map of strategic directions, including the following (CFI 2012a, 8):

- Maintaining the integrity of the CFI merit-review process to ensure that only research infrastructure projects meeting the highest levels of excellence receive support
- Placing greater emphasis on the sustainability of CFI-funded projects
- Developing funding partnerships with the granting councils and Genome Canada to improve application processes and increase the overall impact of federally funded research and technology development
- Engaging with research communities in the United States and Europe to share best practices in the management of large-scale research facilities
- Implementing the 2012 CFI communication strategy to effectively inform Canadians about the value and impact of federal investments in research infrastructure
- Developing new methodologies to better assess and evaluate the impacts of research infrastructure on Canada's research capacity.

Although the Canada Foundation for Innovation drew attention to its presence in 66 Canadian communities and 138 research institutions, in reality the investments are concentrated in a much narrower set of major research universities. Table 6.1 shows examples of its concentration as follows drawn from the CFI website list under *Our Investments* (CFI 2014).

It is not surprising that such concentrations have occurred. It mostly reflects the roles these universities play as established major research universities, their internal capacities to organize and submit proposals,

Table 6.1 Top seven universities receiving funding from the Canada Foundation for Innovation

University	*Funding received ($millions)*	*Number of funded projects*
University of British Columbia	358.9	659
University of Toronto	349.9	772
McGill University	278.3	596
University of Montreal	262.8	475
University of Alberta	257.7	441
University of Saskatchewan	175.7	255
Dalhousie University	86.9	222

Total of $1.77 billion or 35 percent of total CFI funding on 3,420 projects or 39.5 percent of all CFI projects in seven universities.

Source: Calculated from *Our Investments* (CFI 2014).

and their ability to raise the levered money from partners. In spite of this concentration, about two-thirds of the funding was distributed widely.

Lopreite and Murphy's (2009) analysis of the Canada Foundation for Innovation from its birth in 1997 to 2009 posits that the CFI acts as both a patron and regulator of Canada's research infrastructure. They show that the CFI is a *patron* based on its role as a funder and promoter: "Although the CFI is an arm's-length foundation and is independent of politics in the sense of partisan politics, it does have to deal, as all patrons do, with small-p politics among universities in the form of distributional disbursements regionally or among different realms or disciplines of science and the infrastructure each needs. These are of course not part of the CFI's terms of reference but they are a part of how diverse research interests lobby the CFI and seek to change its program structure" (124).

Lopreite and Murphy (2009, 124) argue that the CFI acts as a *regulator* because investments in R&D infrastructure:

> must conform to certain "rules" set by the CFI, including requirements that universities submit strategic research plans and also that funding be levered. In short, universities do not get infrastructure money unless they bring money in the form of funding commitments from other funders, public or private. Universities have widely different capacities to obtain these levered money commitments, which constitute 60 percent of all CFI awards. Aspects of regulation also occur in the realm of defining what is eligible for funding and hence

in defining what in fact "infrastructure" is as capital spending rather than operational spending (CFI Eligibility). Such issues also inherently arise in the CFI's role of developing research "capacity" and in determining what "innovation" means in these funding choices and also in determining what kinds of research capacity-building might occur when CFI funds are allocated. Capacity building involves both complex physical-technical capacity and also human capacity in the form of Highly Qualified Manpower (HQM) and managerial and even entrepreneurial capacity.

The CFI's initial strategy was to develop several types of programs. These included the following: (1) the Innovation Fund to help institutions strengthen their research and training environment in priority areas that they themselves identify; (2) the Infrastructure Operating Fund to help with the incremental operating and maintenance costs associated with infrastructure projects; (3) the Research Hospital Fund to contribute to hospital-based projects that focus on innovative research and training; (4) the New Opportunities Fund to provide infrastructure support to newly recruited academic staff; and (5) the Canada Research Chairs Infrastructure Fund for infrastructure support to the Canada Research Chairs Program. Targeted international funds were established later.

Following a consultation process in 2007 a revised suite of programs emerged: the main competitive funds were the John R. Evans Leaders Fund, the Innovation Fund, and the Major Science Initiatives Fund. As Lopreite and Murphy (2009, 133) point out, some of the basis for the shift in the CFI suite of programs "can also be found in its program evaluation reports which were a part of its commitment to accountability." An evaluation report by Bearing Point (2003, 4–5) used a complex methodology to identify three strategic considerations facing the Canada Foundation for Innovation:

1 Maintaining long-term sustainability will require institutions to convince their provincial partners to supply matching funds, and institutions to find O&M (Organization and Management) support for the long term. This is the most important long-term strategic issue by far.
2 Additional opportunities for CFI to act as a catalyst for pan-Canadian strategic planning related to research infrastructure should be investigated, possibly including opportunities to act as "the Canadian voice" in these matters internationally.

3 CFI and the Social Sciences and Humanities Research Council should continue to investigate ways to encourage involvement in CFI from researchers in the social sciences and the humanities.

The Canada Foundation for Innovation, like all federal agencies, has had to explain its role, with respect to science and research infrastructure, in the context of the knowledge and innovation economy and social innovation cast in terms of health research charities, the environment, and the local-regional economies that host the universities, community colleges, and polytechnics that operate the infrastructure.

As a foundation, the Canada Foundation for Innovation has some implied independence, but its chief funder nonetheless remains the federal government. In its fifteen-year history there have been ten funding infusions, each involving a funding agreement with the federal government (CFI 2012a, 4). These funding agreements allow the government to exercise some influence in the direction and scope of the infrastructure program areas. The government cannot direct the Canada Foundation for Innovation to fund specific projects, due to the merit-review system, but it can nudge the CFI to move in some different directions or shift it by insisting on managerial reforms. Overall, three areas of the CFI's evolution warrant brief further mention: its merit-review process, boundary issues with the granting councils, and the impact on universities from the required research plans.

As noted in chapter 1, standard peer review was not enough for infrastructure decisions for two reasons. The management of universities or research hospitals had to be involved and had to be assessed as to their merits and claims of merit on their own and/or in comparison with other competing project applications. This had also been true in earlier periods when Canadian research funding had to deal with so-called Big Science centred on large and politically visible research allocated according to classical distributional regional politics. Second, research partners in and users of a new facility or equipment were fundamentally involved because they brought much of the money needed to make the projects work. Project reviewers in the merit-review process had to judge (and rank) projects in relation to three main criteria: the quality of the research and the need for infrastructure, the project's contribution to strengthening the capacity for innovation, and the potential benefits for Canada (CFI 2012a, 3). These are quite broad criteria and are imbued with particular acts of judgment regarding "capacity" for innovation and "potential" for benefits. Evidence and judgment can be brought to bear by expert reviewers, but these are as much the art of decision making as science.

In this regard, the first priority statement in the *CFI Strategic Roadmap 2012–17* (CFI 2012b) has a potentially double-edged meaning. As mentioned above, the CFI gives its highest need to that of "maintaining the integrity of the CFI merit-review process to ensure that only research infrastructure projects meeting the highest levels of excellence receive support" (CFI 2012a, 8). This statement could just mean that it is restating its obvious importance as a CFI value or it could also mean that somehow the merit-review system is under pressure internally, from applicants or from the government, all who might want it loosened or made more flexible. In research infrastructure, as in other kinds of infrastructure allocations and decisions, there is typically room for merit-based decisions but also for distributive politics of the "small p" or even occasional "big p" varieties.

The boundary between the Canada Foundation for Innovation and the other granting bodies discussed above is complicated; it is not difficult to see how pressures, overlaps, and gaps could arise. We have seen examples in the above narrative. Given that the Canada Foundation for Innovation is mandated to fund new infrastructure, who and how the resulting projects would fund operations and renewal is less clear. The granting councils fund research, which can and often does cover some of the operational costs of the infrastructure, but seldom covers all the costs and seldom contributes much to renewal costs. In many universities CFI infrastructure facilities are seen as drawing off resources that otherwise would cover other university priorities. The federal government responded in 2012 with the Major Science Initiative that offered a special pool of $185 million competitive funds to support the operations and maintenance in 2012–17 of large-scale infrastructure projects (minimum $25 million capital costs). In 2011 four facilities – Compute Canada, SNOLAB, Canadian Light Source Inc, and Ocean Networks Canada Observatory – were allocated $111 million for four years. The road map also mentioned the need for "developing funding partnerships with the granting councils and Genome Canada to improve application processes and increase the overall impact of federally funded research and technology development" (CFI 2012a, 8), illustrating the awareness of a need to better link operational and infrastructure coordination. It is also worth noting in this context that the birth of the Canada Foundation for Innovation was accompanied in 1999 by a federal government agreement to pay universities indirect overhead costs as a separate financial commitment to universities (Morgan 2009).

Last but not least, we take note of the impact of CFI requirements that universities submit research plans as a part of their project applications.

In this regard, Lopreite and Murphy (2009) looked briefly at two small universities, Saint Mary's University and Brock University, and at Canada's largest university, the University of Toronto. Their analysis concluded that Brock University and Saint Mary's University were able to use this process to improve their impact in CFI competitions and to transform their institutions into more focused research universities; nevertheless, and not surprisingly, the University of Toronto was able to garner an above average share of the funding.

Overall, the Canada Foundation for Innovation continues to garner federal funding support. The Harper 2015 Budget (Minister of Finance 2015, 95–6) commits the federal government to $1.33 billion over six years, and stresses that CFI-funded projects "are felt across the economy," which also translates in politically distributive terms to across the country and its regions.

THE THREE ELEMENTS

We now track with the aid of Table 6.2 the three elements being utilized to understand change and inertia across the four domain policy and governance histories.

Policy Ideas, Discourse, and Agendas

With regard to the three foundational granting councils the central ideas and agenda trajectory show strong support for discipline-based basic or curiosity-driven research, later cast by NSERC as "discovery" research. Quite early on, however, their ideas and agendas focused on support for thematic or priority research, overtly or indirectly implying interdisciplinary research with separate programs of funding. Partly because of these dual kinds of research, all three councils changed the way they described their granting review processes, shifting to the concept of merit review based or some form of dual peer review and impact assessment. Ideas regarding the innovation and knowledge economy and society, which emerged in the 1980s, also had significant influence. SSHRC sought to rebrand itself as a knowledge council, while each of the councils accepted the need to ensure greater visibility for their funding, citing successful research with applied economic or social impacts.

The NCE program forged the idea and discourse of networked and interdisciplinary research and the accompanying requirement for partners and leveraged co-funders. Collaboration and convergence were increasingly cast as conjoined criteria of research success. Support for research that led to spin-off companies and commercialization was central to this model and became increasingly linked to controversies regarding the balance between public interest and business values and the interplay between promoted interdisciplinary research and the traditional role of university researchers to follow their muse. Partly because of these dual kinds of research, all three councils changed the way they described their granting review processes, preferring to talk of merit review or some form of dual peer review and impact assessment.

The Canadian Institutes of Health Research reinforced the network approach by casting its thirteen institutes of applied health research as virtual institutes and as hosts to practitioners of complex multidisciplinary research within and across the institutes. Research ethics emerged strongly as a CIHR regulatory idea and function and diffused quickly to all the granting agencies.

The CFI's formation and investment activities brought strong public focus both to the idea of research infrastructure in universities and research hospitals and to the explicit role of broader merit-review systems. It also made more explicit the notion of attracting and retaining research talent and mandated university research plans.

Economic and Social Power

Although the prevailing economic and social system generally supported the broad initial structure of discipline-based and peer-reviewed granting councils, changing priorities forced the domain to adapt. Medicare in the 1960s triggered the creation of the Medical Research Council of Canada, which inherently introduced interdisciplinarity, impact assessment, and leveraging into the granting system. Biotechnology, communications industries, and health research charities all fit with the networked research model. Business interests were formally incorporated through the NCE program and in the CIHR where health research was recast as support for a health business and not just health.

Politically, the Mulroney era was very receptive to the networked research idea and the Harper government pushed often for business and commercialized research impacts and outcomes, including the establishment in the NCE of "business-led" networks encouraged through new separate funding programs in the NCE. The demonstration effect of the US National Institutes of Health exerted significant influence in 2000 as an exemplary model when the CIHR was formed.

Table 6.2 Policy and governance histories in the granting, universities, and levered-money domain: Three analytical elements

Policy and governance history	*Policy ideas, discourse, and agendas*	*Economic and social power*	*Time, temporal realities, and conflicts*
1960–1980: The foundational granting councils	• Discipline-based, curiosity-driven research • Top-down thematic research • Early basic research ethos • Medicare as growing ethos for more applied health research in MRC • More research on the interface between disciplines • Greater visibility of research for NSERC • "Discovery" research as new NSERC name for basic research • Innovation economy and society ideas • "Leading-edge" knowledge • New core technologies in NSERC mandate • SSHRC cast as a "knowledge council" • More need in social sciences for "merit review" • Flexible programming • Need for supra-governance across the three councils	• Broad political and economic support for creation of the three discipline-centred councils • MRC as the product of early and later health care lobby and growth of research hospitals in the provinces • Influence of new biotechnology and communications technology lobby • Societal-health influence of health charities • Growing provincial influence in concert with growth of universities • Business influence through NSERC need for greater research integration with business • Pressure and power in Harper era for SSHRC to strengthen the "cultures of innovation"	• Growth in number of research faculty and research students in baby-boom growth of universities underpins need for more funding • Changing medium-term nature of thematic research topics • Long-term basis for research in health in birth and growth of medicare • Later/current retirement bulge in senior faculty and need for young replacement faculty
1989–2015: Networks of Centres of Excellence	• Networked and interdisciplinary research as core rationale • Levered expenditure ideas; bring money to get money • Create spin-off companies • Collaboration and convergence • Increased visibility of research • Support for HQP • Networks but with gatekeepers • University public values vs. Private sector commercial values	• Support and push in Mulroney era for networked research • Harper addition of "business-led" NCE • Government belief and pressure on levered-money provisions • And also on value of networks per se even though new forms of bureaucracy created with them	• Life-cycle and sustainability of NCE over 10- to 15-year period and what happens if an NCE is cancelled • HQP cast as permanent human capital need for innovation economy and society • Impact of Internet as permanent technological basis for new networks • Networks as basis for new forms of bureaucracy

Table 6.2 Policy and governance histories in the granting, universities, and levered-money domain: Three analytical elements (*Continued*)

Policy and governance history	*Policy ideas, discourse, and agendas*	*Economic and social power*	*Time, temporal realities, and conflicts*
2000–2015: Canadian Institutes of Health Research	• Excellence in creation of new knowledge • Virtual institutes, not centralized bricks and mortar • Multiple types of interdisciplinary research required within each of the 13 CIHR institutes • Knowledge translation • Investigator-driven and priority-driven research • "Model review" of all 13 institutes • Research ethics role • Support for full spectrum of health research	• As with NCE, CIHR based on considerable bottom-up social power and influence for CIHR institutes as applied and interdisciplinary health research • Influence of US NIH model • Some increased pressure from business to see health as a "business" and not just as a health and medicare "sector" • Government and scientists' support for stronger ethics regimes for health research	• 13 institutes already 15 years old • Sustainability of current institutes vs. plan for new ones • Health science sector vs. health science business as normative and real long-term battle over discourse and projects
1997–2015: Canada Foundation for Innovation	• Support for research infrastructure • Merit-review decision making centred on three main criteria • Leveraged funding with CFI supplying up to 40% of any project • Attract and retain world's top research talent • Innovation and commercialization ideas • Ideas regarding need for well-developed university research plans by each university	• University presidents as direct lobby for infrastructure needs • Martin's power regarding CFI in the context of emergence of budgetary surpluses • Power of big established research universities in garnering large share of project infrastructure funds • Weaker power of some regions and their universities to lever or raise the needed non-CFI funds • CFI as a foundation conferred some greater power and independence than the granting council model	• Research infrastructure as physical asset needing both initial establishment and investment and long-term renewal and change • Retaining top research talent, but for how long and at what cost? • Does CFI as a foundation need extra funding for its long-term viability, or does its foundation status largely make it not much different from other arm's-length granting bodies?

Note: see the chapter text for expansion of abbreviations in this table.

The lobbying power of some university presidents and Universities Canada (formerly the AUCC) and the fit with the research infrastructure focus and ideas of Paul Martin as Liberal minister of finance were pivotal in the formation of the Canada Foundation for Innovation. As we have seen, in the actual allocation of CFI funding the major research universities demonstrated considerable power to leverage and exploit their existing advantages. But at the same time distributive funding dispersed power, allowing some smaller players to seek and find a goodly share of the investment;, however, some university applicants from less affluent regions found it difficult or impossible to come up with the required levered money, and hence were excluded from the benefits.

As noted in the analysis, there is some concern that, despite each moving toward interdisciplinarity and innovation, the three granting councils remain quite separate in their governance and may well need some kind of overall supra-governance reform. Recent changes in programs and their administration at CIHR have raised questions about intra-agency governance, and as NSERC and SSHRC have installed new presidents in 2015 who likewise might undertake agency reforms, there may need to be some kind of supra-governance reform, potentially mediated by Industry Canada.

Time, Temporal Issues, and Conflicts

Temporal issues are present in each of the four policy and governance histories. For example, the emergence and growth of funding needs for the three foundational granting councils arose from the post-war baby-boomer demographics of the 1970s and 1980s. There were more universities established and more students were conducting research under the supervision of a considerably larger body of faculty members (many new hires). All of the above implied and compelled the need for more research funding as a national and provincial priority. More recently, a faculty retirement bulge from the same demographic group has meant that new faculty had to be found, retained, or given research support, this time in an even more competitive global university market.

Time and temporal issues and conflicts are also inherent in the case of the Networks of Centres of Excellence and Canadian Institutes of Health Research. Even though they were both created as permanent agencies, they both have exhibited a ten- to fifteen-year life cycle or staying power at the institute level. Should support for them be ended with assumptions or hopes that someone else will pick up the slack? How should new centres or institutes be chosen as medium-term creatures and objects of funding? We have seen that both these funders, as well as the Canada Foundation for Innovation, have mandates related to supporting highly qualified personnel as a permanent multi-decade effort to meet the changing human capital need for an innovation economy and society. One can also see the Internet as both an initial new technology and now a pervasive underlying one in supporting network and new virtual institute research formation and change. But this is also a technology that is changing some aspects of peer review and how one judges impacts under merit review.

Last but not least, the Canada Foundation for Innovation as a supporter of complex research infrastructure is quintessentially grounded in a world of complex physical asset values that decline rapidly in a world of competitive research and are linked closely to increasingly valuable human knowledge capital.

CONCLUSIONS

This chapter has examined the granting, university, and levered-money domain in the federal S&T and innovation policy and governance system. The granting councils and agencies examined directly spent about $3 billion in 2014–15 and levered about a further half-billion dollars from funding partners. The $3 billion is a vanishingly small percentage of the federal expenditure budget, but it is a the main source for Canada's university researchers and graduate students in the conduct of their disciplinary and interdisciplinary research across science and engineering, health sciences, medicine, social sciences, and humanities.

We have traced a mainly federal story though the four policy and governance histories examined: the three foundational granting councils (MRC, NSERC, and SSHRC); and then the NCE, CIHR, and CFI, with some brief mention of the Canada Research Chairs program as well. But we have also stressed the roles of universities per se as creatures of provincial governments and also the related pressure of direct lobbying in Ottawa from university presidents and through Universities Canada. This and related developments may suggest that in some respects the federal government was partly ceding STI content and influence to the provinces, given how the provinces had their own single integrated granting bodies, and also how provincial network involvement emerged on CIHR funding of the CIHR Strategy for Patient-Oriented Research. But otherwise the CIHR was mainly a federal dynamic rather than one led by the provinces per se. The CFI was partly a provincially driven initiative regarding research infrastructure, but it was key university presidents rather than the provinces that pushed for the CFI, and the big powerful universities such as the University of British Columbia, University of Toronto, and McGill were dominant recipients of CFI grants. But the level of provincial influence must ultimately be gauged as well against the Statistics Canada data cited earlier in the chapter, which show that the provinces are minor players in the linked areas of R&D spending, where the federal government is dominant. We leave, however, to the book's conclusions to provide further comments on the thesis of ceding STI to the provinces after we have explored our remaining domains in chapters 7 to 11.

The levered-money dynamics also say a lot about the Canadian STI space. This new imperative has inextricably linked scholarly research to the larger society and economy. At one level, it has transformed universities. Leveraging implies accommodating different interests and perspectives, sometimes in unexpected ways. Although the granting programs have become de facto regulators as they impose rules about research ethics and university priority setting, the need to match those with the norms and interests of their partners has blunted their impact. Hence, there is a subtle but fundamental change in how we define research disciplines and areas, how we adjudicate and review research, and indeed, what kinds of institutes will be funded and how many and for how long.

Two other linked developments emerge from our account of the four policy and governance histories. One is the enduring presence of funding that is intrinsically bottom-up and curiosity-driven as opposed to funding for research that is top-down and thematic or priority-driven. The second is the testing of new review systems, described variously as peer review and merit review. The more that research is curiosity-driven and emerges from within disciplines the more that peer review is a reasonable descriptor of the review process used, although even here it is not yet peer review as found in the processes for reviewing research submitted to a journal for possible publication. The more that research is thematic or priority-driven the more it is likely to be interdisciplinary and the more that this requires merit review. This is because merit review involves other kinds of expertise focusing on assessing the potential impacts of proposed research on users. In the case of the Canada Foundation for Innovation, merit review also requires diverse expertise regarding infrastructure per se and its proposed or hoped for impacts based on diverse and often broad criteria of outcomes.

The chapter has shown the importance of the changing nature of the sciences and disciplinary and multidisciplinary boundaries due to developments within science and also due to new technologies used in research globally. Competition for funding among and between granting bodies and programs has emerged, which has created pressure to increase the visibility of grant decisions and subsequent research outcomes and success stories locally, regionally, nationally and internationally.

This domain, perhaps not unexpectedly, brings out important aspects of the innovation economy and society nexus, as examined through the changing structure of the granting system across the various funding bodies. The original Medical Research Council first explored the integration of social content in scientific research in the context of hospitals and health care; this naturally led to introduction of research ethics. These were extended massively under the CIHR and its virtual institutes. SSHRC, and the social sciences are social content incarnate. SSHRC extends its granting structure and networks to such critical areas for innovation as education and learning, the environment, communities, and women in the labour force. The Networks of Centres of Excellence and the Canada Research Chairs program both involved in different ways focused support for highly qualified personnel goals, particularly creating and attracting talent. The CFI with its infrastructure focus also funded diverse areas of new capital and infrastructure to foster social, environmental, and health benefits. NSERC had the least obvious "social" content in its mandate and its funding except perhaps for the HQP goal, which it expressed frequently. Nevertheless, impact is an increasingly important metric for NSERC-funded scientists. Innovation is now firmly embedded in the discourse regarding all these entities, with notions of the innovation economy expressed as "cultures of innovation," "commercialization," and also partnership funding on a significantly increasing scale.

Some of the political-economic features of the granting and reviewing bodies are examined in the domain chapters to come, including the emergence and influence of Genome Canada in chapter 10. However, as we see in our next chapter on the industrial S&T and innovation domain, these features can be found widely in the government-supported S&T and innovation system.

7

The Industrial S&T and Innovation Domain

Glimpses into the nature, content, and evolution of the industrial S&T and innovation domain have already emerged in previous chapters, including the content of macro S&T policy agendas, the analysis of government departments and agencies including the National Research Council of Canada and its changing business focus, and trends in the research granting system that partly involves leveraging funds for R&D from business.

The industrial focus in this chapter tends to refer to S&T and innovation policy aimed at Canadian manufacturing or policies impacting on manufacturing. But it also deals historically with the impacts of technology, which often were positive even where there were policies not explicitly labelled as S&T policy. The domain historically is anchored in overarching and changing views and policies about Canada's industrial landscape. Our overall focus on manufacturing is only one strand of Canadian industrial policy; Canada also has significant natural resource industries that are a part of the total economic and social history examined in this book (see chapters 1 and 3), and they are to this day a continuing presence as we see in more detail, for example, in the examination in chapter 9 of the agriculture, food, biosciences, and biotechnology domain.

Canada's industrial landscape and S&T policy were profoundly influenced historically by Prime Minister John A. Macdonald's National Policy that in addition to settling the agricultural West created a tariff wall to encourage and protect the development of Canadian manufacturing, mostly in Ontario and Quebec (Creighton 1955). The protective tariff barrier was largely eliminated by 1995, but the mindset of protecting strategic or important industrial sectors (albeit through other means) remains potent even in today's globalized marketplace. The protective barrier encouraged and supported the development of technology-based farm machinery manufacturing, steel foundries, and an automobile production and assembly industry that to this day remain important regional and national industrial manufacturing sectors.

In addition to offering firms the prospect of avoiding import tariffs in a small but growing captive market, Canadian governments have come to realize that they must do more to sustain and grow the country's industrial capacity. Successive governments have used a mix of macroeconomic stabilization (especially fixing our exchange rate or managing a "dirty float" to provide greater predictability of returns in foreign markets) and targeted subsidies, tax expenditures, loan guarantees, infrastructure investment, and sectoral planning, programming, and institution building (Doern, Maslove, and Prince 2013). In addition to those sectors supported behind the tariff wall, the other major national investment has been in aerospace. A recent Fraser Institute study showed that between 1961 and 2012 more than $13 billion – $5 billion in grants and other non-repayable contributions, combined with $8.5 billion in repayable (but often not repaid) support, equal to about $22 billion in 2012 dollars – has been directed to industrial support, much of it in the aerospace sector (Milke 2013; see Table 7.1).

The heyday of such public investment was in the 1970s and 1980s, before the Canada-US Free Trade Agreement (CUFTA), the North American Free Trade Agreement (NAFTA), and the World Trade Organization (WTO) put restrictions on such programming – more than 45 percent was doled out in the 1980s. Most of the funds are directed to a small subset of firms. The top twenty-five companies using these programs, which represented about half of the funds invested, were regular dippers, drawing an average of twenty-five times from the support programs (range, 1–83 times). The main beneficiary was the aerospace sector: in all, thirteen aerospace firms (out of the top 25 recipient firms) collected about 38 percent of the funds. Pratt & Whitney Canada received the most money over the decades, almost $3.3 billion via seventy-five disbursements; Bombardier and de Havilland were the second- and third-largest recipients, each receiving disbursements worth $1.1 billion over the years (Milke 2013). This goes a long way to explaining how Canada remains a global competitor in the aerospace sector.

Bombardier is in many ways Canada's poster child for such industrial development and the potential for science, technology, and innovation to advance economic and social development. The company started in Quebec in 1942, having invented the snowmobile. Over the years it expanded into a range of product areas. In addition to an expanded

Table 7.1 Type of assistance to business, Industry Canada, 1 April 1961 to 31 March 2012

	2012 ($ millions)	*Nominal ($ millions)*
SUBSIDIES NOT EXPECTED TO BE REPAID: "GRANTS"		
Grants	338	264
Conditional grants	131	120
Contributions	7,924	4,840
Interest contributions	81	64
Other assistance	375	214
SUBTOTAL	8,849	5,501
SUBSIDIES EXPECTED TO BE REPAID: "LOANS" OR LOAN GUARANTEES		
Repayable contributions	1,902	1,258
Conditionally repayable contributions	10,647	6,760
Conditional/unconditional repayable contributions	202	171
Participation loans	45	12
Loan guarantees	474	260
SUBTOTAL	13,269	8,467
TOTAL	22,118	13,963

Source: Milke (2013).

recreational line (including snowmobiles, jet skis, boats, and ATVs), the company moved into transportation. The first move was in 1970, when the company moved into mass transportation equipment with the purchase of Lohner-Rotax, an Austrian manufacturer of snowmobile engines and tramways. It used that investment and a series of acquisitions in the 1980s to become one of the world's largest manufacturers of railway rolling stock; it has more than a hundred thousand rail vehicles in operation in almost every continent and has subway trains operating in more than forty cities worldwide. Its highest profile investment is in the aerospace sector. In 1986 Bombardier acquired Canadair from the Canadian government; shortly afterward it absorbed de Havilland Canada, Short Brothers, and Learjet. Bombardier used these assets to develop a fleet of popular aircraft, including the Dash 8, CRJ series of regional airliners (which compete directly against Brazil's Embraer jets), Learjets, and, currently in trials, the CSeries medium-range jets, which are expected to compete directly with planes produced by Boeing and Airbus.

The aeronautical industry, Bombardier in particular, presents an interesting case study, although it is more the exception that makes the rule. To a great extent, except for the aerospace industry, most of the industrial support and policy measures in Canada have been targeted on building and maintaining domestic capacity rather than on using science, developing technologies, and innovating per se. There have been some recent efforts to shift the focus, but much of Canadian industry remains weakly focused on innovation.

Governance of this domain has been squarely centred on the industry, trade, and commerce ministry, which has gone through a variety of name and organizational changes to reflect its shifting priorities. The Department of Trade and Commerce started in 1892, but it was a relatively modest affair until C.D. Howe, the legendary minister of munitions and supply who marshalled industrial capacity in support of the military effort in the Second World War, was appointed minister in 1948. He used the ministry to rebuild civilian industry after fifteen years of depression and war. In 1963 the Diefenbaker government created an additional agency, the Department of Industry, which in 1969 was merged with trade and commerce as the Department of Industry, Trade and Commerce (ITC). In 1983 it was renamed the Department of Regional Industrial Expansion (DRIE), in 1990 as Industry, Science and Technology Canada (ISTC), and in 1995 as Industry Canada (Doern 1996; Blais 1986). Along the way, the portfolio absorbed Consumer and Corporate Affairs (in 1995), Communications (in 1996), the Foreign Investment Review Agency (FIRA, in1985, which was subsequently renamed Investment Canada), and in 1990 the Ministry of State for Development Activities – variously called Ministry of State for Economic Development (MSED), Ministry of State for Economic and Regional Development (MSERD), and Ministry of State for Science and Technology (MOSST).

The Industry Canada portfolio currently manages a wide range of programs and special operating agencies (SOAs) engaged with the following issues: business regulation, including competition, incorporation, bankruptcy, and consumer affairs; intellectual property (IP) protection, including patents, copyright, and trademarks; trade, including weights and measures, standards, international marketing, and export financing; business and sectoral development and regulation, which includes tourism, the Canadian Space Agency, the Business Development Bank of Canada, the Canadian Radio-television and Telecommunications Commission (CRTC), and the Internet; foreign investment review; science and technology programming; and all of the peer-reviewed granting agencies examined in chapter 6, including the Canada Foundation for Innovation (CFI), Networks of Centres of Excellence (NCE), Canada Research Chairs (CRC), the Natural Sciences and Engineering Research Council of Canada

(NSERC), the Social Sciences and Humanities Research Council of Canada (SSHRC), the Canadian Institutes of Health Research (CIHR), and the National Research Council of Canada (NRC), the country's largest S&T enterprise (examined in detail in chapter 5).

In the context of innovation, Industry Canada now has most of the key levers for incentivizing, supporting, and regulating investments in science, technology, and innovation (STI) in Canadian industry. About the only other major federal player is the Department of Finance, which manages and controls tax expenditures for R&D, which in most years dwarfs the direct outlays for S&T and innovation. In recent years Canada more than any other OECD country has relied relatively heavily on non-directed, tax-based support for STI (see further discussion below).

FOUR DOMAIN POLICY AND GOVERNANCE HISTORIES

The motivations and nature of Canada's efforts to develop and sustain a competitive S&T-based industry encompass four policy and governance histories. In the first instance, our heavy manufacturing sector can be traced to the dirigiste industrial policies and programs of the 1950s–1970s; the automotive, steel, and military hardware production can be traced to that set of policies and instruments. The second history centres on efforts to create the climate for innovative firms to compete in the global economy. Since the mid-1980s the focus has changed dramatically, with most of the effort and energy on developing the framework policies in support of industrial development and a national system of S&T development that supports industrial innovation. The third policy history centres on the Scientific Research and Experimental Development Tax Credit (SR&ED) and dynamics in the relations between small and large firms since the early 1980s. The fourth policy history focuses on policies on clusters, business, and local-city-regional innovation systems.

1945–1980s: Dirigisme, Made-in-Canada, and Made-for-Canada Industrial Policy

This policy history concerns mainly the period between the Second World War and the 1970s, but its underpinnings are traceable to earlier periods. Canada has actively supported infant industries through aggressive trade and industrial policies. The repeal of the Corn Laws in 1846 and the loss of preferred access to the British Empire in the first instance forced the British colonies in North America to band together to create

Canada and to rationalize and generate a more robust and interconnected domestic economy. In 1878 the federal government introduced the above-mentioned National Policy involving Western agricultural settlement and industrial development in Central Canada behind a tariff wall. To finance the developments Canada, for the most part, relied on global capital, first from London and, following the crippling First World War, increasingly from New York.

The results were usually firms and sectors scaled to service domestic demand; few exports were attempted, and for much of the period import tariffs were prohibitively high, effectively cutting off imports and thereby limiting the integration of these industrial ventures into global production systems. After the National Policy was implemented, new industries appeared: electrical equipment and chemicals in the 1890s; cars, aluminum, and steel after 1900; pulp and paper between 1890 and 1914; radio and home appliances in the 1920s; and aircraft in the 1940s.

Canada had three main proactive strategies that it used, in some instances up until present times, which may or may not have produced what we would now cast as S&T policy. In the first instance, it developed a range of regional development plans and programs. Savoie (2003a) asserts this policy effort reflected a tension between "needs versus opportunities." Initially, the consensus was that regional development should be tied to "needs," but there have been frequent efforts to shift the focus more toward optimizing opportunities, which has led to regional competition (both among the provinces and in federal Cabinets). The operation of the federal Industrial and Regionals Benefits Policy (discussed below) has tended to minimize or neutralize those efforts.

From Confederation to the mid-1950s, the federal government had no explicit regional policy – all the policy was directed to national priorities (Savoie 2003b). Partly in response to the Rowell-Sirois Commission (the Royal Commission on Dominion-Provincial Relations, 1937), the federal government as part of demilitarization and reconstruction after the Second World War sharpened its focus on provincial and regional development. Federal and provincial tax sharing was offered and accepted by all but Quebec, and federal infrastructure investments and industrial support were directed to the regions, albeit often with controversial offers of fifty-cent dollars, the first major use of formulaic leveraging, which still dominates federal-provincial relations. In 1957 fiscal equalization was initiated to help provide a fiscal floor for "have not" provinces, which provided them with some increased resources and greater certainty.

The Gordon Commission (the Royal Commission on Canada's Economic Prospects), in its 1957 report, expressed concern about increasing foreign ownership of Canadian industry, which prompted a range of more engaged policies and programs over the coming years. In 1960 the federal government passed the Agriculture Rehabilitation and Development Act, which in 1966 it renamed the Agricultural and Rural Development Act (ARDA), to increase industrial development outside of industrialized southern Ontario. A number of institutional innovations were refashioned, including Industry, Trade and Commerce as the Department of Regional Economic Expansion (DREE), creating Special Areas (growth poles) and the Regional Industrial Incentives Act, together with federal-provincial general development agreements, economic and regional development agreements (ERDAs), and cooperation agreements. These were designed to align federal and provincial priorities in industrial development to improve economic performance.

Eventually these ventures got institutionalized and prioritized by a range of federal regional development agencies, including the following: the Atlantic Canada Opportunities Agency (ACOA), Western Development (WD), and Federal Economic Development Northern Ontario (FEDNOR), all in 1988; FORD-Q, in 1991), renamed Canada Economic Development Quebec (CED), in 1998; CanNOR in the Territories, in 2009; and, finally, in response to the collapse of manufacturing employment in the "Golden Horseshoe," FedDEV for southern Ontario, also in 2009 (Bradford and Wolfe 2012; Savoie 1987; Doern and Phidd 1992).

In 1973 the federal government responded to the Gordon Commission and to the1968 report of the Task Force on Foreign Ownership and the Structure of Canadian Industry (the Watkins Report) by creating the Foreign Investment Review Agency, which for the next twelve years assessed all foreign acquisitions and takeovers based on their contribution to job creation. The combination of coordinated federal-provincial support for domestic industrial development combined with the tariff wall and limitations on foreign ownership created a made-in-Canada and largely made-for-Canada industrial structure.

A second approach to industrial development came in the mid-1960s with the development of the Canada-US Automotive Products Agreement, signed in 1965 and commonly known as the Auto Pact (Fuss and Waverman 1986). By the early 1960s most of the Canadian auto assembly functions were undertaken by one of the big three American car manufacturers, General Motors, Ford, and Chrysler, mostly using parts and components imported from the United States. From 1955 to 1964 the Canadian automotive industry faltered due to its inability to produce at the mass scale required; Canada imported nearly all the automotive products used in the country: in 1964 Canada imported $723 million in American automotive products, while exporting only $105 million to the United States.

With emerging competition from European and Asian car manufacturers, the auto industry and the two national governments sought to rationalize the North American industry to improve its competitiveness. The governments of Canada and the United States, with the support of the big three automakers and the governments of the provinces and states hosting those firms, negotiated a sectoral free trade agreement that led to rationalization. In response, the industry grew in Canada, quickly surpassing pulp and paper as Canada's most important industry. After the Auto Pact was implemented, trade deficits in automotive products decreased until the first trade surplus in 1970. Although the policy generated significant benefits, it did little to support innovation in the sector for a long while. Most of the corporate functions – including design, logistics, branding, and marketing remained or relocated to Detroit. The Auto Pact ended in 2001, with a WTO ruling; however, it had de facto been superseded by the deals under CUFTA, signed in 1987, and NAFTA, signed in 1994 (Doern and Tomlin 1991).

After the 2008 financial crisis, governments intervened significantly in the auto sector, purchasing equity and offering loans on a preferential basis (Waddell 2010). The Ontario government contributed about $1.3 billion and Ottawa about $2.7 billion to the package of loans to parts manufacturers, assistance to encourage consumer purchases, funds for GM and Chrysler to assist them to restructure their operations and a line of credit for Ford. The federal budget 2008 also introduced the Automotive Innovation Fund (AIF), initially providing $250 million over five years to support large-scale R&D projects to build innovative, greener, more fuel-efficient vehicles. The fund was renewed in 2013, initially with an additional $250 million and then topped up by another $500 million in 2014.

Although not directly targeted through the Auto pact, steel is inextricably linked to the auto industry. A bit of a chicken and egg story, neither would have emerged or been sustained in the past without access to the other. Steel tends to "cluster" because the transportation costs of moving this heavy but relatively low-value product discourage long supply chains (Warrian 2010). Instead, steel mills tend to be hubs for manufacturing customers who buy and fabricate steel coils, bars, beams, and pipes into industrial and consumer products. The main steel foundries located in and around Hamilton while the auto fabricators sited in a penumbra of towns and cities around the mills. The first seventy-five years of the steel industry in Canada was largely the story of the emergence and leadership of Stelco; the next twenty to twenty-five years saw the rise of Dofasco which developed a new sort of "learning" steel company; then new market entrants (including IPSCO in Regina, which processed recycled metals) and the advent of "minimills" further developed the industry.

As recently as the mid-1980s many pointed to the steel industry as a leading example of Canadian industrial success – it was the largest and most successful Canadian-owned and -managed industry. Then came trade disputes that restricted access to the US market, a rising dollar that made Canadian production less competitive and the emergence of a wide number of new producers from Asia (first Korea and then China) and Latin America (Brazil) that ramped up competition. This initially encouraged Canadian companies to divert any new investment to the United States, with most extra production occurring south of the border. The industry seemed strong and competitive until 2006, when in just twenty-four months the ownership structure and the familiar names of the Canadian steel industry disappeared. In 2014 only one of the top ten companies operating in Canada is domestically owned; the rest are owned and linked to global ventures led from the United States, the European Union, India, Russia, Brazil, and South Africa (Warrian 2010).

Some might say in spite of these changes while others might say because of the transformation, in 2012 the industry employed directly and indirectly 130,000 Canadians and produced $7 billion annual exports. Warrian (2010) reports that in spite of all the disruption in the industry, there is a strong and viable steel technology cluster comprised of steel producers and their suppliers of material and professional services (e.g., engineering and logistics) centred around Hamilton but extending to other provinces. De-verticalization of the global steel industry expanded the role of steel mills in the supply chain to include services and has encouraged (perhaps forced?) mills to develop networks to coordinate supply to a global rather than local industry.

A third strategy was to exploit the investments and purchases of government by pursing a regional economic benefits policy. Canada has a long-standing policy of requiring suppliers of major defence materiel to spend at least the value of any contract in Canada on Canadian purchases or firms. This policy – formalized as the Industrial and Regional

Benefits (IRB) policy in 1986 and often referred generically as "offsets" policy – requires IRBs be provided either directly through contracts for the purchased materiel or indirectly through contracts unrelated to the purchase. The impact of this policy has ebbed and flowed with the volumes and values of military procurement. Globally, industries serving the military are unique in that governments are for the most part their only customers. Governments therefore are core to building innovative defense firms. Jenkins (2013) argues that the proof is that the successful companies in Canada started with Government of Canada contracts.

Flagship innovative firms in this sector include CAE Inc, which entered the simulation business in 1952 with a contract from the Royal Canadian Air Force to develop a CF-100 flight simulator. Since then the company has gone on to sell more than 1,300 simulators and flight training devices to civil and military customers in more than 190 countries, growing its revenues to $1.8 billion in 2012 and employment to more than 8,000. The company is an acknowledged world leader in simulation equipment, commercial and helicopter aviation training, military virtual air training, and healthcare simulation technology. Another oft cited example is General Dynamics Land Systems Canada in London, Ontario, which is a leading designer, manufacturer and provider of technical support for light armoured vehicles (LAVs), which was initially sourced by the Canadian army but now is sold to the US; the company has a supplier base of over 400 Canadian companies, at least one in each province.

With the aging of the Canadian military equipment, partly due to delays in decision making, a pent up demand for new equipment has emerged. In 2008 the Government of Canada established the Canada First Defence Strategy (CFDS) to provide twenty-year funding and a roadmap for the modernization of the Canadian Forces. The CFDS targets to invest $240 billion in procurement for equipment and infrastructure. In order to rebuild the fleet, the National Shipbuilding Procurement Strategy was adopted in 2010, establishing a strategic relationship with two Canadian shipyards: Irving Shipbuilding Inc in Halifax for combat vessels and Seaspan in North Vancouver for non-combat vessels (Spence 2014).

Enthusiasm for supporting infant industries has been on the wane for more than a generation, but targeted industrial subsidies and buy-local requirements and procurement policies remain resilient. The challenge is that in a globalizing world where firms increasingly succeed based on their capacity to innovate and capture a piece of the global market, domestically oriented policies are hard to sustain. Slowly the policies are shifting to address that new reality. Regional policies now at least talk about science, technology and innovation, even if there is little evidence of them delivering benefits beyond the distribution of major science infrastructure. Direct support for automakers is now tied to innovation goals while the steel sector, partly in spite of and partly in response to policy, has adapted to the new reality with little fanfare and little in the way of federal investment. Finally, the IRB policy for military procurement, the last bastion of directed government investment, is beginning to look more carefully at how the offsets can create innovative, sustainable industry rather than one-off purchases (Jenkins 2013).

1985–2015: Creating the Climate for Innovative Firms to Compete in Global Markets

The core of this second policy and governance history stretches from the mid-1980s to the present, but there are earlier contributing features as well. Beginning in the 1970s, governments began to rethink their approaches to industrial policy in relation to changing macro-economic, micro-economic and S&T and innovation ideas. The Keynesian consensus of proactive and engaged government, implemented in the dark days of the depression in the 1930s and by all accounts successfully pursued through the mid-1960s, began to unravel in the 1970s as evidence began to mount that governments were far from omniscient and all powerful (McFetridge 1985; Harris 1985). Economists and firms began to promote the idea that the state had expanded too far and too fast, and that many of the policies, programs and investments did little to improve socio-economic outcomes. A few voices were highly influential. Milton Friedman and Anna Schwartz (1971) undertook a groundbreaking long-term study of US monetary policy, concluding that proactive, macroeconomic policies designed to moderate the business cycle were seldom beneficial – the leads and lags involved in detecting recessions and then stimulating demand were so long and unpredictable that society would generally be better off without proactive policies.

Rational expectations theorists added another nail in the coffin of core Keynesian stabilization policy, arguing that regardless of what measures government used, rational individuals and markets could and would make choices that would sterilize any positive effects. Economists came to the view that stabilization was misguided. Beginning in the late 1960s the global economy began to gyrate wildly. In quick succession the global system faced stagflation (economic stagnation and inflation combined), the end of the gold standard and fixed exchange rates and the great

Grain Robbery of 1972, which caused an historic spike in grain prices. Then the first OPEC oil crisis in 1973 tripled world oil prices, triggering the most severe recession since the 1930s and precipitating a global commodity price collapse. At the end of the decade, a second OPEC oil-price increase occurred sending oil prices into the triple digit range. By the end of that tumultuous decade, governments everywhere were seeking new policy options.

The tipping point came in 1979 when Margaret Thatcher and her Conservative Party were elected in the United Kingdom and her new government adopted a monetarist solution of high interest rates and lower public spending to choke off inflationary pressures. When Ronald Reagan was elected president in the United States in 1980 he adopted a similar policy. As global inflation dropped and macro-economic stabilization was wound down, the supply side model came into its own. Governments everywhere began to look seriously at the microeconomic foundations of successful enterprise, including the role of policies and programs related to finance, markets, technology and skills. The public choice model as the new "economics of politics" gained influence and – in effect, governments adopted a modified Hippocratic Oath, vowing only to intervene where private markets failed *and* where government action could make things better.

Canada was no exception. In 1984 the newly elected Progressive Conservative government of Brian Mulroney adopted these new policies and practices and governments have continued broadly in this direction (Doern, Maslove, and Prince 2013; Atkinson and Coleman 1989). The Bank of Canada had already mirrored the practices in the United Kingdom and the United States, causing the Canadian bank rate to peak at 19.5 percent in 1982, effectively slowing inflation. Structural reform followed with the new government. Somewhat ironically, the impetus for the most significant reform effort was laid by the Royal Commission on the Economic Union and Development Prospects for Canada (1985), set up in 1982 by the Liberal government of Pierre Trudeau and headed by Donald Macdonald, the former minister of finance in that government. The Macdonald Commission reported in 1985 that Canada needed to foster a more flexible economy, capable of adjusting to international and technological changes. The main recommendation was generally to push toward a market-driven economy, with diminished government oversight and regulation, and specifically to purse a free trade agreement with the United States (Doern 1986).

A number of themes dominated in the Mulroney era, including free trade, deregulation, the strong encouragement of foreign investment, and privatization. The most significant venture of the new government was to start and conclude the first comprehensive trade agreement with the United States. Canada had a chequered history of cancelled and aborted efforts to negotiate with its continental neighbours. In 1855 the British colonies in North America had a Reciprocity Treaty, but that was abruptly cancelled in 1866 and was one of the triggers that led to confederation the next year. Following the Alexander Mackenzie Liberal government failure to negotiate a reciprocity agreement with the United States in the 1870s, John A. Macdonald was re-elected and moved to implement the National Policy. Again in 1911 the Liberals under Laurier campaigned on a platform of free trade with the United States, but were defeated. In the intervening years there were a few bilateral agreements that reduced tariffs and irritants, including the Auto Pact, but Canada mostly relied upon the General Agreement on Tariffs and Trade (GATT) to govern bilateral trade. Throughout the 1970s and early 1980s debate about the appropriate balance continued, but the general view was that multilateral solutions were most appropriate.

In 1986 Mulroney reversed his party's historical opposition to free trade and accepted the Macdonald Commission's challenge and took a "leap of faith" to initiate talks (Doern and Tomlin 1991). After two years of intensive back-and-forth negotiation, the resulting agreement was a game changer, both for the two parties and for world trade policy. In addition to removing most of the tariffs not already removed by the seven previous GATT rounds and the Auto Pact, the agreement included a series of innovative measures: a chapter on trade in services, which was not then in the GATT; a chapter governing technical standards as non-tariff barriers, which under the GATT were dealt with through non-binding "understandings"; chapters on agriculture (which was largely exempt from GATT through Art. 11c), energy (that extended what has become a continental market for energy), and wine and spirits; a chapter establishing government procurement rules; rules providing for labour market mobility in selected professions; an investment chapter that provides national treatment to reciprocal investments; a financial services chapter; and a binding dispute settlement system. This agreement was the prime focus of the 1988 federal election campaign – the result vindicated the leap of faith, as the Mulroney government was re-elected, albeit with a reduced majority.

The exciting thing for trade policy theorists and practitioners is that this was a groundbreaking agreement. Most of the provisions in the agreement were novel and helped to move the goal posts for all subsequent trade negotiations. They set both the agenda and structure (sometimes plus or minus some elements) for most of the free trade agreements in the past twenty-five years. In particular, they were institutionalized in the North American Free Trade Agreement of 1994, the World Trade Organization Agreement of 1995, and a series of bilateral and regional accords and agreements negotiated and implemented around the world.

Since then Canada has continued to negotiate and implement agreements to open up new markets and to manage disputes about trade and investment. The prevailing logic is that trade negotiations are like cycling, if you stop pedaling you will crash. So, the policy is to continue onward and outward. In addition to being a charter member of the GATT and its successor, the World Trade Organization, Canada has comprehensive rules to deal with matters with the United States (CUSTA and NAFTA) and Mexico (NAFTA), a comprehensive economic and trade deal with the European Union and trade agreements with the European Free Trade Association (especially Norway and Switzerland), Israel, Jordan, South Korea, and six states in Latin America (Chile, Costa Rica, Honduras, Peru, Columbia, and Panama). In 2014 Canada had eleven ongoing negotiations with the Caribbean Community and Common Market (CARICOM), the Central America Four (El Salvador, Guatemala, Honduras, and Nicaragua), the Costa Rica, Dominican Republic, India, Israel, Japan, Morocco, Singapore, Ukraine, and with eleven Pacific nations as part of the Trans-Pacific Partnership, not yet fully approved. Canada also was exploring potential discussion with Turkey, Thailand, and MERCOSUR (Argentina, Bolivia, Brazil, Paraguay, Uruguay, and Venezuela, with its associate countries Chile, Peru, Colombia, and Ecuador).

As Canada's national borders started to open, governments and industry in Canada became increasingly concerned about the interprovincial barriers to flows of goods, services, people and capital. Interprovincial trade and commerce is under federal jurisdiction (s. 91.2), and no tariffs existed; nevertheless, a multitude of regulatory, institutional, policy, and program provisions enacted by provinces in support of their local interests effectively restricted the free flow of economic activities between and among the provinces. There was a prevailing view in federal circles that the economy was balkanized, so that after NAFTA it was easier for firms from the United States or Mexico to trade in multiple provinces in Canada than for a domestic firm to do so.

As an early priority, the newly elected Liberal government of Jean Chrétien convened the provinces to negotiate what came to be called the Agreement on Internal Trade (AIT). The agreement, implemented in 1995, provided binding rules that enshrined the principles of reciprocal non-discrimination among Canadian persons, the right of entry and exit, no obstacles to interprovincial trade (unless justified by legitimate objectives) and reconciliation and transparency (Doern and MacDonald 1999). It directly addressed regulatory issues in eleven specific policy areas, including procurement, investment, labour mobility, consumer-related measures and standards, agricultural and food products, alcoholic beverages, natural resource processing, energy, communications, transportation, and environmental protection. To add teeth to the agreement, the signatories created a permanent secretariat and a dispute resolution process for governments, individuals, and the private sector.

This agreement has helped, but it has not been enough for some provinces. Several adjacent or neighboring provinces have negotiated bilateral or multilateral agreements that address specific irritants or opportunities between the provinces (Public Policy Forum 2014). In 2007 a new Ontario-Québec Trade and Cooperation Agreement) was created that incorporated aspirational economic development goals (e.g., innovation corridors, a high-speed rail link, and coordinated green development) and addressed several specific regulatory issues (e.g., mutual recognition of labour qualifications for professionals and tradespeople and harmonized rules for financial services). Concurrently, the provinces of Nova Scotia, New Brunswick, and Prince Edward Island pursued a Maritime Economic Cooperation Agreement, which enshrines in provincial statutes many of the same goals of the other agreements but which has a particular focus on economic advancement through labour and product mobility and consistent regulatory systems.

In addition, after more than a decade of discussion among western premiers about the potential to strengthen the Western Canadian economy through greater liberalization of markets, British Columbia and Alberta, both with right-of-centre governments, signed in 2006 the Trade, Investment and Labour Mobility Agreement (TILMA) to reduce impediments to the free flow of people, investment, and products through the mutual recognition and reduction of barriers to market access. In 2010, with a new right-of-centre government in Saskatchewan, TILMA was extended to incorporate Saskatchewan and renamed the New West Partnership Trade Agreement (NWPTA).

Trade liberalization has been paired in most agreements with new rules governing foreign investment. Canada has long-standing concerns with the scale and scope of foreign ownership of Canadian industry and resources. Given that much of the capital used to develop our infrastructure, our trade, and our industry comes from global markets (in earlier years from London, from the United States after the Second World War, and most recently from Asia), the problem is real and will not likely go away. Perhaps surprising to some, but Canada holds more assets abroad than foreigners hold in Canada. Foreign investment in Canada in 2010 amounted to about $549 billion, while Canadian investment abroad was valued at $593 billion, for a net advantage of $44 billion. Nevertheless, the concern is real and long-standing. We have already traced some of the concerns to a succession of reports, starting in 1957 with the Gordon Commission and followed up with the previously mentioned Watkins Report of 1968, but later studies collectively made the case for the Foreign Investment Review Agency (FIRA). Foreign takeovers were assessed based on their contribution to job creation, Canadian participation in management, competition with existing industries, new technology, and compatibility with federal and provincial economic policies.

This agency was to many in industry the "bogey man" as it slowed and sometimes turned back foreign investment. With the move to more liberal trade, new and more binding investment rules came into use. In 1985 Mulroney renamed FIRA as Investment Canada and changed its mandate, directing it only to look at "significant" takeovers; as of 2014, the threshold for review was $354 million. The declaratory provisions of Investment Canada have only been invoked four times.

In support of the view that investment is actually encouraged is the spate of privatizations starting back in the 1980s. In 1984 the federal government held sixty-one different Crown corporations and during the Mulroney mandate they divested twenty-three, including Canadair and de Havilland Canada to Bombardier in 1986, Teleglobe in 1987, Eldorado Nuclear to Cameco in 1988, Polymer Corp to Nova in 1988, Air Canada in 1989, Petro-Canada in 1991, and the Canadian National Railway Company in 1992. PetroCan was finally private in 2009. While the government was divesting assets, it was also prying open doors for more incoming investment and more secure markets for Canadian investment abroad. Canada's stated view is that investing abroad can help companies gain access to overseas markets, reduce input costs, secure access to key resources, acquire new technologies and provide better support to foreign customers.

But the government acknowledges that the risks of investing offshore can be high: political instability, weak legal institutions, uncertain regulatory regimes and the possibility of expropriation all raise concerns. In response, the Canadian government (along with most OECD member states) has negotiated Foreign Investment Promotion and Protection Agreements (FIPAs). These bilateral agreements seek to gain national treatment for reciprocal investments, subject to transparent and judiciable exceptions and legitimate objectives. As of 2015 Canada has negotiated and implemented twenty-six FIPAs (9 with east European countries, 9 with Latin American counties, 4 with Asian countries, 4 in the Middle East and one in Africa). Fourteen more FIPAs have been signed or negotiated but have not yet come into force, and eleven negotiations were ongoing in 2014. These agreements have sparked significant controversy in Canada (especially the recent deal with China); nevertheless, these agreements are becoming ubiquitous, with an estimated 2,700 bilateral investment treaties signed globally as of 2014 (Dattu and Morgan 2014).

Along the same lines, governments everywhere, including Canada, have moved to reform their regulatory landscapes (Doern, Prince, and Schultz 2014). Governments are learning from each other about new and better ways to develop, manage, review, and renew legislation and regulation. The first move along this line came in the 1980s, as conservative governments in the United Kingdom, the United States, Australia, New Zealand, and Canada came to power on platforms of reducing the role of government, including by reducing regulatory red tape. In Canada, Erik Nielsen, the deputy prime minister in the Mulroney Conservative government in the spring of 1986, developed a federal regulatory reform strategy that, for the first time, called for all new regulations to be subject to economic and social cost analysis and committed the government to greater public involvement in the regulatory process. To demonstrate its commitment, the federal government later that year created the Ministry of State for Privatization and Regulatory Affairs (MSPRA), which worked with the Office of Privatization and Regulatory Affairs to promote the government's regulatory objectives.

The main focus in the 1986–91 period was to review and amend federal telecommunications, transportation, and environmental regulations, all of which had knock-on effects on the provinces. Following this period, the federal government moved authority for regulatory reform to the Treasury Board Secretariat and put a new plan into place that commits the government to annual regulatory plans, regulatory impact analysis (RIA) statements for all new regulations, public consultation and information delivery on all draft regulations, and review of all regulatory efforts over set periods (statutes every 10 years and regulations and regulatory programs every seven years). The 2012 federal budget added several elements to regulatory policy, introducing a "one-for-one" rule requiring agencies not add to the "burden" of regulations, especially for small and medium-sized firms.

These fundamental changes to trade, investment, taxation, and regulations precipitated a landslide of concomitant reforms. One of the earliest and most controversial was the replacement of the narrowly levied, hidden, 13.5 percent federal manufactures sales tax with an almost universal value added tax, called the Goods and Services Tax (GST). When introduced in 1991, the government argued it was needed to make Canadian manufacturing more competitive in domestic and international markets. But most citizens were not convinced – in part because the tax was so continuously visible in every day purchases. The Liberal majority in the Senate in a highly unusual move tried to challenge this tax measure, but the government initiative was sustained. The GST legislation finally passed but left a bad taste for many. Over the intervening period, Ontario and the Atlantic Provinces officially harmonized their provincial sales taxes with the federal system and Quebec has implemented and collects its own parallel value added tax called the QST. All other provinces levy their own provincial sales tax but they are not administered on a value added basis, which diminishes the value of the GST. British Columbia harmonized in 2010, but then withdrew from the agreement in 2013 after a public referendum rejected the measure.

With these major, game changing policies in place, the focus has shifted to filling in a more explicit and effective contemporary S&T policy framework. We have seen key aspects of these kinds of past efforts in chapters 4, 5, and 6. In 2007 Industry Canada finally delivered an S&T strategy, entitled *Mobilizing Science and Technology to Canada's Advantage*. Since then the Harper government has commissioned several studies of the state of science and technology as discussed in chapters 2 and 4.While all these reviews were reasonably well argued and structured, few have made much of an impact on the federal superstructure or on the performance of Canadian firms in the context of science, technology and innovation.

In essence, the climate for innovative firms in Canada to compete is built, but there is little that is noteworthy or exceptional about the overall choices Canada has made. While recognized globally and within Canadian industry as a good climate for science, technology and innovation, industrial effort and national performance have not responded in any appreciable way. Conventional wisdom says if there is a strong science and technology R&D base acting as an "ideas pump" into the economy, wealth and prosperity should flow. The Canadian dilemma is that this is not occurring. Canada has about 20 percent more researchers per capita than the OECD average (but about 20 percent fewer than the US) and publishes approximately 4.5 percent of all basic research in academic journals (OECD 2010), yet productivity is lagging both expectations and key comparator countries (Castle and Phillips, 2011). OECD data for 2009 show that the average Canadian works about 8 percent longer than an American, harder than the OECD average (almost 13 percent more than the average for the OECD), and more than all of our key comparator countries. Despite this extra effort, we generate a GDP per capita that is about 19 percent below the US average and barely equal to the average of the OECD countries.

1986–2015: The SR&ED Tax Credit Policy

The main thrust of both of the above two policy histories did not have S&T at their centre, although both had some S&T impacts on industry, especially developments since the mid-1980s. Our third policy history concerns a policy that was centred on S&T. In an effort to encourage more domestic firms to invest in research, the Scientific Research and Experimental Development Tax Credit (SR&ED) was established in 1986 by the Mulroney Conservative government (Doern 1987, 1995a; Bernstein 1986). By 1995 the annual payments to Canadian companies reached about $1 billion and in 2014 stood at about $3.4 billion per year. Unlike earlier business incentives based on tax deductions, the SR&ED was initially trumpeted as a credit payment to smaller R&D-oriented Canadian firms that did not yet have taxable corporate income. We need to understand the SR&ED in relation to what had preceded it and what its core features as a tax incentive are, how the program has changed, and the complex set of multi-agency governance institutions and dynamics involved (Phillips and Castle 2011). We also need a sense of the changing business and related R&D lobby initially at its formation and now. Moreover, it is essential to see the changing notions of how to ensure that the SR&ED credit remained a framework incentive rather than one that treated different sectors differently either via actual policy or in interpretative guidelines intended to flesh out what scientific research and "experimental development" meant in very diverse types of industrial and corporate research situations. The cumulative ability to evaluate the effectiveness of the SR&ED is influenced by later or parallel developments such as S&T incentives by the provinces and the trend in the last decade to significantly reduce corporate taxes (Lee 2012; Phillips and Castle 2012; PricewaterhouseCoopers 2011).

Prior to the SR&ED there had been from the early 1960s onwards an array of R&D tax incentives and grants for corporations (Lalonde 1992; Department of Finance 1983). One of the policies introduced by the Trudeau Liberals in 1983 was the Scientific Research Tax Credit (SRTC). It provided a mechanism for research and development firms in a non-taxable position to transfer their unused tax incentives to outside investors. The gaming of this incentive by firms and their tax advisors produced a sudden and quite massive revenue drain from the federal treasury that exceeded $2 billion in the first ten months of its operation (Doern 1987; Bernstein 1986). That practice was stopped and the program entered limbo.

One of the early initiatives of the Mulroney Conservative government, carried out in concert with its deficit reduction focus, was to end the Liberal SRTC program which was not only bleeding the treasury but was also raising concerns about possible tax fraud and corruption. The SRTC was ended in 1985 and was replaced by the SR&ED. It was intended to be especially favourable to smaller Canadian R&D firms without taxable revenue but which would receive tax credit payments for expenditures on scientific research and experimental development. The 1984 election campaign had seen the Conservatives promise to be a much better supporter of R&D for business than the Trudeau Liberals had been.

The SR&ED operated fairly smoothly and very much out of the public eye from the early 1990s onwards and was often cited by ministers as one of the most successful and generous incentives among OECD R&D and tax systems (McKenna 2011). Gradually two kinds of criticism emerged. One centred on the view by some industrial sectors that the definitions of what constitute "scientific research and experimental development," particularly the latter, were too narrow or too complex or simply ignored basic realities of some industrial sectors (Doern 1995a). This created pressure for, in effect, a sectoral (mini-industrial policy) approach rather than a "one size fits all" incentive for Canadian firms.

A related more recent criticism is that the program, buoyed by specialized tax lawyers and R&D consultants, has wasted money because of the potential for almost anything to be called and interpreted as SR&ED and hence losing its intended focus (Expert Panel on Federal Support to Research and Development 2011). Moreover, the greater involvement of larger corporations with taxable income was significant; these firms had greater lobbying capacity and worked to influence the interpretation of the guidelines through dialogue with Revenue Canada and the Department of Finance and through litigation. The SR&ED also presents a significant challenge to small and medium-sized enterprises. SMEs, particularly start-ups and small firms, are often beset by cash flow pressures. A reimbursement system of tax credits might help between financial years, but adds pressure in the intra-financial year that not all firms can withstand. Reformers have called for "direct investment" in part to offset this problem. But, direct supports are inherently directional, as we see further below.

Before dealing with these later evolving criticisms and dynamics it is essential to have a brief sense of SR&ED's governance structure. In relation to federal institutions, it involved: the Department of Finance (as the home base of all federal tax and related fiscal and macroeconomic policy but also of S&T tax incentives); Revenue Canada (as the normal revenue collector but under the SR&ED also a program "spender" of tax credit payments and now for the first time with its own S&T personnel to issue guidelines and interpret individual corporate claims which were not subject to evaluation as to whether they were meeting the purposes of the program), and Industry Canada (as the main S&T and innovation policy advocate as per our discussion above and in earlier chapters). Other institutions involved include the Auditor General of Canada (as auditor of the programs) and the Tax Court of Canada (regarding tax cases/appeals brought to the court).

The private interests involved in pressing for the SR&ED and for later amendments included the following: R&D (and later innovation) interest groups including initially the Canadian Advanced Technology Association (CATA) and the Information Technologies Association of Canada (ITAC) but also the Canadian Federation of Independent Business (CFIB), formed in 1971 with political roots forged in a campaign against high corporate tax rates; individual firms; and professional advisers and rent-seekers, including lawyers, accountants, and R&D consultants. Larger corporations were also involved both directly and via national business lobby groups such as the Business Council on National Issues (BCNI), now the Council of Chief Executives (CCE), and the Canadian Chamber of Commerce (CCC).

By 2011 the SR&ED was described by the Jenkins Report (EPFSR&D 2011) as a program whose nature and purpose is to "encourage Canadian businesses of all sizes and in all sectors to conduct R&D in Canada that will lead to new, improved or technologically advanced products or processes." It offered two types of incentives (ibid., 65):

- Income tax deductions – all allowable SR&ED expenditures, including capital, may be deducted from taxable income in the year incurred at the taxpayers discretion …
- Investment tax credits – earned as a percentage of qualified SR&ED expenditures, investment tax credits can be used to reduce the amount of income taxes otherwise payable.

The eligible participants are corporations, proprietorships (individuals), partnerships, and trusts. The allowable costs include: salaries and wages, overhead expenses, capital equipment expenses, contracts and payments to third parties such as postsecondary institutions. Eligible SR&ED projects include those on: experimental development, basic and applied research, and support work (such as in engineering design and operations research) carried out for the purpose of achieving technological advancement or advancing scientific knowledge (ibid., 66).

The SR&ED program was assessed in 2007 through a benefit-cost analysis conducted for the Department of Finance (Parsons and Phillips 2007). It concluded that the SR&ED's public benefit exceeded its full costs. The program, in contrast to grant programs, however, "is not subject to assessment of the quality of any particular business project" (ibid., 66). Interestingly, the Jenkins Report had to rely on 2007 program data since more recent information was not available. The annual value of the credit in 2007 was $1.29 billion for small businesses (19,806 in total), $1.82 billion for large firms (2,599 firms), and $136 million for others (1,310 entities; ibid., 67). By then there were also a range of provincial tax credits that had become part of the overall tax incentive mix, not to mention general corporate tax rate reductions under both the Martin Liberals and the Harper Conservatives.

The Jenkins Report (EPFSR&D 2011, xii) examined the SR&ED program and recommended that the federal government "simplify the … program by basing the tax credit for small and medium-sized enterprises (SMEs) on labour costs" and that it "redeploy funds from the tax credit to a more complete set of direct support initiatives to help SMEs grow into larger, competitive firms." The report noted that the current cost base "is wider than that used by many other countries, includes non-labour costs, such as materials and capital equipment, the calculation of which can be highly complex. This complexity results in excessive compliance costs for claimants and dissipates a portion of the program's benefits in fees for third-party consultants hired to prepare claims" (ibid., E-3). The report also stressed that "Canada's program mix is heavily weighted toward the SR&ED program and, during our consultations, we heard many calls for increased direct expenditure support. As well, many leading countries in innovation rely much less than Canada on indirect tax incentives as opposed to direct measures … (and therefore it is important) to ensure a greater focus on promoting the growth of firms, the portion of the credit claimed by SMEs that is refundable – that is, paid regardless of whether the firm generates taxable income – should be reduced, such that part of the benefit would depend on the company being taxable" (ibid.). If some part of the SR&ED credit "spending" is reduced, our research interviews suggest that it would be transferred to direct grant programs such as the long established NRC's Industrial Research Assistance Program but this is by no means a certain outcome. The SR&ED review in the Jenkins Report and in other forums was also merging with a policy debate about whether SMEs are too broad a target for support and whether they should be replaced (as seen in chapters 1 and 4), by incentives directed at "start-up" innovation firms, a much smaller subset of firms requiring greater policy and institutional detective work and precision.

Overall then the story of the SR&ED policy history is one which emerges from an earlier short-lived S&T tax incentive debacle under the Liberal's SRTC where sudden and very public gaming the system led to its quick demise. Its replacement, the SR&ED, was initially aimed at small Canadian R&D firms without taxable income and paid as a tax credit with cheques issued by Revenue Canada acting as a "spending" as well as a revenue agency. The program has been assessed as having net benefits but the Jenkins Report shows it has morphed into a broader incentive for larger firms as well as SMEs and that makes it more difficult and complex to assess at both the program and individual project levels and in relation to start-up firms with cash-flow problems.

1990–2015: Clusters, Business, and Local-City-Regional Innovation Systems

Our fourth policy and governance history examines industrial S&T in relation to policies driven by business in urban clusters or in regional or national innovation networks and systems. In this world of policy innovation, as introduced in chapter 1, innovation is cast as complex forms of non-linear S&T in contrast to earlier theories of S&T as a linear continuum that follows a fixed set of stages. We trace it further here, especially in the context of what it says or reveals about industrial and manufacturing research and innovation. Its tentacles and ideas have already been seen in the analysis in chapter 6 of research granting and levered funding where clusters, networks, and business involvement were strong features. Innovation policy, theory, and discourse have been in political-economic ascendancy for the past twenty-five years. The concepts of national innovation systems, local-city or regional innovation systems, and clusters have emerged internationally and diffused to most countries (Porter 1990; Freeman and Soete 1997; de la Mothe and Paquet 1998; Nelson 1993).

These ideas about innovation, networks and applied commercialization often imply that one cannot actually have determinative policies but rather only strategies for recognizing and fostering individual and firm performance through networks of interaction, co-funding, and collaborative networked science (Guston 2000). Indeed knowledge is sometimes seen more broadly as "knowledge transfer," where knowledge consists of services rendered and transferred through teaching, research, and service or other kinds of "tacit knowledge" (Wolfe 2009; Pavitt 1991; Polanyi 1967).

Some of the emphasis in Canadian policy on innovation and networks or clusters can be traced to the early years of the Mulroney Conservative government (de la Mothe 2000). But as the examination in chapter 2 of prime ministerial agendas has shown, the innovation policy approach was picked up more centrally by the Chrétien Liberals and given emphasis in Industry Canada and other micro-economic studies as early as 1994, soon after the Liberals took office (Industry Canada 1994a). Innovation policy was always linked to the earlier S&T policy developments but was intended to signify a Liberal effort to push harder towards realizing greater applied development of new products and processes throughout Canadian industry. By 2000, innovation policy was the central theme of the 2002 federal policy paper, *Achieving Excellence* (Industry Canada 2002).

Network and cluster theory and practice first emerged in bodies such as the National Research Council, particularly regarding how it saw the changing role of some of its research institutes located and strongly engaged in different Canadian cities and communities (Doern and Levesque 2002). Within the NRC, innovation policy, discourse, and related institutional change emerged first in its Quebec-centred biotechnology research institute and then, at a more meso-level, in its institute on construction. The long established NRC Industrial Research Assistance Program was also restructured to offer advice and funding on a more networked basis. The NRC in some ways was seeking to show that it was not just an

Ottawa-centred national agency, but one which was actively fostering local innovation systems with its competitor-partner institutions across Canada. Part of the challenge was, as seen in chapter 5, that the NRC was characterized in a range of federal reviews as being still insufficiently focused on recognizing and supporting businesses with real, viable commercialization capacity and potential.

Gradually, strategies for growing or fostering clusters became an explicit aspect of innovation policy in other competitor countries (Phillips and Castle 2010, 2012; Niosi 2005; Phillips et al. 2012). Institutionally they create what Etzkowitz and Leydesdorff (1997) refer to as the "triple-helix" of interacting university-industry-government relations. They also create often-legitimate opposition centred on what this means for the independence of academic researchers and for science and research as a public good (Atkinson-Grosjean 2006). These issues and non-linear complexities create the not inconsiderable dilemma of how to assess performance and outcomes of such clusters and networks and indeed whether they can be evaluated vis-à-vis industrial and commercialization impacts or indeed broader impacts.

For example, Kristian Behrens' (2013) study for the C.D. Howe Institute examines manufacturing clusters in Canada, arguing overall that they have a weak effect on Canadian productivity. Gaisford et al. (2010) came to a similar conclusion with respect to biotechnology clusters. Defining clusters as "the spatial concentration of interrelated industries, specialized service providers, and sophisticated customers," Behrens (2013, 2) cites other international research (Duranton 2011), concluding that "most clusters do not live up to the hype" and that "there is indeed little solid evidence that clusters make regions generally prosperous." The study examines the spatial structure of Canadian manufacturing so as to investigate whether changes in that structure between 2001 and 2009 are associated with aggregate sectoral performance. Behrens (2013, 3) argues methodologically, "Whereas most cluster studies have focused on the local impacts of industrial concentration by looking at specific regions, a broader national perspective is useful ... [because] when some industries cluster in a region, this might come at the expense of other regions or other industries. Given that, in the short and medium run, specialized workers and other production factors are mostly in limited supply – especially in high-tech sectors and for highly educated workers-gains from clustering in some locations might well be dampened by losses in other locations."

Among the key findings of the Behrens (2013, 4) study are the following:

- High-tech manufacturing industries are not systematically among the most strongly clustered industries
- Over the 2001–09 period, the spatial clustering of manufacturing industries generally decreased in Canada.
- Over the 2001–09 period, there was no strong link between the clustering of manufacturing industries and their economic performance in terms of either output per worker or wages.

A recent discussion paper (Mowat Centre 2014) on a *Federal Economic Agenda for Ontario* diagnoses Ontario's innovation policies and performance. It reiterates that "innovation in all its forms will be crucial" to Ontario, but that "Ontario's record in business innovation has been weak" (ibid., 11). In addition to declining productivity growth, the policy challenges under the innovation rubric are centred on the fact that "Ontario experiences low levels of commercialization; there is a lack of alignment between federal and provincial programs designed to support innovation; and Ontario firms are more likely to be SMEs – which usually invest less in R&D than larger firms – than those of our peers" (11).

Another recent study (Galvin 2012) focused on "cluster-related innovation" in the City of Toronto, with an empirical focus on two clusters, aerospace and textiles/fashion. Its analytical focus was to zero-in on the city's approaches and draw on the interacting roles of leadership (business and city), local government finance, and multilevel governance programs and activities. Of considerable interest to this chapter's industrial and manufacturing focus, the two clusters both had historically long periods of state intervention and support as industrial sectors, aerospace with federal and Ontario support, and textiles with tariff protection. While state intervention in aerospace continues, support for the textile sector now involves the city seeking to reorient it as a "fashion" industry and therefore a creative industry cluster. Galvin's conclusions centre on the weakness of city-level budgets, even for a mega-city such as Toronto, and weak coordination of multi-governance support activity. Although clusters have potential, Galvin (2012, 255) concludes that an innovation policy based on clusters "suffers from a lack of conceptual clarity because it is embedded in a range of other closely related policies described and promoted in other ways including economic development policy, prosperity policy, competitiveness, and industrial sector policies."

Our interviews with senior Industry Canada officials suggest that the federal government now is seeking to find some kind of different balance between the supply and demand for innovation in a changing manufacturing and service economy. They stress the weak productivity record as a crucial context and reality, emphasize Canada's supply-side emphasis and record on innovation policy and propose a policy mix with more demand for innovation coming from within firms and sectors, albeit incentivized by demand-inducing policies for innovation. Indeed, the Harper 2015 budget plan (Minister of Finance 2015), downplayed cluster policy per se and explicitly linked its pre-election budget initiatives on "supporting world class advanced research" as a subset of the larger section of the budget on new tax support plans aimed at "supporting the manufacturing sector." It stressed, moreover, that the government's "commitment to manufacturing is long standing and strong" (82), possibly because of the government's concerns that it was seen also as one focussing on energy and the oil sands as the key driver of the economy. But such shifting of current policies and discourse will only be possible to assess over the medium term and it will be difficult to attribute cause and effects to any impacts due to the embedded nature of this policy approach and the multiple dynamics involved across industrial and economic sectors.

THE THREE ELEMENTS

We now track with the aid of Table 7.2 the three elements being utilized to understand change and inertia across the four domain policy and governance histories presented in this chapter.

Policy Ideas, Discourse, and Agendas

Our four policy and governance histories reveal quite profound changes in ideas, discourse and agendas. Initially, S&T and innovation ideas were scarcely mentioned but dominant policies nonetheless produced important technology impacts in the Canadian industrial structure. The industrial policy space has seen a clash and shift of purposes, reflected in nation building and then enthusiastic dirigiste industrial policy in the period after the Second World War to the late 1970s and now a more hands-off approach.

From the mid-1980s onward, as the world changed and national boundaries became less influential, S&T and innovation policy ideas were submerged under the strong shift from macro-economic stabilization policy to micro-economic pro-market growth ideas, which proscribed a more limited role for the state. Gradually but profoundly the

Table 7.2 Policy and governance histories in the industrial S&T and innovation domain: Three analytical elements

Policy and governance history	*Policy ideas, discourse, and agendas*	*Economic and social power*	*Time, temporal realities, and conflicts*
1945–1980s: Dirigisme, made-in-Canada and made-for-Canada industrial policy	• Dirgiste, infant industry strategies dominated • Nation building at the root of most policies • Emergence of more nuanced ideas and understanding of the role of government spending • Markets and private sector not held in high regard • Technology impacts often present without explicit S&T ideas in policy discourse	• Policy driven by reconstruction and uneven regional economic prospects and interests • Emerging competition from the US and abroad pressed firms and industries to adapt or perish • Some industries lobbied but also simply adapted without policy or funding or tax incentives	• Government sought to exert fiscal authority with new medium-term levered conditions • Regional development plans and agencies also somewhat more medium term in focus • Auto Pact solidified the auto and probably steel sectors, reached zenith in 1970s • Procurement and IRBs critical for aviation and military hardware producers • Shift to innovation in later years also cast in medium- and long-term ways
1985–2015: Creating the climate for innovative firms to compete in global markets	• Government macro-economic stabilization replaced by micro-economic growth policy ideas and programming directed and constrained by public choice theory • Market-based view in ascendancy	• Globalization of trade and investment after 1985 pushed for greater policy engagement and coordination among governments and market interests internationally • Greater power of business overall	• Free trade efforts 1985 to present, but also some provisions for exemptions and phased-in adjustment • Foreign investment reform and policies to encourage longer term investment • Privatization of many Crown corporations • Deregulation, but also increased social regulation • Greater medium-term policy focusing post-2007 on federal S&T strategy

1985–2015: SR&ED tax credit policy	• Initial focus on Canadian SMEs S&T and "experimental development" as explicit further purpose to be judged by Revenue Canada • Morphed to include support for large firms • "Start-up" innovating firms compared with SMEs emerges as better idea re main target for support	• SR&ED system propelled by failure of earlier system that produced big sudden losses to federal revenues • Gaming the system by accountants, lawyers, and consultants problematic in different ways • Revenue Canada establishes its own S&T staff to assess claims • Revenue Canada, normally a revenue collector, in this case was a "spender" • System pronounced as generous in global comparative terms	• Challenge to find the right balance of inputs and results, fine tuning in 2012 to present • Over its 30-year history, the SR&ED functioned well under the radar of macro-policy coverage and scrutiny • Individual company claims not easily evaluated in part because of the secrecy features of tax collection system • Interacting but hard to differentiate combined effects of SR&ED and general steady lowering of corporate taxes
1990–2015: Clusters, business, and local-city-regional innovation systems	• Non-linear nature of clusters and S&T and innovation idea • National vs. local-city and regional innovation systems • Initial focus on collaborative and network-based S&T relations as valued in and of themselves • Triple-helix systems including universities • Knowledge economy and systems • Importance of tacit knowledge	• Support in early Mulroney era • Strong support in Chrétien era overall, including in the granting system (chapter 6) • Experimentation with clusters in some NRC institutes • Demonstration effect of city-regional systems in US and Europe • City support for approach conceptually, but often with constrained city budgets	• Problems of attribution, and cause and effect claims regarding clusters • Good or plausible impact claims for one cluster or city-region may later produce adverse impacts in other clusters/regions in other parts of the country • Underlying temporal features regarding productivity slowdown or prevent positive impacts

Note: see the chapter text for expansion of abbreviations in this table.

new dynamics of risk and opportunity in the twenty-first century yielded science and technology based innovations that both threatened existing operators in traditional value chains and offered tantalizing rewards for those firms and entrepreneurs able to work in increasingly fluid, fast and frictionless global markets.

The changed industrial S&T and innovation space is in many ways the battle ground for a vibrant and compelling set of alternate world-views. Some see disorder and chaos in the emerging world system and harken back to simpler times, when it was enough to be able to assemble the means of production (capital, labour, and land) and to work hard. Now creativity, the intangible factor that sparks invention and innovation, is not equally accessible. This is frightening and challenging to many who look to governments to slow the process, foster adjustment, and control the outcomes. This world view sees the state as the solution and the market as the problem. In contrast those thinkers and doers attracted by the excitement and rewards of the intangibles of capitalism, creativity and innovation see the state as the problem and the impersonal, global market as the solution.

Ideas, discourse and agendas also emerge in other basic, often more particular ways as the second and third policy histories reveal. The SR&ED sought initially to be a focus for federal support of smaller Canadian firms that did not yet have taxable income but gradually morphed into a program that supported larger firms. It also revealed ideas about basic tax-centred support but in a context where other corporate tax incentives were also changing.

The clusters, business and local city/regional innovation policy paradigm of contextualizing policies ultimately flowed from the global context but became assimilated to the Canadian context, where policy makers had come to accept the non-linear nature of S&T and innovation and recognized the value of networks and the fostering of complex networks that conformed with the Canadian geography.

Economic and Social Power

The structure of economic and social power in this domain reveals the importance of globalization, even before that term of power and policy took hold in the 1980s. Thus, the first policy history, even though cast as dirigiste in nature, showed that Canada's industrial landscape under nation-building and the national policy had been dominated by large firms working at the margins – the margins of the resource economy, the margins of global markets and the margins of profitability. The same is true in many ways in the recent and present context, where globalization of power is both a problem and an opportunity. The increasingly interconnected input and output markets and the emergence of global value chains and networks threatens those firms and sectors that are not connected to secure supplies, ample and appropriately priced markets, global finance, and the latest technology.

But in the dirigiste era, and to some extent today as well, regional interests lobbied to produce regional agencies. While their mandates and power remained constrained, they nonetheless developed policies to deal with regional socio-economic needs and opportunities, including more recently explicit S&T and innovation plans and programs. Overall in the policy and governance histories, business power is dominant but there is significant power-sharing with political parties and prime ministers in each of the Chrétien-Martin and Harper eras. The influence of cities and local interests emerged in the past few decades, involving both advocacy and use of the clusters and networks, only limited by civic budgetary capacity. The SR&ED history reflects business influence but also the capacity of various market interests to game the system and change its key features.

The economic and social power story thus far tends to view the industrial domain as an isolated realm focused on manufacturing. But since the end of the gold standard in 1971, manufacturing has been whipsawed between volatile energy and resource prices and deep plunges and dizzying heights of the Canadian dollar, with the result that the industry has had periods of abundance and periods of pain. Obviously natural resources are also a part of the overall story of economic and social power as shown in earlier chapters.

Time, Temporal Realities, and Conflicts

The four policy and governance histories are reasonably representative of much of the direct and indirect S&T and policy innovation effort in the industrial space in the past fifty years or so. In the dirigiste era, in temporal terms, policies had considerable staying power, including the use and experimentation with conditional spending, the continued presence of regional agencies (albeit with changing mandates), but also with programs such as the Auto Pact and procurement incentives. From the mid-1980s onwards, international free trade agreements, starting with CUFTA, NAFTA, and the WTO, were pivotal and led to further bilateral agreements and to internal trade agreements. Privatization and key aspects of deregulation were cast as medium- and long-term change and have had considerable staying power.

In more particular ways, the SR&ED policy was designed to be a permanent feature and for years it has remained so, in part because it functions well below the radar of the macro-policy agenda. Clusters and related city-region innovation system policies are still broadly favoured even though there are difficulties in attributing positive impacts to them, given the non-linear and networked nature of these innovation structures.

In temporal terms and in various cycles of change, one of the most profound difficulties in this policy domain is that the ideas, policies, governance arrangements and outcomes possess the capacity to continuously alienate, disenfranchise and frighten the average citizen and threaten the comfortable future for many small and medium sized firms in Canada. Canada will continue to be challenged by global markets that never stop determining the value of its effort, and impersonally inflicting great harms on those persons or communities found wanting and, perhaps more challenging, immense rewards earned by those lucky or prescient enough to be in the right place at the right time with the right product or service at the right price.

CONCLUSIONS

This chapter's analysis of the industrial S&T and innovation domain shows a complex trajectory of policy and change, including features where manufacturing but also technological change was fostered without S&T policy being explicitly stated or involved. This was especially the case in the dirigiste and industrial strategy policy history both in the sense of the National Policy but also when industrial policy and industry departments were created in the period after the Second World War and up to the late 1970s.

Macro-economic and then micro-economic growth policies in different ways then sought to create a climate for innovation by Canadian industry. Policies increasingly moved away from demand management and toward micro-economic measures driven by trade and globalization, including efforts to privatize and deregulate, but also to foster a more export-oriented Canadian manufacturing industry. The domain story then shifts or is complemented by more explicit business tax incentives such as those featured in the SR&ED analysis. Finally, we have explored the somewhat divergent policy track defined by clusters and local-city-regional innovation systems.

Some aspects of our interest in the innovation economy and society nexus emerge in this chapter. Dirigiste industrial policy was initially seen as nation building via infant industries. Regional policy initially and now has been cast as addressing society's employment needs and economic opportunities. Clusters and local-regional innovation systems have now embraced both the economic and social aspects of development, including the involvement of cities, universities, community colleges, and in some cases, social economy actors. Tacit knowledge, the lubricant that makes the system work, is fundamentally a socio-economic construct and process. But finally and importantly the chapter shows that in spite of a long-standing focus on S&T in industrial manufacturing, involving repeated restatements of the goal and repeated program innovation, performance has never been very strong when compared with other countries.

We noted above that this domain analysis does not examine Canada's resource industries. However, the industry story in a deeper historical sense reveals the tension at the heart of Canadian S&T and innovation policy. Canada developed its resource base using foreign capital and accessing global markets. In this sense Canada was one of the first globalized economies. For an extended period of time nation builders downplayed our historical resource-based roots and projected an image of Canada as a vibrant manufacturing and services-producing economy. In spite of all the efforts to develop capacity that is disconnected from Canada's resource roots, much of our industry that is thriving is also inextricably connected to the primary sector – providing inputs, processing outputs, financing, developing new production technology, and moving and marketing those products. Although there have been efforts to show the interconnectivity and mutual future prosperity from further engagement between some of the resource sectors – e.g., agriculture in the 1980s and oil and gas in the 2000s – and the mining and forestry sectors, that message is still often subsumed into a misleading two-category world of old economy–new economy, where we lament with Harold Innis (1977 [1930]) our past and possible future in the stables trap, as mere "hewers of wood and drawers of water." We return to these issues in later domain chapters and in the book's conclusions in chapter 12.

8

The Intellectual Property, Invention, and Innovation Domain

The intellectual property (IP), invention, and innovation domain has grown in visibility and importance in the context of federal and international innovation and trade policy after many decades of functioning out of sight and out of mind in national politics and in economics and S&T agendas. Increasingly, the value generated in markets and visible in the stock valuations of firms is intangible value, variously described in terms of intellectual property, but also reputation, brand value, and good will. With the rise of the knowledge-based economy, governments around the world have been alerted to the need to consider intangible asset management in concert with the management of tangible assets, and in research this has led to a focus on the role of intellectual property, particularly patents, in innovation. Firms and countries now assiduously measure and report the value and share of their operations or trade generated by intellectual property. In that context, a series of ongoing debates and processes about the appropriate values, metrics, and methods of valuation have kept the issue of intellectual property in play in the policy system over the past forty years.

Intellectual property has existed since before the industrial era. What would now be called "trade secrets" abounded and were guarded, cultivated, and exploited by medieval guilds. Patents – as state grants of limited monopolies – emerged as a precursor to the Industrial Revolution. The earliest patents were issued in Venice as early as 1450, as an incentive to encourage the transfer, reduction to practice, and scale up of glassmaking technologies. The English patent system evolved after 1688, when all existing Royal letters of patent for monopolies were revoked – largely due to misuse – and patents were henceforward offered only for inventions. The Canadian patent system also traces its roots to the US Jeffersonian model centred on its access-incentive philosophy (Fox 1969).

IP policy in Canada can be traced back to early patent laws that pre date Confederation and then emerged formally as Canadian law in 1869 (Fox 1969); it has been amended periodically since then. The discussion on intellectual property has historically been divided into two fields, "industrial property" and "copyright." Industrial property IP rights include protection by means of patents, trademarks, and industrial designs. Copyright gives authors and other creators of works of the mind such as literature, music, and art, rights to authorize or prohibit, for a certain period of time, certain uses made of their works; so-called neighbouring rights also supply rights to performers such as singers and musicians. More recently a range of other claims and rights have broadened the debate, including plant breeders' rights (PBR), animal pedigrees, and traditional or Indigenous knowledge. IP rights in force internationally include 8.66 million patents, 23.96 million trademarks, and 2.71 million industrial designs (World Intellectual Property Office 2013). This chapter examines these rights in a range of applications but overall focuses on patents.

The central framework policy trade-off in intellectual property as a whole is between that of protecting creations and inventions of the mind in order to facilitate commercialization and a desire to disseminate widely the inventive insights of such creations for the broader good of society (Castle 2009; de Beer, Gold, and Guaranga 2011; Doern 1999b). Different patent systems arising in different jurisdictions (and at different times in the evolution of those systems) offer different balances between the goal of making inventions accessible and the goal of rewarding inventors. Especially regarding patents, the rules both protect the patent holder (provided that a patent is renewed through the periodic payment of fees and efforts are made to "work" the patent), and also provide for detailed information about the invention to be made public so that others might use those insights to develop further inventions. The general practice is now to offer twenty years of protection in exchange for full disclosure of the inventive step, albeit often a few years after the patent is filed. Patents are rights conferred on inventors of new products, technologies, business processes, or compositions of matter that are judged to be an inventive step, in short, one that is novel, useful (effective), and non-obvious. Patents can be filed in multiple markets, but the rights embodied are only accessible in the market where the patent is granted. Fundamentally, however, a patent gives the patent holder a right to exclude others from using rather than a right to use. What patent holders get is the right to sue others, but only inasmuch as they can afford to do on their own.

Other IP mechanisms offer different balances. Trade secrets – such as the recipe for Coca-Cola – are universal and perpetual, as long as no one either discloses them or reinvents them. Copyrights and trademarks offer different applications, both in terms of reach and duration: copyrights are universal by treaty and extend to up to seventy-five years beyond the life of the creator, while trademarks are for the most part universal and perpetual but have to be renewed frequently by payment of a fee.

The desire for disclosure, and hence use of new knowledge, creates the case for states to explore options for intervention, but it does not in itself make the case for *how far* to intervene or *what instruments* to use to intervene (Doern and Sharaput 2000; Trebilcock 2011). In some cases states will promote the use of formal IP rights, while in others cases states will provide subsidies for private action or simply have the state undertake the research itself and make strategic decisions about how new knowledge is released and to whom. These options can and do work side-by-side in modern economies.

Governments approach IP patent rights as a framework-oriented policy domain issue. The challenge is to define one overall system to foster intellectual creativity and inventiveness rather than create niched approaches for different industrial sectors such as the biosciences, chemicals, computers, or pharmaceuticals. This way, patents especially can each be treated consistently within a legal jurisdiction, and the international Patent Cooperation Treaty is an intergovernmental approach to extrajurisdictional harmonization. Yet, in practice, the complexities and particular demands presented by innovation in different sectors means that sectoral trade-offs are inevitable. They have certainly driven key aspects of the politics of patents given that key sectors internationally and in Canada have economic and political power and thus have driven patent and related agendas.

In policy and regulatory terms, the core IP regulatory institutions (and related interests) include the following: the Canadian Intellectual Property Office (CIPO), an advisory Patent Appeal Board, the Copyright Board, the Trademarks Opposition Board, Industry Canada, Canadian Heritage, Foreign Affairs, Trade and Development, and Big Business such as the pharmaceutical industry. Also important are the IP professionals, such as patent agents, patent examiners, and technology transfer officers, who serve both the regulatory system and the inventors (CIPO 2015; Doern 1995b). As suggested above, these core players centre on the protection and rights function but by no means exclusively.

The secondary set of institutions and interests, typically with a more social focus, include other federal departments and agencies such as the Patented Medicines Prices Review Board, Health Canada, and Agriculture and Agri-Food Canada (and their regulatory units), whose clientele may have major or periodic concerns about patents; the network of federal research organizations such as the Networks of Centres of Excellence (NCE), the Canada Research Chairs (CRC), the Natural Sciences and Engineering Research Council of Canada (NSERC), the Social Sciences and Humanities Research Council of Canada (SSHRC), the Canadian Institutes of Health Research (CIHR), and Genome Canada; provincial governments; small and medium-sized enterprises (SMEs); university, educational, library, and research institutions; and individual inventors or creators (Doern and Sharaput 2000).

This chapter explores the intellectual property, invention, and innovation domain through four policy and governance histories. In the first instance, Canada is inexorably pushed and pulled by its global engagement, both involving Canada's membership in the World Trade Organization (WTO) and the World Intellectual Property Organization (WIPO) and because of the large number of major sectors and firms that are intensely connected to other markets. Our first policy and governance history shows how this system has pushed Canada, and other countries, to extend and refine its IP system, in terms of amendments to the law and changes in institutional practice, over the past thirty years. These changes focused on the protection side of the IP policy trade-off. One of the main areas of significant change, and thus our second policy history, centres on how patents and other IP rights are now available for living organisms but are also a realm of continuous controversy.

Our third policy and governance history explores the limits of patents. In spite of Canada's apparent enthusiasm for extending and strengthening IP rights, there remain some significant concerns raised about the limits of patents on humans, the counterclaims of Indigenous peoples for recognition of traditional knowledge, and the practice of patent trolls in the information and communication technologies (ICT) sector. Moreover, there is significant support for an "open source" model of development, both in the life sciences and ICT sectors. Our final policy history shows how IP rights are a means to an end, and not an end in themselves. Canadian policy is also fixated on trying to figure out how to convert intellectual property into socio-economic benefits and impacts.

FOUR DOMAIN POLICY AND GOVERNANCE HISTORIES

Patent and IP policy in Canada is inextricably interconnected with the global system of intellectual property, globalized economic activity, the changing landscape of inventive activity, changing social interests, and ultimately with the broader innovation policy agenda.

1980s to Late 1990s: IP and Patent Policy Emergence in the Federal and Global Trade and Innovation Policy Agenda

There is not at present or historically an IP policy or set of policies that one could conveniently or easily point to. IP policy involves statute law, common law, and policies vested in departments and programs. Canada's patent policy and laws in part emerged or more accurately re-emerged and congealed to some extent in the 1980s Mulroney era with some important carry-over into the Chrétien era following the completion of the GATT trade negotiations that led to the formation of the World Trade Organization and the Agreement on Trade-Related Aspects of Intellectual Property Rights (TRIPS), administered by the WTO (Doern 1999b). Intellectual property has become an important, and increasingly conspicuous layer of the overall *innovation policy and strategy* context of this period where federal policies and funding programs are supportive of the growing importance of intellectual property, especially patents, as a crucial feature of commercial originality and development.

It is important to keep in mind, however, that patents constitute invention and are not yet innovation in actual markets where products and production processes are bought and sold. Patents, rather, register and legitimate claims to possible innovation opportunities, and often act as important signals in markets that attract important collaborators such as venture capitalists. Innovation based on patent portfolios often occurs when firms or inventors acquire investment capital to fund the subsequent regulatory approval process for products but also the manufacture, marketing, and sales stages of the full innovation and commercialization process (Corbin 2010).

Although Canada does not have a formalized single IP policy framework, there exists a body of policies, common law, regulations, and the Patent Act that jointly enable the cross-industry use of patents. Having this framework in place, and defending Canadian interests in intellectual property, has in the past decade become a core focus for corporate interests, such as the Council of Chief Executives (CCE). In Canada and abroad, however, there are growing concerns about the trade-off that is being made between "core protection" versus missed opportunities for greater economic gains, a concern centred on striking a balance between what has been called the "fortress mentality" with respect to intellectual property and the open IP movement – a balance that must be considered sectorally and overall in the public goods versus private property debate (Castle 2009; and see further policy histories below).

The essence of Canada's modern patent policy and law, influenced by US pressures and by key sectoral business interests within the US economy, was that Canada's system should offer the maximal scale and scope of protections possible. The first battle was over the length of patent protection, which Canada set at seventeen years from the granting of the patent and the United States wanted extended to twenty years from the date of filing. Canada argued in international trade litigation that the terms were effectively equal, but in the end complied with the US model for patents filed after 1 October 1989. Basic economic logic, and the most cursory glance at the intrasector development and intersectoral comparisons suggest that these periods of protection probably ought to vary greatly by field or sector depending on the varying cost structures, investments, and payback periods (Doern and Sharaput 2000). This also suggests that countries could have different views about what kinds of protection across sectors would make the most sense given their national state of development, comparative advantage, and strategies for development. Thus, the market economics underlying patent protection suggests the suitability of many periods of protection and that these periods could change over time.

The political and institutional logic differs from the market logic, however, and stipulates that longer protection periods and indeed one defined long period, now twenty years, are better. This view was driven in the United States and in Canada by firms such as those in the national and global pharmaceutical and biotechnology industries, which sought out and achieved maximum effective protection for an expanding range of fields of invention (a main battleground was over patents for living organisms, which is discussed in our second policy history). Their desire for maximum periods was driven by factors in the 1980s such as high up-front costs in R&D, and longer and more expensive drug approval processes that eroded the *effective* protection available.

Canadian and US changes to laws and trade regimes were undertaken against a backdrop of international negotiation on the international enforcement regime for intellectual property and patents. Under the General Agreement on Tariffs and Trade, in effect from 1948 to 1995, the international regime for intellectual property fell well short of a harmonized regime (Sell 1998, 2010; Doern 1999b). Indeed, IP issues were largely outside the GATT purview. Moreover, international IP organizations such as the World Intellectual Property Organization did not contain any effective dispute resolution process; WIPO regularly reported on disputes and sometimes mediated disputes but, like the GATT enforcement process, was voluntary and sporadic (Doern 1999a; Trebilcock 2011). The Uruguay Final Round Act of 1994 included, for the first time, a comprehensive agreement on TRIPS that seeks to balance the conflicting values inherent in intellectual property and between developed and developing countries (Sell 1998, 2010; Marcellin 2010).

During the Uruguay Round of the trade negotiations, the issue of mandates in intellectual property between WIPO and the proposed World Trade Organization generated considerable dispute. The developing countries preferred WIPO as the lead institution because it had facilitated diverse IP policies and institutions in developing countries. The United States and Europe, but especially the former, preferred a stronger WTO mandate because it wanted better dispute settlement and enforcement of harmonized IP rights, especially regarding key developing countries whose IP regimes were either weak in law or weak in implementation.

The United States also applied sustained pressure on Canada and its policies that gave preference to generic drug manufacturers. This issue was prominent in a range of successive free trade agreements (FTAs), including the North American Free Trade Agreement (NAFTA), and GATT negotiations (Doern and Sharaput 2000). In the late 1980s and early 1990s, in particular, there were few if any effective counter-pressures from those interests/countries with different interests. Developing countries mounted some opposition to stronger IP rights but were eventually worn down by more powerful forces. Consumers in some overall sense had a vested interest in less monopolistic practices but, at both national and certainly at international levels, they were weak, diffused, and virtually voiceless on patent issues. Perhaps the only exception to this was in the health sector where health ministries and NGOs were often a surrogate representative of consumer, patient, and disease-specific interests (Doern and Prince 2012).

In the later part of the Chrétien era and the Martin years, the issue of patenting of higher life forms (as opposed to microbial life forms which were patentable in many countries) and the broader economic and ethical issues inherent in regulating this aspect of biotechnology emerged. Somewhat surprisingly, decisions to extend the patent scope to include living matter were not the result of legislative action but were, instead, the result of private litigation in the courts regarding the intent of patent laws, such as in the Abitibi, pioneer, Harvard oncomouse, and Schmeiser cases (see more below in our second policy and governance history).

Debate turned on what constitutes an *invention* and the degree to which the manufacture or the composition of matter was under the control of the inventor as opposed to control by the laws of nature. These concerns raise serious ethical issues about patenting life forms, and similar issues attend other genetic testing products and bio-life processes (Castle 2009).

Emerging counter-pressures and arguments about the limits of patents and property rights appeared due to the need for public science that keeps patenting as an option but not a default, as contrasted with private interests in science that bias innovative activity toward patenting. All patents are issued by nation states, and national patent systems basically focus on the protection side of the core protection versus dissemination trade-off. As the Doha trade negotiations began in 2001, and as criticisms emerged since then, the focus has turned more to the dissemination side of this policy equation, that is, to the dissemination, public goods, and related public science and public knowledge side of the trade-off (Dutfield 2003; Castle 2009).

At the level of the World Trade Organization, this shift in focus to dissemination transpired through changes to the TRIPS Agreement that were intended to create greater *access to medicines*, especially bio-health products by WTO member countries with insufficient or no manufacturing capacity in the pharmaceutical sector. The concern was that a strict interpretation of the TRIPS Agreement would exclude those countries from making effective use of the compulsory licensing allowed under it. Under the provisions of the Agreement on TRIPS (Art. 31) compulsory licensing or government use of patents is allowed without the authorization of the patent owner. One of the conditions under which this can be done is when such use is predominantly for the supply of the domestic market. TRIPS generally prevents WTO members with manufacturing capacity from issuing compulsory licences. In 2003 WTO members agreed to a waiver stipulation regarding this provision. Canada was the first country to announce that it would implement this waiver, and in May 2004 Canada's legislative framework was given parliamentary approval for Canada's Access to Medicine regime. It was implemented through amendments to the Patent Act, and a year later its regulatory provisions came into force (Mills and Weber 2006). Alas, due to industry opposition and other Canadian and international bureaucratic inertia, the policy has not had much actual effect (ibid.) and thus the rules and policy are still being reviewed by the Harper government.

Both the WTO's waiver (and later permanent amendment on this matter) and the Canadian regime to implement it were motivated by wholly desirable broader foreign and international development and health policies. The policy of access backed by legislative and regulatory change had to ensure that Canada was still complying with its overall international obligations under the TRIPS Agreement and also NAFTA, was respecting the integrity of its own patent law, and was responding to the competing industry and NGO interests involved in this issue.

Access to medicines is one issue in the much larger debate over the patent regime. Other analyses are often biotechnology-specific and draw attention to the actual unwieldy natural bio-world (which would not be patentable) in contrast to practices and pressures in patent law to create "discrete immutable biological 'objects'" (Carolan 2010). Similar biotechnology-focused criticism centres on the need to reign in the scope of patents in the interests of the garnering of public trust and obtaining contributions to public benefits and knowledge (Caulfield 2009).

More broadly, there are concerns about excessive patenting, the ease of getting patents for questionable inventions, and the rise in some fields of patent thickets that are harmful rather than conducive to innovation (Heller and Eisenberg 1998; Canadian International Council 2011; Castle 2009; Corbin 2010). These concerns are valuable from an analytical standpoint, and if they have influenced practice, they have not yet led to major legislative and regulatory change in Canada or internationally.

The role played in the past by requirements for *compulsory licensing* is one feature of this trade-off. These requirements, which now are seriously restricted, once were based on the belief that inventions made and patents granted should actually benefit consumers and society. Thus, if the inventor was not going to turn the invention into a product or process actually available in the market, then others should be able to through the payment of a licence fee to the patent holder. Licensing certainly exists in the current system, but persons or firms that want to be able to use the patented item or process then have to negotiate with the patent holder and, in the end, the patent holder may decline to license. Such issues have not gone away – they are, for example, at the centre of disputes about patented genetic diagnostic tests. The history of Myriad Genetics Inc and its commercial strategies has exercised governments and health policy advocates around the world (Gold and Carbone 2010).

Early patents were often awarded to inventions of an engineering nature, but today a significant proportion of patents deal with other science products and processes, including computer hardware and software, business processes, genetic and genomic applications (Doern and Prince 2012). Thus, CIPO patent examiners have to come from a variety of technical and science backgrounds including biology (CIPO 2015;

Doern 1995b). The Canadian Intellectual Property Office functions increasingly in an era of high volumes of patent applications with national governments, including Canada, seeing patenting performance as one important international indicator of innovation. Canada still struggles in this and other patent races in the global rankings.

The CIPO must be differentiated from the role of the Commissioner of Patents and Registrar of Trade-Marks. The CEO of CIPO is also the Commissioner and the Registrar. In the latter capacities CIPO's head official is a statutory person. The role as a statutory person is important because all of the regulatory powers reside in this legally defined role. This position is a statutory appointment to ensure that decisions on patents and trademarks and other IP rights are based on independent objective judgments and not on political considerations. In other respects as CEO, however, the head of CIPO functions within the jurisdiction of Industry Canada. As a result, if legislative policy changes were being considered to the Patent Act they would ultimately be the responsibility of the minister, although it is possible for the Commissioner to comment on amendments to the legislation for the minister's consideration and approval. All of the other previously noted aspects of Industry Canada's mandate mentioned in previous chapters apply in the IP patent realm as well. Of particular importance is the fact that Industry Canada's Intellectual Property Policy Directorate and its Patent Policy Directorate are also engaged in international IP and trade negotiations and monitoring (Industry Canada 2011).

CIPO's statutory base flows from six Acts, one for each type of intellectual property, under which there are also separate regulations (CIPO 2011b). The Patent Act is underpinned by a set of Patent Rules. The patent regulatory system also has the above-mentioned Patent Appeal Board. Crucially, the courts are heavily involved, largely because there is no patent police force to protect patent rights. Enforcement occurs mainly through private litigation, a process that occasionally creates the above-mentioned complex patent thickets patrolled by patent trolls, especially in the aggressively litigious United States (Canadian International Council 2011; Bessen and Meurer 2008; and see more below).

The Canadian Intellectual Property Office has also been involved in numerous consultations to address specific issues, but most of these are very technical and administrative in nature (CIPO 2011b). In the Harper era these politically sensitive consultations have centred mainly on new copyright legislation issues rather than patents (Geist 2011). As we see, however, from the policy-regulatory histories that follow, other pressures are building on the patent side regarding where the protection versus dissemination trade-off should occur in the interests of both creativity and innovation. Furthermore, there is debate about what an invention is and the limits to property rights where the boundaries of property in patents are increasingly blurred (Dutfield 2003).

1980–2006: The Fight to Extend Property Rights to Living Matter

As mentioned already, the original focus of patents and other IP rights was on product and process innovations in the mechanical and industrial world. Most inventors of compounds – be they chemicals or foods – tended to use trade secrets, trademarks, and commercial brands to exploit their innovations. But given that a large and growing part of the Canadian and international economic and industrial landscape involves manipulating and using living matter, it was inevitable that pressures would build for extending, or developing new property rights into this realm.

By the 1930s the US patent system granted rights for asexually reproduced plants, and in the 1940s a number of European nations granted plant breeders' rights for new varieties developed domestically. Plant breeders' rights were extended internationally in the 1970s through negotiation of the International Union for the Protection of New Varieties of Plants (UPOV). During the 1980s rights were revised and extended as new biotechnologies enabled new inventions. In 1980 the US Supreme Court, in a 5–4 decision in *Diamond* v. *Chakrabarty* overturned a ruling of the Patent and Trademark Office and granted a utility patent for claims related to a genetically engineered micro-organism that degraded multiple components of crude oil, based on defining the new organism as a new composition of matter. This decision was shortly followed by grants of utility patents for genetically modified (GM) plants and for genetically modified animals. During that same period, the technology and science became more refined, enabling researchers to identify and isolate specific genes from various microbes, plants, animals, and humans. Many researchers applied for and (mostly) were granted utility patents on those genes or gene constructs.

This effort was internationalized in the 1990s, with the negotiation of the World Trade Organization in 1995 and the TRIPS Agreement, which requires all member states (159 in 2014, collectively representing more than 96 percent of global production and trade) to either grant full patent protection to all patentable materials (including plants and other higher life forms) or to provide a "sui generis" (i.e., purpose built) system of protection (such as PBR for plants and pedigrees for animals). All developed countries had until 2001 to comply with TRIPS, while developing countries had until 2006 to comply.

In Canada the events surrounding life patents have differed considerably from those of the United States, and they have progressed more slowly (see Table 8.1). In 1982 the Canadian Patent Board, following suit with the Chakrabarty case in the United States, extended the scope of patentability to micro-organisms by granting a patent to the Abitibi Company for a "foam flotation activated sludge process" (CIPO patent no. 1,131,371 issued 7 Sept. 1982). In 1983 the impact of this new ruling was tested for plants when Pioneer Hi-Bred filed an application for a Canadian patent for a new soybean variety. The application was ultimately rejected by the Supreme Court of Canada in 1989 because of its subject matter and a failure to satisfy public disclosure. This decision drew a response from Parliament, which moved to encourage the innovation of new and useful plant strains by enacting the Plant Breeders' Rights Act in 1990, albeit only enacting the limited provisions of the 1978 International Convention for the Protection of New Varieties of Plants, which had established the International Union for the Protection of New Varieties of Plants. In 1991 UPOV negotiated a new agreement that extended breeders' rights, first, by allowing patents and PBR to be used on the same varieties (in UPOV 1978 only one form of protection was allowed per variety), and second, by extending the breeders' right to import, export, stock, and condition reproductive materials, which effectively allows for end-point royalties collected in the handling system rather than only seed royalties.

On 21 June 1985, following the Pioneer Hi-Bred application, Harvard University filed a Canadian patent for the oncomouse. On 24 March 1993, the patent examiner granted a patent for the cell culture, process, and use of the oncomouse, but not for the mouse itself. Upon review, the decision was upheld by the Commissioner of Patents on 4 August 1995. Harvard appealed the ruling to the Federal Court of Appeal and won a favourable judgment, but a final decision by the Supreme Court of Canada in 2002 rejected the claim, determining that the mouse as a multicellular organism was patentable.

The issue about the patentability of higher organisms in Canada appeared resolved until the Supreme Court of Canada judgment in *Monsanto* v. *Schmeiser*, whereby the court upheld Monsanto's patent on the Roundup Ready gene (see chapter 9). But they added a small codicil to the debate, affirming that because the gene is expressed in all of the

Table 8.1 Comparative history of US and Canadian life patents

United States	*Canada*
24 Dec. 1970 Plant Variety Protection Act passed	19 June 1990 Plant Breeders' Rights Act passed
31 Mar. 1981 Single Cell Life Patent issued to Ananda Chakrabarty for oil slick–consuming bacterium (USPTO patent no. 4,529,444)	7 Sept. 1982 Single Cell Life Patent issued to Abitibi Co for a foam flotation activated sludge process (CIPO patent no. 1,131,371)
22 June 1984 Harvard files first oncomouse patent titled "Transgenic Non-Human Mammals" (USPTO patent application no. 623,774)	21 June 1985 Harvard files for Canadian patent titled "Transgenic Animals" (CIPO patent application no. 484,723)
1985 Ex Parte Hibberd – USPTO extends patent protection to plants, seeds, and plant tissues under Patent Act s. 101	22 June 1989 Pioneer Hi-Bred is denied the first Canadian seed patent (filed 18 May 1983)
12 Apr. 1988 Harvard granted world's first animal patent for the oncomouse (USPTO patent no. 4,736,866)	24 Mar. 1993 CIPO patent no. 1,341,442 granted to Harvard for oncogene, process, and use, but patent on oncomouse itself rejected
11 Feb. 1992 Harvard granted 2nd oncomouse patent protecting cell cultures derived from animals (USPTO patent no. 5,087,571)	3 Aug. 2000 Federal Court of Appeal overturns Commissioner of Patents and trial division court, ruling oncomouse is patentable
20 July 1999 Harvard granted 3rd oncomouse patent for testing for oncogene expression (USPTO patent no. 5,925,803)	5 Dec. 2002 Supreme Court of Canada overturns Federal Court of Appeal, determining the oncomouse cannot be patented

cells of the plant, the patent owners have de facto rights to control all potential uses of the plant. In effect, because the transgene was universally expressed rather than expressed in only specific sites (such as the roots, leaves, or seed), the patent owners can lay claim to the entire living organism.

The end result is that there are now literally tens of thousands of patents for biotechnology tools (e.g., *Agrobacterium tumefaciens*, and polymerase chain reaction), for isolated genes, and for constructed single and multicellular organisms derived from the technologies.

1983–2015: Finding the Right Balance in Patents

As emphasized above, Canada has had a long and varied debate about what is the right balance between protection and dissemination or disclosure with regard to intellectual property. This has involved wrestling with compulsory licensing of patented drugs, engaging with claims of traditional and Indigenous knowledge, considering if or how to structure research systems in response to increasing patent claims, and addressing the emergence and predatory practices of "patent trolls," especially in the telecommunications and ICT sector.

Canada's unease with complying fully and completely with private ownership of intellectual property goes back to 1923, when the Patent Act was amended to provide for compulsory licensing for patented drugs if the medicine's active ingredients were manufactured in Canada. A compulsory licence is a statutory licence that gives the licensee the right to manufacture, use, or sell a patented invention before the patent expires. Licences could be granted without the consent of the patent holder, and the licensee was required to pay a royalty. In 1969 the Patent Act was again amended to permit compulsory licences to allow generic drug producers to import a medicine's active ingredients and process them into final form for sale. The Commissioner of Patents was authorized to issue compulsory licences to import; royalty rates were set at 4 percent of the net selling price of a drug in its final dosage form.

Beginning in 1983 the government began to consider changes to the 1969 policy on compulsory licensing in order to generate growth in the pharmaceutical industry. A Commission of Inquiry on the Pharmaceutical Industry (the Eastman Commission) was established in 1984 and in 1985 made recommendations that in 1987 informed Bill C-22, to amend the Patent Act. The Amendments offered brand-name drug manufacturers receiving a Notice of Compliance (NOC) to approve marketing after 1986, ten years of protection against compulsory licences to import and seven years' protection against compulsory licences to manufacture in Canada. Drugs already in the market by 1986 had modified protections. Drugs invented and developed in Canada were protected from compulsory licences to import and subject to compulsory licences to manufacture within seven years of the NOC only if the inventor failed to make the drug in Canada for the purpose of completely or substantially supplying the Canadian market. Bill C-22 also set in October 1989 the term of a patent at twenty years from filing date. Before that, inventors had seventeen years exclusivity from the date of issue, sometimes extended for drugs that took an unusually long time for review and to receive their Notice of Compliance.

International trade negotiations caused the change. The Drunkel Text in 1991 of the WTO Agreement on TRIPS, particularly Article 31 and its provisions on "use without the authorization of the right holder," would make Canada's compulsory licensing regime for pharmaceutical products non-compliant. Although the WTO took another four years to finalize it, the text of the North American Free Trade Agreement included the provisions of the then-draft TRIPS Agreement as Article 1709(10). In 1992 the Canadian federal government introduced Bill C-91, the Patent Act Amendment Act, which eliminated compulsory licences for pharmaceutical products issued after 20 December 1991; compulsory licences in existence before that date continued in effect. Canada, however, reserved in the bill two exceptions. An "early working" exception allows a person to use a patented invention to obtain regulatory approval to sell an equivalent product after the patent has expired (s. 55.2(1)). The "stockpiling" exception allows a person to use a patented invention for a period of time before the patent expires in order to manufacture and store a product intended for sale after the expiry of the patent (s. 55.2(2)).

Following these changes, Canada's policies were challenged at the World Trade Organization by the European Union and the United States. In March 2000, a WTO panel ruled against the EU challenge of Canada's early working exception but ruled in favour of the EU challenge of the stockpiling exception. Canada implemented that ruling in 2001. Canada also lost a challenge from the United States that Canada should have extended all patents issued before 1989 to twenty years, from the then seventeen-year term.

The extension of private ownership claims through the World Trade Organization Agreement and its related TRIPS Agreement revolutionized worldwide ownership of intangible assets, especially in non-OECD countries. Because of the binding nature of the WTO Agreement, every country that has signed onto the WTO now assumes the benefits and burdens arising from the TRIPS Agreement. The agreement provides that member states must provide any invention either patent protection or, in the case of living organisms, some form of sui generis protection such as plant breeders' rights.

The TRIPS Agreement also permits an *ordre public* (public policy) provision to incorporate non-economic values into the patent system. Japan and some member states of the European Union have adopted related measures. On a case-by-case basis, patents can be refused should the commercial exploitation of the invention violate public order or morality. European Directive 98/44 on the legal protection of biotechnological inventions, for one, explicitly states that processes to use human embryos for commercial purposes or processes to clone human beings violate public policy and morality. In practice, the public policy provision is usually invoked by a third party in an opposition procedure after the patent has been granted. Others provide a broader range of opportunities to challenge patents, including an opposition procedure that provides a forum for raising challenges, typically in terms of novelty and inventiveness, but also with respect to public policy. Australia, the European Patent Office, France, Germany, India, and Japan have opposition processes, and a US report has recommended establishing an opposition procedure in the US patent system. Even in the absence of a more open opposition procedure, the United States has established the Court of Appeals for the Federal Circuit to increase uniformity in appeal decisions.

Along the way, numerous controversies have emerged. The longest and most involved debate has been about the above-mentioned patents granted to Harvard's oncomouse. These patents have raised issues of the morality and ethics of patenting higher life forms (leading to an extended review and debate in Europe), the appropriateness of patents on higher life forms (the subject, as mentioned earlier, of legal wrangling in Canada between the Harvard regents and the Canadian Commissioner of Patents), and the impact on freedom to operate and follow-on research. In an unrelated issue, Myriad Genetics Inc, of Utah, holds patents on two genes implicated in breast cancer (known as *BRCA1* and *BRCA2*) and have patented gene tests on them. There is increased concern that these types of patents constrain both research and public use.

There has been significant concern in Canada about the commercial practices of patent holders (especially Myriad Genetics), with the result that some provinces have ceased to use the contested technology, others have conformed to and duly licensed the technologies, while still others decide to use the technology without concluding licences. A consortium of European researchers successfully challenged those parts of the Myriad patent on *BRCA1* that restricted its use as a diagnostic test in research and health institutions, winning a favourable decision at the Opposition Division of the European Patent Office. The same consortium challenged the claims related to both genes in the US courts, and in 2013 the US Supreme Court invalidated the patent on the genes.

There have been various fights about IP rights and their use in the agri-food world (see chapter 9). Although farmers groups around the world have expressed concerns about plant patents, the debate has tended to focus on a few high-profile examples. Numerous groups have disputed – and at times formally challenged – utility patents issued in the United States and the European Union for apparently new cultivars that subsequently are determined to be traditional varieties from developing countries. High-profile cases include the following: Mexican yellow bean (patented as "Enola" by POD-NERS, in the United States and subsequently challenged and overturned in US courts by the International Center for Tropical Agriculture); Indian turmeric (a patent for its wound-healing properties was issued in the United States, challenged, and subsequently cancelled); the Indian neem tree (a series of patents were issued in the United States and the European Union assigned to W.R. Grace for this traditional Indian tree with industrial and pharmacological properties; and Basmati rice (patented in 1997 by RiceTec and challenged in the United States and subsequently amended). Some of the patents have been challenged and struck down in Europe.

These and other proven or alleged acts of bio-piracy have raised debate about the concept of invention, the definition of prior art, and the ethics of using traditional knowledge or genetic resources. The overall tenor of the debate is that the IP regime for plants is inimical to the interests of traditional farmers. This issue was most recently engaged in a widely watched and cited case involving a Monsanto civil patent prosecution of Percy Schmeiser, a Saskatchewan farmer and activist. In his defence, Schmeiser challenged the validity of the gene patents, challenged the right of private ownership of living organisms, and raised concerns about the economic, social, and ethical effect of private patents on self-propagating life forms. Although the Supreme Court of Canada sided with the plaintiff, Monsanto, the case continues to focus attention on the Web and among NGOs.

A subset of concerns relates to traditional knowledge (TK). A large body of legal scholarship and a host of international negotiations are founded on a normative set of principles that vest Indigenous people's traditional knowledge and related plants and animals with intrinsic value, above and beyond what might be measureable in any explicit applied use. These principles are derived from a variety of sources, including philosophy, common law, human rights, and sociology (Crookshanks and Phillips 2013). For the most part, this style of scholarship has generated a range of valid and informative arguments in favour of extending and affirming collective rights to traditional knowledge. An extensive literature examines the definition of rights for traditional knowledge and the framing of claims to benefits from those accessing those rights.

These arguments and efforts to deal with them suggest benefits that could involve monetary compensation in the form of fees, research support, royalties, and salaries; or non-monetary benefits, such as in-kind support for institutions and communities, information, transfer of equipment, software and knowhow, training, joint R&D, capacity building, and local employment. Canada is a party to these agreements and is engaged with its Indigenous, First Nations community. This has led to the articulation of best practices in research involving the First Nations, Inuit, and Métis peoples of Canada in the 2010 Tri-Council Policy Statement: Ethical Conduct for Research Involving Humans (chapter 9). In spite of this effort, the Aboriginal community exhibits low awareness, poor preparation, and weak processes for dealing with research involving their populations (Phillips et al. 2013).

In the face of rapid changes in the volume and distribution of private investment in R&D and a surge in proprietary claims across all realms, governments everywhere have been considering the potential impact on the economy. Many micro-economic studies indicate that in discrete areas the extension of property rights has accelerated investment, invention, and commercialization, but there is significant ongoing debate. There are many who conclude that while patents may incentivize investments in commercialization technologies, in many instances, especially for basic research and process patent inventions, there is little compelling evidence that patents have much impact on the scale and scope of effort (Castle 2009). Instead, many scientists are concerned that private rights to fundamental processes, specific genes, or whole organisms could block other users. Heller and Eisenberg (1998) called this the "tragedy of the anti-commons," and there is anecdotal, but not systematic evidence for the anti-commons argument.

There are also concerns that follow-on inventions may be reduced, as licences to use intellectual property are either strategically withheld to limit competition or the cost of negotiating and enforcing arrangements exceeds capacity or is greater than the value that would be generated by access. Thus, there is some debate about whether private IP rights are the only or the best method of supporting and encouraging research into basic science and scientific processes.

Other strategies that might be more open and transformative and accelerate public-interest, public-good science include the following: more public investment in upstream science, public, private, or public-private consortia such as research centres that, with the Rockefeller Foundation, brought us the Green Revolution or the Human Genome Project spearheaded by the United States and the UK's Wellcome Trust that led to sequencing the human genome (see chapter 10); the dynamic random-access memory (DRAM) and Hybrid Cube Consortia that advanced DRAM technology; public research programs or subsidies to private companies to undertake directed research; crowdsourcing, which is increasingly used to fund and advance public goods research in development; and X-prize competitions.

These are examples what is called "open innovation," centred on a mix of research networks and collaborative structures in which industry seeks external, rather than developing internal, intellectual property (Chesbrough 2005). In spite of all this experimenting with new models to enhance upstream discovery research, governments in Canada have increasingly sought to move downstream to applied, translational research (Castle and Phillips 2011), effectively offsetting the enthusiastic but limited private investments being targeted to upstream topics.

One recent vexing concern is the emergence of so-called non-practising entities (NPEs), often labelled "patent trolls," who aggressively litigate, especially in the US legal system and culture. Trolls purchase patents, often from bankrupt enterprises, and then sue other companies they claim are infringing on the rights they have purchased. These trolls generally have no research capacity, have no manufacturing base, and have no interest in working the patent themselves, hence the NPE designation. Their interest resides solely in exhausting rents available through the patents they own. The biggest concern is that these non-practising entities quite often sue widely, launching cases against apparent non-copiers and, sometimes, whole industries. As free riders, these trolls have had some success in extracting rents from the ICT sector, in particular.

The challenge is that the scope of patents has grown massively in recent years. The number of patents filed in the United States increased six-fold between 1983 and 2010 (Canadian International Council 2011). The sheer volume of patenting leads to overlapping and interlocking claims that sometimes are assigned to different owners in different markets or different applications, creating a potential patent thicket ripe for pruning. Mueller (2014) notes that Patent Freedom, a consultancy, reports that trolls in 2013 filed more than 3,170 infringement lawsuits in the United States (nearly six times higher than the number in 2006). According to Patent Freedom, based on the largest patent holdings, the top five NPEs are Intellectual Ventures, Interdigital, Round Rock Research LLC, Wisconsin Alumni Foundation, and Rockstar Consortium LLC.

From 2009 through mid-2014, Apple Inc was the defendant in 212 lawsuits brought by NPEs, followed by Samsung (172), AT&T (171), Hewlett-Packard (160), and Sony (148). Bessen and Meuer (2014) estimate that targeted firms accrued $29 billion of direct costs in 2011.

Although a small subset of very large firms accrued over half of direct costs, most of the defendants were small or medium-sized firms. Bessen (2014) estimates that in 2011 nearly 2,900 small firms were sued for patent infringement by trolls; the median defendant's annual revenue was $10.3 million. RPX Corporation (2015) estimates that in 2014 more than 60 percent of all patent infringement cases in the United States were filed by non-practising entities. Companies with less than $100 million in revenue accounted for over 60 percent of the unique NPE defendants.

Canada has some patent troll activity, but of a different order and style from the activity in the United States. The main trolls in Canada include Rockstar (which bought the lion's share of the IP assets of Nortel when it was wound down), Conversant Intellectual Property Management, Wi-Lan Inc, all located in Ottawa, and Dovden Investments, registered in the United Kingdom. Rockstar, Conversant, and Wi-Lan do the vast majority of their litigation in the United States, the majority in the Eastern District of Texas, US District Court, which is the court of preference for patent and tort cases (partly because it is perceived to improve the likelihood of success). Dovden is the most active NPE in Canada, launching forty-two patent infringement actions as of March 2015, which accounted for about one-third of the patent actions started in Canada during that period (Fancourt-Smith 2015). Concern about trolls has fluctuated depending on what cases are in play. The Canada International Council (2011) raised it to the national level in 2011 at a meeting in Ottawa, but it has since slipped below the radar as a pressing public issue.

1995–2015: Finding Better Ways to Generate Benefits from Intellectual Property

As discussed in chapter 6, Canada differs from many other OECD countries in that government (particularly the federal government) is a relatively large investor and funder in science and technology, especially for basic and early applied research – through a combination of grants and funding for postsecondary education and research, for infrastructure projects, through special operating agencies (SOAs) such as Genome Canada, through intramural research, and to firms through tax expenditures. In 2013 the federal government invested more than $14 billion directly or indirectly through tax measures to support R&D, much of this undertaken by universities and public labs. While a large absolute number, it represents only about 1.62 percent of GDP in 2013 equal to just over half of the national targets of comparative OECD member states.

Evidence suggests, however, that the optimal gains from this research are not realized due to delays in decision making within institutions and among actors within the global policy regime. Graff, Zilberman, and Bennett (2009) report that the "research sieve" culls up to 99 percent of all potential agri-food innovations before they get to the market; they identified that 560 crop biotechnology traits in the United States at the "proof of concept" stage led to only 383 early stage field trials, 47 advanced trials, 14 regulatory applications, five market introductions, and just two sustained products in the market. Phillips McDougall (2011) calculated that the full discovery, development, review, and commercialization process for successful global market introduction takes about US$136 million and 283 months (27 percent of that in regulatory review); one specific concern they noted is that the cost, duration, and uncertainty in the system has risen in the past decade in almost all of the key adopting countries (see Dewar 2014). These studies focus on agri-food R&D; however, there is significant evidence that these trends are evident in many other sectors.

As we have seen, Canada's basic challenge is that it invests too little in R&D (see chapters 1 and 7), few innovations are successful (only about 1 percent of them), and the cost and time to get to market is high, rising, and becoming less certain. Most of those costs and uncertainties do not arise directly from the science – rather, they arise in decision-making systems in government, universities, the private sector, industry associations, research teams, and by individual producers and consumers. Effective, efficient, and timely decisions are critical to ensuring new technologies reach the marketplace to benefit Canadian industry and global consumers. Canada's small market size compounds these difficulties.

The United States tends to set the pace in commercialization, with a large and responsive market, a well-documented entrepreneurial class (which according to Zucker, Darby, and Bewer (1998) includes many university faculty researchers), a relatively large and aggressive venture capital industry (raising approximately 30 times the funds as the Canadian industry), and engaged and competitive municipal governments (with municipal development bond programs). Canada has tried to follow the US lead, but with rather less success.

In 1981 the implementation of the US Bayh-Dole Act transferred ownership of any university-based inventions resulting from US government support to the universities and required them to undertake efforts to commercialize those inventions. Previously, the results of federally funded research generally were owned by the US federal government. The implications of this legislation have been far reaching, with university researchers and university administrators now taking a much greater interest in undertaking inventive activities. The key decision in the United States through the Bayh-Dole Act to devolve ownership of intellectual property to universities and researchers directly (and emulated in analogous ways in most other OECD countries) had a major impact on how their devolved dollars influence inventive activity. Now that more people see themselves as engaged in and possible beneficiaries of invention, they are assigning greater value to the inventive portion of their research and to whatever they might invent. Before the Bayh-Dole Act became effective on 1 July 1981 US federal agencies (like most government funding organizations) kept tight control over any intellectual property resulting from research they funded. The Bayh-Dole Act effectively devolved to universities the ownership for all research undertaken with US federal government funds.

Essentially, the Act requires any US university using federal funds to have written agreements with faculty and technical staff requiring disclosure and assignment of inventions and specifies that universities are then obligated to disclose for each new invention to the federal funding agency and to elect whether to assume responsibility for that invention. If the university chooses to retain title, it must file a patent application and share with the inventor(s) a portion of any revenue received from licensing the invention, and it must provide the government a non-exclusive, non-transferable, irrevocable, paid-up right to practise the invention.

This one piece of legislation turned the university research world on its head. The US-based Association of University Technology Managers (AUTM), representing more than 350 universities, research institutions, teaching hospitals, and government agencies in the United States and Canada, reports that a survey of 198 US universities, hospitals, and research institutes shows that these institutions undertook more than US$27.8 billion in federally sponsored research, disclosed almost 16,800 inventions, and filed more than 13,750 patents during 2004. They also reported that they had more than 11,000 licences and options generating more than US$1.4 billion in licensing income in 2004 and during the year spun out more than 460 start-up companies. Canadian firms reported proportionally smaller but still significant sponsored research, disclosure of patents, licensing, and revenues.

Canada has tried to emulate this US breakthrough, but has found the Canadian system to be less responsive. In 1985 the government passed the Public Servants Inventions Act to govern and manage exploitation of inventions by public servants in line departments or special agencies, such as the National Research Council of Canada of Canada. In some ways this was just the opposite of what the United States did with the Bayh-Dole Act. A major challenge in Canada is that the bulk of federal research efforts is undertaken by universities, but those institutions are under the authority of the provinces and have generally been slow to respond to federal interests. Given that most of the funding for universities flowed until the 1990s through block grants, it was hard to incentivize those institutions to focus on technology transfer and commercialization. This was especially true in the times of restraint that spanned from the mid-1980s to the late 1990s – money talks, and the smaller flows from the federal government simply dampened any enthusiasm.

Then in the late 1990s, as the federal government realized budgetary surpluses, it recapitalized the university system, but in a new and more directing way (see chapter 6). Rather than on top of the block grants, the new capital flowed through new initiatives in NSERC, SSRHC, CIHR, and the NCE program that imposed as a condition a new focus on dissemination of results and through a number of new research agencies such as the Canada Foundation for Innovation and Genome Canada, that increasingly had as their target the realization of socio-economic benefits from their investments.

As the funds started to pile up, universities and faculty responded. Universities created new industry liaison or technology transfer offices (TTO). The earliest reported technology transfer office was created at the University of Toronto in 1980. In the 1980s another twelve institutions opened offices; fifteen more opened in the 1990s and another dozen by 2006. Meanwhile, colleges and faculties changed peer- and merit-review systems to acknowledge patents and other knowledge mobilization, and individual scholars responded to the incentives and bid on these new grants.

The simple linear view of innovation, somewhat paradoxically, drives this model. The difficulty is that most technology transfer offices, at best, break even and most cost more than they generate for the institutions involved (Bird 2009). Smyth's (2011) analysis of Canadian data from 1998 to 2008 shows that while the total investment in university research has increased seven-fold, the proportion of patents actively licensed by universities is declining, and the number of spin-offs has fallen to half what it was a decade ago. IP management costs of technology transfer offices are virtually equal to the licensing revenues they generate, and more concerning, costs are trending upward – particularly in respect of litigation – while revenues are relatively flat. These data likely understate the problem, as they fail to account for the probable increases in costs of enforcement over the entire life cycle of an organization's intellectual property, and none of these reports incorporate the full cost of securing proprietary positions through patents.

Recognizing that current models of technology transfer are imperfect, a number of organizations are experimenting with new models of IP management. These models rely on patent and other IP protection to implement creative licensing. The common feature in all these approaches is that they help to facilitate multilateral IP transactions. In centralized systems, like a patent pool, an agent assembles and licenses a bundle of IP rights; in decentralized systems, such as a clearinghouse, an agent merely provides a venue for trade in IP rights. Historically, patent pools have been discouraged as they have been viewed as fundamentally anti-competitive, but ICT patent pools have been tolerated when they are used to develop common technological standards and biotechnology-patent arrangements have been successfully used by social entrepreneurs for philanthropic purposes. The Medicines Patent Pool, a partnership formed under the auspices of UNITAID, provides a "one-stop shop" for clearing patent rights related to antiretroviral medicines for treating HIV.

One reason for disappointing financial returns on investment in acquisition-oriented IP management strategies is the significant expense of acquiring and enforcing rights, especially patents. De Beer et al. (2011) assert that these costs can be entirely eliminated by choosing to forego any IP protection, instead freely revealing knowledge and technology directly into the public domain. This is qualitatively different than the open source movement. The open source approach taken by some software developers involves a creative commons system of licensing copyright-protected works, and several open source biotechnology initiatives all depend, fundamentally, on acquiring IP protection. The novelty of such open source systems is that intellectual property is then licensed to require rather than restrict access to the protected content or technology. The free-revealing approach offered by de Beer et al. (2013) "sidesteps" the IP property system entirely. It would both involve foregoing proprietary claims and developing "strong community norms" to ensure inventive steps are not purloined by others.

The Harper government's 2015 Budget (Minister of Finance 2015) contains the latest statutory efforts to seek the difficult "balances" characterized above while strengthening IP rights. It indicates that new amendments to the Patent Act, Trade-marks Act, and Design Act will be proposed to provide "intellectual property agents with statutory privilege for confidential communications with clients, enhancing Canada as a place in which to invent and market inventions" (ibid., 113). This IP modernization will bring "Canada's framework in line with other common law countries such as Australia, New Zealand, and the United Kingdom" (ibid.).

THE THREE ELEMENTS

We now track with the aid of Table 8.2 the three elements being utilized to understand change and inertia across the four domain policy and governance histories presented in this chapter.

Policy Ideas, Discourse, and Agendas

The ideas that ultimately defined intellectual property as a core part of science, technology, and innovation policy in Canada and elsewhere emerged in other policy spaces and flowed across to this domain. In the first instance, the effort in the 1980s to late 1990s period to redesign the GATT and to entrench trade-related aspects of IP rights reflected the replacement of neo-classical growth theory with knowledge-based growth theories, underpinned by intellectual property. But in a deeper historical context, IP underpinnings showed the continued influence of historic Jeffersonian US ideas about invention. While the protection versus dissemination trade-off at the core of intellectual property (especially regarding patents) were present, key interests globally and nationally ensured the primacy of the protection idea, largely via the entrenching of twenty years of protection.

Also still inherently present in the intellectual property, invention, and innovation domain were ideas and agendas centred on how far the state should intervene and using what policy instruments and whether IP policy was framework policy or sectoral policy. Patents give the patent holder the right to exclude others from using but only through the patent holder being able and willing to sue others for infringement.

The global process of extending property rights to living matter and in most cases multicellular organisms was a logical extension of the trade-related IP agenda. Bioscience innovations like gene sequencing, gene amplification, and gene splicing led to living matter being open to manipulation. Hence, protection and property rights were a logical extension, albeit with concerns about exactly how far to go.

The last two histories highlighted the fundament conflict between public good and private greed embodied in property rights and trade-offs regarding dissemination. The capacity to exclude others offered by patents and other IP rights may be necessary for innovation but not sufficient for realizing the optimal socio-economic benefits of invention and innovation. Thus, dissemination ideas such as compulsory licencing, traditional knowledge, and TRIPS-related non-economic values in the patent system emerge to some extent.

Economic and Social Power

The extension of IP property rights, for the most part, is a reflection of the distribution of economic and closely related political power. The focus in the Uruguay Round of the WTO on IP rights was a clear projection of US hegemonic interests. The dominant position of drug, biotechnology, and other technology companies in the US economy focused the government's attention. The sheer size and capacity of those firms and industries ensured that when the US government secured agreement to engage in negotiations, they were ready, able, and willing to advance their own (and by extension US) economic interests. Although there were dissenting voices inside and outside the United States, especially in less developed countries, the US-led initiative was stronger and dominated.

Within Canada similar shifts in power occurred with multiple federal agencies and laws, centred in part in Industry Canada, leaning strongly toward the primacy of IP protection. Secondary IP social institutions were weaker and on the defensive overall, especially in the first policy and governance history examined.

The expansion of property rights to living matter, including multicellular organisms, was not the result of any explicit or visible public policy choice. Rather, it was in almost all cases the end result of private litigation and the role of the courts. Firms in pursuit of their own self-interest used the legal systems available to them to extend IP property rights and advance their own interests.

Governments in our third policy history seriously and continuously sought ways to moderate and mitigate the impacts of expanding private rights, and public policy for the most part was caught in the middle, trying to find a compromise. Public goods advocates pressed for compulsory licensing, collective rights, traditional knowledge, and wider diffusion of the technologies. Meanwhile patent trolls, among others, acted as pure

Table 8.2 Policy and governance histories in the intellectual property, invention, and innovation domain: Three analytical elements

Policy and governance history	*Policy ideas, discourse, and agendas*	*Economic and social power*	*Time, temporal realities, and conflicts*
1980s to late 1990s: IP and patent policy emergence in the federal and global trade and innovation policy agenda	• Neo-classical growth theory replaced by knowledge-based growth theories • IP as underpinning growth • Continued influence of Jeffersonian US ideas about invention • Ascendancy of protection side of IP trade-off vs. dissemination • 20 years of patent protection • How far to intervene and using what instruments? • IP as framework vs. sectoral policy	• US projected hegemonic interests in Uruguay Round of WTO and weakening of WIPO • Dominant position of drug and biotechnology companies in US economy gave them the pen at the WTO TRIPS table • OECD consensus that trade should encompass domestic policy was stronger than the concerns in less developed countries • Multiple federal agencies, laws, and programs lean toward protection idea overall • Secondary IP social institutions weaker and on the defensive	• Serendipitous match of the negotiating cycle with the emergence of new economic drivers (biotech and ICTs) • Re-emergence of IP after decades of being out of sight and out of mind policy-wise • Access to medicines concerns emerge, but with slow progress • CIPO's gradual acquisition of relevant technical expertise to assess patent applications (e.g., from engineers to biologists) as patent content and volumes change
1980–2006: The fight to extend property rights to living matter	• New concepts of living matter as simply engineered DNA constructs • Plant breeders' rights internationally and nationally	• Private litigation and the courts rather than public policy drove the extension of IP rights to living matter • Tens of thousands of patents for biotechnology tools	• Biotechnology and ICT create real opportunities • US developments move faster than those in Canada

1983–2015: Finding the right balance in patents	• Conflict between public good and private greed • Continued debate about • protection vs. dissemination ideas • Compulsory licensing • Traditional knowledge • TRIPS non-economic values in patent system	• Public policy caught in the middle, trying to find a compromise • Patent trolls as rent seekers in response to overlapping and interlocking claims: strong in the US litigious culture, but also a factor in Canada	• Conflicts between those looking back (traditional knowledge and Indigenous peoples) and those looking forward (patent trolls) • Slow progress in Canada regarding Aboriginal community and traditional knowledge issues • Long-running Harvard mouse case
1995–2015: Finding better ways to generate benefits from intellectual property	• Humanitarian and public good ideas • Ideas about ownership of university-based research compared with US	• Budget cuts followed by targeted recapitalization of research universities • Universities as provincial public institutions and systems • Poor commercial returns • Slow decision processes in multiple institutions • Technology transfer offices emerge, but limited revenues earned	• Impatience about returns on investments • Canada's progress slowed by long-term weakness in level of R&D • Follow-on invention may be reduced as licences to use IP are strategically withheld

Note: see the chapter text for expansion of abbreviations in this table.

rent seekers hoping to exploit the friction created by the profusion of new IP rights, often exhibiting overlapping and interlocking claims that created opportunities to hold out for a higher share of use-benefits. Patent trolls were especially strong in the United States but also were emerging in Canada.

Finally, the commercialization agenda of governments is inextricably tied to a sense that public research investments should generate measureable benefits. In a somewhat perverse way, the budget cuts to universities in the 1980s and early 1990s created the circumstances that allowed targeted recapitalization at the end of the 1990s, conditional on further effort to work their creations. Universities have responded, but so far they have not demonstrated much success, delivering for the most part poor commercial returns. Nevertheless, the creation of technology transfer offices, vice presidents of research, and peer-review and merit-review processes and incentives have enabled individual faculty to engage in innovation and commercialization activities. Along the way, the academy has fundamentally changed by a combination of factors but only one of which was intellectual property.

Time, Temporal Realities, and Conflicts

To a great extent, these policy and governance histories highlight the importance of time in policy systems. The serendipitous match of the Uruguay Round negotiating cycle with the emergence of both new economic drivers (biotech and ICTs) and new economic theories undoubtedly contributed to the "success" of the WTO in extending its reach through the TRIPS Agreement to extend private property rights to all the member states. This also weakened and changed WIPO as a global IP institution. The conditional nature of, and mutual self-referencing among the Bretton Woods institutions – the International Monetary Fund, the World Bank, and the WTO – makes it increasingly difficult for countries to remain outside the scope of their rules. So while countries do in theory have a choice about joining, in practice the costs exceed the benefits of an independent existence. At present, 160 countries are members of the World Trade Organization, and most were also members of the other two organizations.

The emergence in the 1980s and 1990s of biotechnology, genomics, and ICT created real technological and market opportunities that drove IP rights, especially the expansion of rights to living matter and multicellular organisms. In many ways the search for balance between incentives for creating intellectual property and the generation and distribution of the benefits of innovation is caught in a seesaw battle between those looking back (e.g., proponents of traditional knowledge and the rights of Indigenous peoples) and those looking forward (e.g., patent trolls and venture capitalists).

Time also played a major part in driving the policy of commercialization. In the R&D game time is money, and Canada's weak R&D record as traced in earlier chapters again has adverse consequences. Delays and uncertainty reduce the socio-economic returns to investments. Governments and private investors have exhibited significant impatience about returns on R&D investments, which suggests the tinkering with commercialization models is difficult but far from over.

CONCLUSIONS

The intellectual property, invention, and innovation domain is now inextricably embedded in the science, technology, and innovation policy space. In addition, this is one policy space where the federal government has a lead role but not always a controlling one, given the wide array of players and interests, internationally and nationally. Intellectual property is a very old science, technology, and innovation domain in the sense that initial laws emerged very early in Canada's history, influenced by both US IP values and history and also British and colonial history. At present the domain involves statute law, common law, and policies vested in multiple departments and programs. Canada's patent policy and laws, as we have seen, emerged or more accurately re-emerged and congealed in the 1980s and 1990s. Although the IP domain overall comprises several types of IP rights, including trademarks, copyright, designs, and trade secrets, our analysis of the domain has necessarily had to focus mainly on patents, the volumes and complexity of which have grown massively in global and national terms. The four policy and governance histories bear witness to the central place of the patent story and overall to the greater emphasis on the protection side of the IP policy equation.

Our interest in the innovation economy and society nexus shows that such a nexus is central to the history of the IP domain in that it is ultimately driven by a core protection versus dissemination values trade-off. The natural temptation is to treat the protection idea and value as the economic dimension and the dissemination idea of invention and knowledge as the social dimension. There is some truth in this and our policy and governance histories suggest overall the power of the economic imperative defined in this way. But the protection side of the IP value trade-off also at its heart is designed in a social and even cultural sense to reward and encourage inventors and creators, many of whom are individuals working on their own or in small teams in various institutional and non-institutional community settings. And the dissemination side of the IP value trade-off is also replete with economic innovation nexus attributes, given that complex chains of learning and distributed knowledge can produce economic gains of crucial kinds in a modern economy.

In the next three chapters we now turn to three domains each of which are anchored by and embedded in three transformative international technologies which Canada had no choice overall but to live with and seek to gain from and manage in complex, fast-moving ways. We begin in chapter 9 with the agriculture, food, biosciences, and biotechnology domain, followed in chapter 10 by the genomics, life sciences, and technology domain, and in chapter 11 by the Internet, communications, and social media domain.

9

The Agriculture, Food, Biosciences, and Biotechnology Domain

This chapter examines the Canadian agriculture, food, biosciences, and biotechnology domain. Its story must be set in the fundamental reality that agriculture and food occupy a unique place in the hearts and minds of people. As the first and until recently the most important social and economic activity for mankind, agriculture and food gets a special place in the science, technology, and innovation (STI) space. Canada is no exception. Agriculture led the settlement and development of this country; the big migrations to and within Canada were largely driven by agriculturally based opportunities in Western Canada and on migration from areas in Eastern Canada where farming faced serious limitations regarding land, geology, soils, and crops (Russell 2012). Until the Second World War agriculture employed the largest portion of the Canadian workforce. Farmsteads were home for a plurality of Canadians, and Canada's global position in markets was anchored on huge exports of grains that generated large trade surpluses. In that context Canadian science, technology, industrial, and innovation policies were firmly anchored on the land. Reflecting its core importance, the British North America Act 1867 assigned constitutional authority to agriculture jointly to the federal and provincial governments.

This is not unexpected. The history of Canada is inextricably entwined with the exploitation of natural resources – cod, wood, and beaver pelts – in the early colonial period and after 1867 with the development of agriculture on the Great Plains area in the West (Innis 1956). The National Policy of 1876 was founded on a few core goals: the opening of the West through a transcontinental railway, the settlement of immigrants on farms in the West, and a high tariff wall to provide incentives for industrial capacity to develop in Central Canada to exploit the emerging market in the West. Agricultural settlement of the West was the lynchpin for making this policy work (Fowke 1973). But mere wishing was not enough. A trickle of farmers moved into the region before 1905, using largely European technologies and seeds – legend has it that Red Fife wheat came to Ontario in 1842 as a few seeds filched from a shipment from Ukraine in the Glasgow harbour – but these were not well suited to the harsh growing conditions in Western Canada. Farmers had to contend with only about a hundred frost-free days a year, limited natural rainfall or water for irrigation, and wide seasonal weather variations. Farming remained a precarious occupation.

In the mid-1880s the Dominion Department of Agriculture was founded, with a plan to develop a set of experimental farms to spur innovation in agriculture. William Sanders was appointed director, and he was sent on a fact-finding mission to the United States and Europe to investigate how they undertook research there. In 1885 legislation was passed creating five experimental farms, one each in the Maritimes (Nappan, NS), Ottawa (for Ontario and Quebec jointly), Manitoba (Brandon), the Northwest Territories (Indian Head), and British Columbia (Agassiz). The Act directed these entities to investigate livestock breeding, the diseases of domestic animals, and dairying; the testing of cereals and other field crops, as well as other plants; the study of seeds, fertilizers, plant diseases, and insect pests; and such other experiments or researches as might benefit agriculture (Minister of Agriculture 1925). Over the past 130 years a lot has changed but these experimental farms largely remain active and important in the Canadian agri-food innovation system.

The key breakthrough that changed the face of Canada was the development in 1906 of the Marquis wheat variety from a cross of Red Fife and Hard Red Calcutta. This new variety matured more quickly (7–10 days earlier than Red Fife), had a shorter stalk, a higher yield, and a high gluten content that would deliver the largest loaf after baking. By 1918 Marquis was planted on more than twenty million acres in the Great Plains of Canada and the United States, capturing 80–90 percent of market share. Virtually every variety of wheat grown in the past century can trace its lineage back to that original cross.

Ever since then, STI policy has been a core mandate of the federal system. Although agri-food policies in many countries turn inward and become parochial, Canada has not had that luxury – none of Canada's core crops or husbanded animals is indigenous to Canada's part of North America, and almost all of its produce is either marketed in internationally interconnected markets or at least priced there. Canada's STI policy directed to agri-food and the biosciences and biotechnology is still pulled between domestic needs and global pressures. The biosciences are any of the branches of natural science that deal with the structure and behaviour of living organisms and biotechnology. Biotechnology is defined by the OECD (2010, 7) as "the application of science and technology to living organisms, as well as parts, products and models thereof, to alter living and non-living materials for the production of knowledge, goods and services."

In Canadian institutional and governance terms, this domain is centred in the agriculture ministry, which regularly changes its name. Currently called Agriculture and Agri-Food Canada (AAFC), this ministry is responsible for more than a dozen key acts and a half-dozen or so operating agencies, the most important of which for our story include its Research Branch, with twenty centres across Canada; the Canadian Grain Commission (CGC), which regulates quality; and the Canadian Food Inspection Agency (CFIA), the lead regulator of food safety and efficacy in concert with Health Canada (HC). Although the Farm Credit Corporation (FCC) and other agencies play important supporting roles, they mostly reflect rather than set strategic direction. Beyond the agricultural ministry, Industry Canada (IC) and its framework policies and special funding agencies – especially the National Research Council of Canada (NRC), the Natural Sciences and Engineering Research Council of Canada (NSERC), the Networks of Centres of Excellence (NCE), the Canada Foundation for Innovation (CFI), and Genome Canada – are the largest public funders of basic biosciences and biotechnology, and the Department of Foreign Affairs and Trade Development (DFATD), in conjunction with the Canadian International Development Agency (CIDA)/International Development Research Centre (IDRC), is tasked with linking Canadian capacity with global policy and capacity.

THREE DOMAIN POLICY AND GOVERNANCE HISTORIES

The more recent history of Canada's agriculture, food, biosciences, and biotechnology policy effort can be told in three overlapping and interlocking policy and governance histories. The heyday of Canada's effort in agricultural S&T and innovation policy was probably the development in the late 1960s of canola, a low erucic acid and low glucosinolate type of rapeseed, and then the transformation of that crop into the world's first large-area genetically modified (GM) crop, with a suite of highly competitive herbicide-tolerant (HT) traits. Canada led the world in this period, generating one of the fastest-growing and highest-value oilseeds yet to emerge.

Since then, the story is murkier. Events beyond Canada and beyond agriculture have forced (some would say encouraged and enabled) the Canadian federal government to rethink its earlier leadership role. Instead of focusing on specific agri-food innovation outputs and outcomes, the government has been attempting to sort out how to come to grips with increasingly globalized STI, public pressures for more socially responsible policies, and an ever-changing constitutional agenda. The three policy and governance histories show the complexity of the system.

1960s–2005: Global Innovation Leadership in the Development of Canola

The world view in the 1950s to 1970s, exemplified by the Malthusian musing of the Club of Rome, was that in spite of pockets of abundance, food and other resources were unlikely to meet the rising demand from a larger and richer world population. New and better sources of food and fibre were urgently needed. In the 1950s a few farsighted and innovative individuals in the federal research system – in the Agriculture Canada research system and the National Research Council – responded to that pressure and began a generation-long program of R&D targeted to create a new edible oilseed that could complement Western Canada's cropping system and meet growing world needs. Rapeseed, a small-area oilseed crop mostly used industrially, was targeted as Canada's solution but needed work to lower erucic acid in the oil (which was implicated in heart defects in animal trials) and glucosinolates in the meal (which were anti-nutritional properties in the feed).

The scientific and industrial story of the emergence of canola as Canada's Cinderella billion-dollar crop has been told elsewhere (NRC 1992; Phillips 2003; Phillips and Khachatourians 2001). One key theme of the early development was the vital role of a few scientific pioneers (especially Keith Downey at AAFC and Burton Craig at NRC) in mobilizing and coordinating the diverse work of scientists and plant breeders from universities (especially the University of Alberta, the University of Manitoba, and Guelph University), industrial oilseed crushers (Canamera, in particular), and groups of innovative producers all coordinated through various provincial development commissions and the Rapeseed Association of Canada, which was later renamed the Canola Council of Canada (CCC).

By 1968 technical breakthroughs in seed evaluation and breeding had delivered a double-zero variety (with low erucic acid and low glucosinolates) through the efforts of public researchers, which was then branded and trademarked as canola by the Canola Council of Canada and actively adapted and adopted by producers, prompted by active extension work of the provincial canola growers' associations. The culmination of this collective effort came in 1985, when canola was granted "generally regarded as safe" (GRAS) status in the United States, opening up global markets.

When biotechnology – the ability to manipulate the genome of specific species through gene splicing and silencing – emerged in the late 1970s as a potential transformative technology, scientists and industry around the world focused on translating the new tools to overcome barriers in plant and animal breeding. As it turned out, canola was an ideal crop for early work: it was relatively easy to manipulate, as a new crop it offered significant opportunities for improvement, and Western Canada, in particular, offered a large and growing market for seed.

Unlike in the earlier venture, when the effort was largely led by individual public sector scientists without any formal sanction and support, the adaption and adoption of biotechnology in Canada was precipitated and supported by governments (Doern and Prince 2012). In 1983 the federal government announced its National Biotechnology Strategy (NBS), which established the National Biotechnology Advisory Committee (NBAC) to advise federal ministers on how to advance Canada's capacities in R&D, human resources, regulatory matters, and the economy. By 1991 the Canadian federal government had acceded to the International Union for the Protection of New Varieties of Plants (UPOV) and implemented plant breeders' rights on new varieties, which provided a commercial incentive for investment. The Saskatchewan government, meanwhile, developed Ag-West Biotech Inc, a number of temporary industrial incentive programs, and provided subsidies to attract Plant Genetic Resources Inc from Ghent to Saskatoon.

Around the same time key industrial actors approached the federal government to encourage restructuring of its agri-science efforts. Research talent was relocated from the Central Experimental Farm (CEF) in Ottawa to a refocused NRC Plant Biotechnology Institute (PBI), new talent was recruited to expand capacity in the Saskatoon AAFC research centre, where Plant Genetic Resources Canada, Canada's gene bank with more than 110,000 accessions, was relocated.

Canada engaged with the international science and regulatory community to proactively assess and revise its regulatory system. In 1983 the United

States had articulated a vision for a coherent "risk analysis framework" (RAF), which became the new international benchmark for risk assessment and evaluation of all new technologies. In 1993 the Canadian federal government adopted the Principles of the Federal Regulatory Framework for Biotechnology. In response, the administration of the Seeds Act was amended to accommodate new breeders and new traits and regulations, and guidelines were adopted to create capacity within existing regulatory agencies (especially Health Canada and the AAFC) to categorize, assess, and manage the risk associated with "plants with novel traits" (PNT). This unique Canadian trigger ensured all plants with novel phenotypes would be assessed, regardless of the technology used to develop the variety. The Canadian regulatory effort came to be viewed among global regulators, at least, as the "gold standard" of quality and performance.

Multinational companies (MNCs), based in the United States, the United Kingdom, and the European Union, in partnership with producers through the Canola Council of Canada and with government through various research contracts, invested heavily in the development of a range of new canola varieties: three new HT types, a hybrid system, a few novel edible oil profiles, and a number of industrial ingredients (e.g., pharmaceutical proteins) were developed, assessed by the Canadian regulatory system, and in most cases, introduced into the market with the full cooperation of the growers' associations, the oilseed trade, and government regulators (Phillips and Smyth 2004). The result was exciting to many. GM HT types came to dominate the market; by the early 2000s, virtually all the seed used had one of three proprietary HT varieties, and all of the varieties were managed and owned by private breeders, not public breeders.

The introduction of GM canola in Western Canada triggered two of the most notable litigations over ownership and control of GM crops. In 1998 Percy Schmeiser, a Saskatchewan farmer and political activist, was sued by Monsanto Canada Inc for patent infringement for allegedly acquiring and planting Roundup Ready™ canola, a patented variety, without licence. After six years of litigation, appeals, countersuits, and controversy, the Supreme Court of Canada ruled 5–4 in favour of Monsanto and levied damages against Schmeiser.

A second case in 2002–04 involved a putative class action lawsuit between Larry Hoffman and Dale Beaudoin (representing organic canola producers and supported by the Saskatchewan Organic Directorate) and Monsanto Canada and Aventis Cropscience, the owners of Roundup Ready and Liberty Linked GM HT canola. The claim was that the widespread adoption of GM canola had damaged the market prospects for organic canola producers because foreign markets were concerned about the co-mingling of GM and organic produce. After a few initial skirmishes, the Court of Queen's Bench dismissed the case, concluding that the application did not identify reasonable cause of action or an identifiable class.

The prevailing opinion is that neither case resolved the fundamental differences between those using biotechnology and those opposed to its use. The award of damages in the Schmeiser case was generally viewed as trivial, with the result that it had little demonstrable effect on other producers who might choose to use unlicensed seeds. Moreover, the case elevated Percy Schmeiser into an icon of the conflict between advanced technologies and farmer control (some viewed him as an activist David against the goliath Monsanto): in 2000 he was awarded the Mahatma Gandhi Award in India and in 2007 the Right Livelihood Award, colloquially called the alternative Nobel Prize. The failure of the organic case, in contrast, presents a lost opportunity to fundamentally grapple with the still-real issue of adventitious presence in the grain and oilseed trade (Kalaitzandonakes et al. 2014).

As these cases wound down, the policy focus shifted. Governments moved away from being proprietors and promoters, with the result that there was a loss of focus and direction in federal biosciences and biotechnology policy. This was partly due to the inability of government scientists to keep up with Monsanto and other companies where researchers possessed a different skill-set applied to plant breeding. In-fill and me-too GM varieties of canola, corn, and soybeans continued to be developed and commercialized, but no new large-area GM crops have been developed for Canadian use, and the rest of the biosciences and biotechnology policy, program, and regulatory system, potentially involving second- and third-generation GM crops, GM animals, and GM microbes, largely stalled, with the result that any GM product inventions migrated to the United States for commercialization.

1995–2015: Groping for New Models of Research Management and the Shift to Partnerships

Four overlapping and interlocking developments have precipitated a major period of experimentation in the search for new models to manage research in this S&T and innovation policy domain.

First, as we have seen in chapters 1 and 2, federal governments after 1980, regardless of their political orientation, shifted focus from macro-economic stabilization through contra-cyclical fiscal and monetary measures and instead invested most of their time, energy, and resources in the implementation of growth-oriented micro-economic policies, programs, and incentives to make markets more efficient and to support and sustain innovative ventures. This revolution started in the economics domain, as shown in chapters 1 and 3, but it migrated into the policy and administration fields, where it sparked developments such as the new public management (NPM) paradigm (Aucoin 1997). In this new way of thinking, government was advised to pull back from spending and doing and use more nuanced policies and incentives to mobilize others to achieve policy objectives.

Second, the 1980s and 1990s transformed the trade policy landscape that biosciences and biotechnology S&T and innovation policy needed to conform to. As chapter 2 has shown, the Canada-US Free Trade Agreement (CUFTA) in 1988, the North American Free Trade Agreement (NAFTA) in 1992, and then the World Trade Organization (WTO) in 1995 established new norms for national policy that could affect international trade and commerce. Collectively, these agreements narrowed the scope for national governments to protect domestic production through subsidies, tariffs, and non-tariff trade barriers such as sanitary and phytosanitary measures and technical barriers to trade. These agreements bound member states to a new set of investment rules that were intended to stop predatory or discriminatory market practices and an expansion of enforceable intellectual property rights (IPRs) for both mechanical and biological inventions (see chapter 8). These measures were particularly influential in agriculture, as for the first time this important sector was fully incorporated in the authorities of the treaties. These agreements prohibited many traditional policies or made them actionable, but they also directed government toward a set of "green light" measures deemed benign in terms of trade and commerce: the key ones for biosciences and biotechnology policy were investments in R&D and environmental goals.

Third, with these pressures as a backdrop, the Canadian agri-food industry has been evolving in a way that has reduced the centralizing tendencies and accentuated regional and sectoral priorities. Partly due to trade policy and partly due to the basic evolution of the domestic productive capacity, agriculture in Central Canada has become increasingly dominated by supply-managed and domestically focused products and producers, while Western agriculture is dominated by farmers producing for export, including grains, oilseeds, pulses, cattle, and hogs.

This shift has stressed the agri-food policy system and fractured the pan-Canadian farmers' federations and the pan-provincial marketing institutions (such as the Canadian Wheat Board (CWB) and the grain co-ops). The result has been that policy and programming are more trilateral, between the federal ministry, sectoral groups, and coalitions of provinces with a position in the sectors involved. The strong export orientation in the West has led to a sharpened focus among producers and provincial ministries on innovation; in Central and Eastern Canada policies and programs need to balance more between stabilization and innovation, which in some ways blunts their effects.

Fourth, in response to the razor-thin vote in the 1995 Quebec referendum on sovereignty, the federal government has been seeking new ways to manage federalism. The Chrétien government tried a mixture of carrots and sticks to develop national policies and strategies, all the while seeking to project the Canadian federal presence to Quebecers. The Union des producteurs agricoles (UPA) in Quebec was a key player in the referendum and a force to be reckoned with. Given the divergence of interests between UPA members and other Canadian farmers, it was only natural that policies and programs needed to become more specific to regional and sectoral interests. The formalized federal-provincial policy frameworks – the Agricultural Policy Framework (APF) and the Growing Forward programs – provided significant provincial autonomy. This approach was consolidated after the Harper government was elected in 2006 as the federal government unilaterally withdrew from areas of provincial authority and practised a form of "one-off" federalism (McGrane 2013), devoting more energies to federal areas of authority, including the North, trade, and national defence.

These kinds of pressures initially were evident in the policy system after the election of the Chrétien Liberal government in 1994. In the face of large and unsustainable fiscal deficits and a relatively weak productivity and growth performance in Canada relative to the United States and other OECD nations, the federal government, first, cut drastically across its departments and agencies, including in Industry Canada and the agricultural ministry and, second, began to retool the micro-economic policy base for innovation and growth. It also began to search for new management models to leverage and optimize its remaining investments in R&D.

Industry Canada undertook systemic changes that had a knock-on effect on all sectoral ministries, including agriculture. As with most agencies, funds were cut mercilessly to contribute to the fiscal plan to balance the budget (Doern and Kinder 2007). Once the plan started to have effect, Industry Canada had the opportunity to recapitalize the Canadian STI system. Instead of simply putting money back from where it was taken (e.g., in line ministries, in NRC, in the standard grants programs, in the granting councils, and in universities), the department, as chapter 6 previewed, took another approach involving the Networks of Centres of Excellence, the Canada Foundation for Innovation, the Canada Research Chairs, and Genome Canada.

In the agricultural and biosciences and biotechnology space, this signalled the end of unilateral federal leadership. Internationally, this conformed to a rising appreciation in the science policy community that the scale and nature of science and R&D was changing. In the recent past, most of the successful science could be attributed to individuals or small groups of scholars and practitioners in academia, government labs, or private companies. The Manhattan Project to develop the atomic bomb in the Second World War offered a new and enticing model – well-financed, large-scale research teams with international partnerships. The Green Revolution that delivered semi-dwarf, high-yielding varieties of wheat and rice suitable for more intensive irrigation and fertilization in less developed countries (LDCs) in many ways exemplified a first attempt to use this model in agriculture. Although the credit is often given to Norman Borlaug, the lead wheat breeder at CIMMYT (International Maize and Wheat Improvement Center) and winner of both the Nobel and World Food prizes, the effort actually involved a good number of member states, individually and through the Food and Agricultural Organization (FAO) and the United Nations Development Programme (UNDP). The Ford and Rockefeller foundations were involved as funders, as were a team of international scientists in universities and public labs. The result was a steep adjustment in world productivity and the transformation of India and China from food-deficit to more sustainable agro-economies.

The value of large-scale global effort was again highlighted in the early 1990s with the Human Genome Project (HGP; see chapter 10), an international scientific research project that sequenced and mapped the genes in the human genome both physically and functionally. Originally a US-based and -funded effort, the HGP team expanded over the years to more than 220 investigators and partners, enabling them to sequence the human genome in a decade rather than the planned fifteen years.

Reforms cascaded through the federal systems after 1995. Agencies sought ways to cut lower priority activities and then to leverage their remaining efforts through partnerships. In 1995 the Agri-Food R&D Matching Investment Initiative (MII) was created to encourage private funding to match, leverage, and lock in AAFC funding for specific research. The MII matched private sector investments in agricultural innovation to a maximum of $35 million annually. The MII was popular with private investors and industry, and by 2000 over nine hundred projects had participated, with a total value of $60 million. In the first few years there was concern that this shift in focus undercut the measurable quality of the basic science being undertaken; however, more recent evidence suggests that some of the partnerships, especially those involving Agriculture and Agri-Food Canada, another order of government, a producer-based organization, and a private venture (public-private-producer partnerships, or P4s), were more likely to produce quality science that would have a good chance of commercialization than research supported by a single actor (Boland 2014). In spite of this, seven of AAFC's research centres were closed over the decade and more than nine hundred staff were let go.

In 2000 Samy Watson was appointed deputy minister of Agriculture and Agri-Food Canada, and he brought in a new management philosophy. In 2002 he introduced a fundamental structural shift toward "horizontal, matrix management" and away from traditional hierarchy (Doern 2004). Then in 2003 he led efforts to establish the first of a series of five-year federal-provincial strategic plans, called the Agricultural Policy Frameworks (APFs). The five priorities of the APFs included food safety and quality, the environment, organizational and sectoral renewal, science and innovation and business risk management. To support these priorities, AAFC's Research Branch was restructured into four program areas: environmental health, sustainable production systems, food safety and quality, and bioproducts and bioprocesses.

In addition, an Office of Intellectual Property and Commercialization (OIPC) was established within the Research Branch to manage contracts and commercialization, and an International Scientific Cooperation Bureau was created to facilitate collaborations with foreign partners. Most importantly, the Agriculture Policy Framework formalized the 60/40 funding ratio that recent federal and provincial governments had informally worked out in the agri-food policy area. In 2006 AAFC announced a Science and Innovation Strategy to target the work with provinces and industry to accelerate agri-food innovation and the Agricultural Bioproducts Innovation Program (ABIP), which solicited and funded joint projects between AAFC, industry, and academics.

In 2007 Industry Canada developed and delivered the Harper government's 2007 science and technology strategy in the document entitled

Mobilizing Science and Technology to Canada's Advantage (see chapter 4), which delegated the task of setting priority areas to the newly formed Science, Technology and Innovation Council. The STIC advised the federal government to invest strategically in the following four priority areas: the environment, including water, natural resources and energy, health and life sciences, and information and communication technologies (ICTs). Agricultural research was noticeably absent; as a result, the main granting councils and other federal agencies retrenched from direct R&D in this area, which put increasing pressure on the remaining federal agencies focused on agri-food R&D, including Agriculture and Agri-Food Canada and the National Research Council. Late in 2014 the government renewed the S&T plan, adding agriculture and pairing it with the environment as a priority; if and how this will affect specific programs and initiatives was not clear in mid-2015.

In 2008, with the end of the Agriculture Policy Framework, Agriculture and Agri-Food Canada negotiated a new five-year federal-provincial agreement, this time called the Growing Forward Agricultural Policy Framework (GFAPF). The focus shifted more explicitly to support science and innovation, including an Agri-Science Clusters Initiatives program to support industry partnerships in ten priority sectors and joint public-private funding for innovation focused on existing market gaps, a foresight effort and measure to match with venture capitalists. In 2010 the AAFC Research Branch was formally tasked with providing sectoral leadership. Industry responses were lukewarm, with many concerned that the targets excluded many smaller-scale ventures, that regulatory change was not part of the package, and that there were simply too many initiatives for the budget and capacity of the ministry and its partners.

GFAPF 2 emerged in 2013, targeted on increasing competitiveness in domestic and international markets and increasing the adaptability and sustainability of the sector. Clusters and public-private-producer partnerships were the instruments of choice. Nevertheless, income stabilization remained a pressing concern, with a range of measures largely disconnected from the S&T and innovation policy agenda.

Meanwhile, as chapter 5 has shown, the National Research Council has gone through its own transformation. In the agricultural space, the centre in Saskatoon was key to the development of canola – it created critical partnerships and relationships with multinational companies that were vital to the translation of biotechnologies into agricultural applications. Since the mid-1990s there has been a sense of uncertainty about what the Plant Biotechnology Institute (PBI) is about and how it should direct its efforts. In 2010 John McDougall was appointed president of the NRC after twelve years as CEO of the Alberta Research Council (see chapter 5). Shortly after arriving, he started to restructure the NRC, which at that time was the largest single producer of R&D in Canada. By 2012 most of its sixteen national institutes had been decommissioned and begun the process of retooling.

In the context of the Plant Biotechnology Institute, the retooling entailed both significant cuts in the budget and staff and, partly at the prodding of the provincial government, the establishment of the Canadian Wheat Alliance (CWA), an eleven-year agreement between the NRC, AAFC, the Saskatchewan Ministry of Agriculture, and the University of Saskatchewan Crop Development Centre; in 2014 it was announced that KWS SAAT AG, a German seed company, and Syngenta AG had joined the Alliance. The NCR's specific portion of the Alliance's work plan is called the Canadian Wheat Improvement Flagship, and it involves NRC's expertise in genomics technologies, development, and plant biotechnology. The total investments locked into this partnership are in the range of $100 million, mostly from the public treasury.

Given the demise of the large-scale farm co-operatives and the modest fall-back in multinational investment, industry money has been relatively scarce. Yet federal programming increasingly relies on leveraging external funding, especially private capital, in its programs. Producer check-off organizations (sometimes called P4s) have emerged as a major player in the absence of other actors (Phillips, Ryan, and Boland 2013). Across the Prairies, for instance, the provincially mandated sectoral associations, along with the Western Grains Research Foundation, jointly raise and invest in the range of $40 million annually to support market development and R&D. P4s are not limited to the Prairies, but span the Canadian agri-food landscape. There are some notable successes in this space. In addition to the Canola Council of Canada, which raises about $12 million annually and was instrumental in the development of both canola and GM canola, the Saskatchewan Pulse Growers also raises and invests more than $12 million; both organizations can be credited with being part of all the major global crop breeding successes in their species in the past twenty years. In addition, Vineland, in Ontario, operates with a $9 million budget from public and private sources that it has used to successfully advance horticulture research in southern Ontario, while the Centre de développement du porc du Québec (CDPQ) operates with a $5 million budget made up of a producer levy, public and private contracts, and royalties (Phillips, Ryan, and Boland 2013).

This fifteen-year experiment with new management structures, new policies, new strategies, and new organizations, all linked to and supported by private or producer partners, is still being tested. Some of these institutional innovations have generated some interesting results, but there has yet to be any definitive analysis that justifies the experiment. Undoubtedly large-scale, globally networked science has a role to play in agri-food STI, but how much is needed, where it could be most effective, what differential impacts it delivers, and when and where other models may be more competitive remains to be determined.

1993–2015: Anti-globalization, New Models for Social Engagement, and GM Wheat

Over the past generation we have fashioned what Ulrich Beck (1992) would call a "risk society," where consideration of hazards drives consideration of new technologies and products. Some date the impetus for the modern application of risk analysis to the advocacy of the precautionary approaches in 1962, when Rachel Carson published *Silent Spring*, which documented the detrimental effects of artificial pesticides on the environment, particularly on birds. Formal structured risk analysis is now pervasive in our society. All manner of products and services are analysed through one form or another of risk analysis. Almost all the new technologies we use to produce the goods and services we consume are evaluated for their effect on the health of the environment or people. New plant varieties, pig barns, slaughtering facilities, chemical operations, manufacturing plants, and many activities and systems in the workplace all require some form of review of the risks they pose to the environment or the people around them. Similarly, the products and services of those enterprises are frequently reviewed and assessed in the context of their potential benefits. The foods we eat, the drugs we ingest, the clothing we wear, and the machinery we use are monitored, assessed, and managed to control for potential harms.

This system was formalized largely in the United States and made almost universal though technical and trade agreements. The US Environmental Protection Agency (EPA), the first dedicated risk management agency in the world, was created in 1970 with an explicit mandate to regulate chemicals and other pollutants, protect human health, and safeguard the natural environment – air, water, and land – upon which life depends. Although risk analysis was not yet a formally recognized process, the EPA undertook much of the early work and completed its first risk assessment document in December 1975 and in 1976 issued its first directive on procedures and guidelines. Then, in 1983, the US National Research Council published *Risk Assessment in the Federal Government: Managing the Process* (commonly referred to as the "Red Book"). This book, for the first time, laid out the context for a systemic, comprehensive process for risk analysis. Under the Red Book paradigm, risk assessment is largely a science-driven process that quantitatively evaluates the probability of risk. Risk assessment is largely, but not entirely removed from emotional factors that influence risk perception. Risk assessment flows in a logical, stepwise fashion from problem formulation through characterization of toxic effect or hazard and exposure analysis, culminating in a risk conclusion. The characterization process is iterative and recursive; if any step yields unexplored concerns, the process loops back to develop or incorporate new science to refine the risk assessment. This process is done in stages, with different levels of potential hazards assessed proportionately.

In practical terms, the risk analysis framework is commonly divided into three distinct phases – risk assessment, risk management, and risk communications – with different actors tasked with different aspects of the analysis. Risk assessment, in the Red Book, was standardized in the late 1990s through the *Codex Alimentarius*, and consisted of four steps: (2) hazard identification, (2) hazard characterization, (3) exposure assessment, and (4) risk characterization. Over time, the US National Research Council expanded on its risk assessment principles in a series of subsequent reports, including *Science and Judgment in Risk Assessment* (NRC, 1994, also known as the "Blue Book") and *Understanding Risk: Informing Decisions in a Democratic Society* (US NRC 1996, known as the "Orange Book") to more explicitly consider the appropriate communications in the system.

At about the same time as the administrative and scientific enterprise was framing the risk analysis framework, NGOs and individuals were organizing nationally and linking up internationally to project criticisms and concerns about the overly technocratic nature of the global economic and political system. Over time, this angst coalesced around a general concern about globalization, and given its increasing globalization, the agri-food R&D system was an early and frequent target for criticism. The movement emerged from the shadows into the spotlight in 1999 in Seattle, when a coalition of anti-globalization NGOs, trade unions, farmers' groups, and anarchists disrupted the meetings of the World Trade Organization. The Internet and social media have provided a new medium and forum for debate (see chapter 11) with the result that controversies seldom remain confined to specific issues or regions, but now are mostly advanced internationally if not globally. In many ways we now have a "glocal" model, with global messaging paired with local advocacy. Interestingly, Canada was the test bed and nursery for this model, which has been exported to the world. A number of leading anti-globalization and environmental activists cut their eye teeth in Canada on Canadian issues and moved to the global stage, including Greenpeace, David Suzuki, Naomi Klein, and Douglas Coupland.

The difficulty is that while governments and social interests were becoming more organized and interconnected, the agri-food industry was becoming more disconnected. Up to the 1990s, each province had an influential set of co-operatives and farm lobby organizations, and they all tended to align with the Canadian Federation of Agriculture (CFA) and project their views into what some have argued was a corporatist system with federal and provincial governments (Wilson 1990). As markets have become more global and technologies advanced, the farming industry has become more heterogeneous. Farmers have less in common with each other and more in common with producers in other regions of the world or firms in other stages of the global agri-food supply chain.

The net result is that many of the umbrella coordinating institutions in Canadian agriculture have lost power and influence: the major grain pools have restructured themselves into public companies; the national and provincial umbrella organizations have either collapsed or lost members and influence; and the Canadian Wheat Board is in the process of being privatized. In their place sectoral and regional groups have emerged, but they only represent segments of the system and have demonstrated little capacity to coalesce to project an industry-wide perspective.

The agri-food supply chain has changed dramatically. Multinational companies have entered at discrete points, especially at the input end and in the logistics and processing stages. The retail sector has been transformed by big box enterprises, such as Walmart, entering the grocery business. In response the Canadian (and also global) retailing sector has become more concentrated and has fought back with own-label products. Non-branded and even some traditionally branded food processors have been squeezed.

Governments and social interests have organized and want to be involved more in decisions in the agri-food system; however, the sector, both at the farm level and across the supply chain, is less able and willing to engage. In the absence of other options, regulators, citizens, and consumers are forced to project their values and interests at the products themselves, putting retailers in the line of fire.

The federal government recognizes the importance of communications in the risk decision system, but it has so far been unable to figure out how to do it any better than anywhere else. Industry Canada, and Agriculture and Agri-Food Canada have each tried a number of models. During the various rounds of 1980s bilateral and multilateral trade negotiations DFAIT constructed Sectoral Advisory Groups on International Trade (SAGITS), involving industry and other interest groups, to assist with framing government policy and strategy. In August 1986 fifteen industry-specific SAGITS were commissioned (each with up to two dozen members) to provide advice in the contexts of the Canadian-US trade negotiations of 1986–8 (Doern and Tomlin 1991). They were also tasked with providing advice on consumer interests, but their confidential deliberations make it impossible to judge how they did on that measure; regardless of how they did, social interests generally dismissed them as co-conspirators in the globalization agenda.

Environment Canada (EC) and Industry Canada tried a different tack, creating a series of national advisory bodies to provide expert opinions to governments on matters of policy and strategy related to environment and technology, respectively. Emerging from the UN-sponsored Brundtland Commission Report, *Our Common Future* (1987), the federal government created in 1988 the National Round Table on the Environment and the Economy (NRTEE or National Round Table), a twenty-five-year venture (abolished in 2012) in convening diverse voices to contribute to national policy on the environment (see chapter 2). The National Round Table focused on a number of topics relevant to agriculture, including water, the carbon cycle, and climate change policy, but few of these had any measurable impact on the agri-food innovation agenda.

Industry Canada first complemented and then replaced national umbrella advisory councils – especially the Economic Council of Canada, 1963–92, and the Science Council of Canada, 1966–93 – with a series of sectoral advisory councils, including a number directly relevant to biosciences and biotechnology innovation. The National Biotechnology Advisory Committee, an expert advisory body, operated from 1983 to 1998, at which time the Government of Canada renewed the National Biotechnology Strategy and, in response to changed global thinking about how the public should be engaged in risk adjudication, replaced it with the Canadian Biotechnology Advisory Committee (Doern and Prince 2012; Peekhaus 2012). The CBAC involved both experts and laypeople and was tasked both to undertake reference studies for government and to engage citizens, scientists, and industry in a national dialogue on the appropriate use of biotechnology.

The federal government in parallel tasked the Canadian Biotechnology Advisory Committee to review the regulatory system for GM foods and made a reference to the Royal Society of Canada (RSC), which struck an expert panel to assess the science underpinning the regulation of GM foods. Both the CBAC and the RSC processes produced a range of reports on the regulation, support, and uses of biotechnology in the food, health, and environmental areas, but the promised national dialogue never really got going, and there is little evidence of direct uptake and use of either the CBAC's or the RSC's specific advice. In support of a more integrated effort, Industry Canada borrowed an idea from the Human Genome Project and provided Genome Canada, a new special operating agency, with about $1 billion to advance genomics science through large-scale projects that were required to be paired with applied analysis of the ethical, economic, environmental, legal, and social aspects of genomics (termed GE3LS work – analogous to the ELSI (ethical, legal, and social implications) and ELSA (ethical, legal, and social aspects) initiatives in the United States and the European Union, respectively, and discussed in chapter 8). In spite of these efforts to funnel discussion into forums that might find compromises, such did not occur. These ventures surely triggered new discourse on the human control of technology, nature, and the economy, but they were unable to find a way to advance any noticeable related policy.

Social interests overall were generally not impressed with these efforts and placed responsibility clearly at the door of Parliament and the political system. The original scoping out and setting up of the regulatory system for GM plants with novel traits was done relatively quietly and without any overt parliamentary dialogue. Initially, this was not a point of contention, but as the products of biotechnology started to emerge, demands for more democratic dialogue escalated. A key trigger was the release of two controversial studies in 1999. In the United Kingdom, Arpád Pusztai from the Rowett Research Institute in Aberdeen, Scotland, announced that he had found that feeding genetically modified potatoes to rats affected their stomach lining and immune systems (Ewen and Pusztai 1999). Then John Losey and his colleagues at Cornell University released the results of a laboratory test that showed larvae of monarch butterflies that consume transgenic corn pollen were harmed (Losey, Raynor, and Carter 1999). This sparked global concern about potential human health and environmental risks. In the European Union this concern led to a freeze on further development that has not yet fully thawed.

In Canada and the United States academic researchers and industry reworked the results of those two studies, ultimately dismissing the Ewen and Pusztai results and determining that while the Losey group's findings were correct in the lab, they had limited application in open field cultivation. Although the scientific controversies subsided, the political controversy has remained salient ever since. These and other more recent similarly contested scientific studies (e.g., Séralini et al. 2012) are cited by some critics as they work to stigmatize and push back on the use of biotechnology in agriculture. The end result is that the global agri-food system has fragmented into GM-using and -consuming countries and regions and GM-free areas.

In Canada social interests have worked with back-bench and opposition members of Parliament (MPs) and in provincial legislatures to trigger debate and new legislation to slow and contain the use of new biosciences and biotechnology. The federal government, for example, held three major consultations beginning in 1993 to discuss the labelling of novel food derived from genetic engineering. Guidelines emerged and in 1999 the government through the Canadian General Standards Board established a multi-stakeholder committee with more than 120 representatives from federal and provincial ministries, farming, food manufacturers, grocery distributors, and consumer groups to develop a voluntary labelling standard. In 2004 the committee created the world's first national standard, which covered both positive and negative claims for both single and multiple ingredients (Canadian General Standards Board 2004). To date it remains the only validated labelling standard in the world – all the other labelling rules involve arbitrary thresholds that are not based in either scientific risk assessment or a collective decision of the sector.

Nevertheless, this has not satisfied many individuals and NGOs. In 2002 the House of Commons Standing Committee on Health undertook hearings on the labelling of GM foods. A series of bills followed. In 2009 a private member's bill (Bill C-474), designed to amend the Seeds Act to require consideration of economic impacts for GM seeds as part of the assessment, passed second reading and went to committee for discussion; it was defeated at third reading, late in 2009. In 2013 Bill C-257 was introduced to amend the Food and Drugs Act to introduce mandatory labelling, but it died on the order paper.

One agro-biotechnology that generated significant controversy and pushback was Roundup Ready™ (RR) wheat, discussed earlier in this chapter. In the early to mid-1990s the Monsanto Company used the biolistic transformation method to insert a gene in wheat that would make the plant tolerant to glyphosate, a broad-spectrum non-specific herbicide sold under the trademarked name Roundup™. Monsanto did much of the early discovery and transformation work in the United States in conjunction with several universities and then entered into research alliances with a number of breeding organizations to develop commercial varieties using proprietary germplasm. Some candidate varieties from that R&D effort entered confined field trials in Canada and the United States in the late 1990s. In the following years, Monsanto assayed the varieties and developed the evidence for regulatory review, ultimately selecting in 2003 a single transgenic variety it would ask the US and Canadian regulators to review prior to release. Monsanto submitted its regulatory portfolio to regulators in both countries to assess and rule on the safety of its product. As commercialization approached, a number of large buyers (e.g., Barilla) and large sellers (e.g., Canadian Wheat Board) expressed reservations about the potential impacts of the new technology on international markets and trade.

In Canada a coalition of more than two hundred groups joined with farmers' organizations (including the National Farmers Union and the Canadian Wheat Board) and petitioned the government to intervene. In July 2003 Greenpeace Canada petitioned the auditor general of Canada to engage in the decision process. As the global debate heated up, Monsanto announced on 10 May 2004 that it had suspended all further research and commercialization efforts, in all countries, and that it would withdraw that transgenic variety from all the regulatory channels but the United States. Monsanto sustained the review in the United States in anticipation that the market at some future point might be more receptive to the technology. Ultimately, the product died and private capital moved away from intensive wheat research in Canada and abroad for almost a decade. Similar stories of truncated or stalled development emerged about the EnviroPig™ produced in Guelph (Ontario), the Newleaf™ bt-potato in Prince Edward Island, and the Aquabounty™ salmon in Newfoundland.

In recent years the government has tinkered further with how it commissions and uses advice on science and public policy. As part of the 2007 S&T strategy, the Harper government decided to wind down all remaining arm's-length expert advisory bodies. The Canadian Biotechnology Advisory Council and the National Round Table on the Environment and the Economy were dissolved and replaced by one single advisory group, the previously mentioned Science, Technology and Innovation Council. This group of eighteen members is explicitly designed to include "leaders" from the public, private, and academic sectors. They consult with experts and have an expert staff, but their advice is fundamentally grounded more in their experiences and interests than in dispassionate, expert deliberation. As noted earlier, the Science, Technology and Innovation Council has tended to ignore (or at least not engage) with the biosciences and biotechnology as they relate to food and agriculture, in spite of some of their own evidence that suggests current investments in R&D generated significant socio-economic and commercial outcomes.

About the same time the federal government started to reduce its reliance on internally generated scientific advice (see chapter 4). To compensate, the government has used or set up specific, time-limited advisory groups (e.g., the Competition Policy Review Panel of 2007–08 that produced *Compete to Win*) or empanelled specific expert panels from the Canadian Council of the Academies,), the Canadian Academy of Engineering, and the Canadian Academy of Health Sciences. The Jenkins panel (i.e., the Expert Panel on Federal Support to Research and Development) on the state of industrial R&D in Canada in 2010–11 is an example of the type of advice the government has received (see EPFSR&D 2011).

Canada is arguably one of the more active countries in trying to find new and better ways to engage citizens in agri-food policy and S&T and innovation policy discussions. But in spite of a wide range of experiments, it is hard to say this has led to greater satisfaction. The federal government, the agri-food industry, and social interests continue to search for a way to engage that generates truly reflexive socially responsible policy. Instead, these processes have too often generated a dialogue of the deaf with frequent skirmishes and forays in the media and on the Internet. Both local and global controversies can and do emerge and for a time can dominate debate (e.g., climate change, organic vs. GM futures, global food security, labelling and regulatory integrity), but any consideration of the impacts on domestic or global S&T and innovation policy and strategy is for the most part blunted and overshadowed by more salient aspects of each topic. Phillips (2012) assessed the use of democratic engagement processes developed and used in Canada and abroad to open the policy dialogue to citizens and consumers. The array of tools developed and tested was impressive. Somewhat disconcertingly, the evidence to date suggests that the application of those tools leads to one of two outcomes: truly engaged processes tend to be undemocratic, while democratic processes tend to be relatively unengaged. Clearly, this area of policy development will remain active for years to come.

THE THREE ELEMENTS

We now track with the aid of Table 9.1 the three elements being utilized to understand change and inertia across the three domain policy and governance histories presented in this chapter.

Policy Ideas, Discourse, and Agendas

This S&T and innovation policy domain has been profoundly affected by some of the same big, global ideas that have transformed the macro-policy landscape both in Canada and abroad, as examined in the industrial S&T domain in chapter 7. Our first policy history regarding the canola success story showed how new technology deliberately and purposefully pursued as transformative technology was pivotal as were related ideas tied to the need for new and better sources of food and fibre. In this process government leadership as an idea was itself respected and exploited and, indeed, supported by strong agri-food producer organizations.

By the time our second policy history begins, newly emerging ideas, some managerial in nature, were evident. New public management ideas and paradigms, reinforced by budgetary constraint themes, were interacting with ideas regarding the value of networked globalized science. NAFTA and WTO free trade ideas were now entrenched, with traditional methods of farm support via tariffs and subsidies winding down. Instead, to intervene, governments had to search for "green" policy options (e.g., R&D support) as possible normative entry points for some kinds of policy and support interventions.

In the GM wheat policy and governance history ideas, discourse, and agendas included a mixed array of changes beginning or being further entrenched. Post-1995 referendum impacts led to efforts to reconstruct ideas/concepts about federalism, including stricter adherence to the federal-provincial division of powers and a shift to a greater role for provinces. This also helped bolster the earlier national consensus of agriculture as a corporatist enclave. The reframing of the world under trade and regulatory policy as the "risk society" was leading to the search for an engaged, reflexive, socially responsible, and participatory society.

Table 9.1 Policy and governance histories in the agriculture, agri-food, biosciences, and biotechnology domain: Three analytical elements

Policy and Governance History	*Policy ideas, discourse, and agendas*	*Economic and social power*	*Time, temporal realities, and conflicts*
1960s–2005: Global innovation leadership in the development of canola	• New technology deliberately and purposely pursued as transformative technology • Ideas tied to need for new and better sources of food and fibre • A green revolution advanced for Canada; made in and for Canada oilseed option • Government leadership as idea itself respected and exploited, supported by strong producer organizations • Explicit idea advanced to focus and relocate research talent	• US-based and -driven technology imperative • Key role of commodity check-off organizations as coordinators and facilitators • Influence of federal labs, including AAFC Prairie Regional Lab and refocused NRC/PBI; Saskatoon-based partnership built on previous public leadership team (Downey) • Shift from public IP ownership in days of rapeseed to extensive ownership of all IP from genes, technologies, to seeds • MNCs attracted to industry for first time; new supply chain relationships emerging • Canada led way in developing national discourse on technology with NBAC, CBAC, Genome Canada, GE3LS, and other ventures, but ultimately overcome by new pressures	• Long-term strategic effort mobilized industry, government, universities, and producers in purposeful partnerships with measureable outcomes • Canada in the vanguard in setting stable medium and longer term norms, rules, processes, and teams; PNT system and CFIA emerge and acknowledged as gold standard • Gradually triggered new discourse on human control of technology, nature, and economy • IPRs and the Schmeiser litigation became a focusing mechanism for dissent

Table 9.1 Policy and governance histories in the agriculture, agri-food, biosciences, and biotechnology domain: Three analytical elements (*Continued*)

Policy and Governance History	*Policy ideas, discourse, and agendas*	*Economic and social power*	*Time, temporal realities, and conflicts*
1995–2015: Groping for new models of research management and the shift to partnerships	• New public management ideas and paradigm and budgetary constraint dominant themes • NAFTA and WTO free trade ideas entrenched with traditional methods of farm support (tariffs and subsidies) winding down; green policies and R&D last remaining ideas/realms for acceptable policy interventions • Rise in appreciation for networked and internationalized science • Rise of P3s and P4s as idea for delivery	• Neo-classical refocusing on incentives for investment, and government gave up setting goals or leading • End of unilateral federal leadership • Emerging division in national agricultural policy and power system: Central Canada dominated by supply-managed and domestically focused products; West dominated by export-dependent products (grains, oilseeds, pulses, cattle, and hogs) • Shift toward the components rather than industry-wide action plans • Federal systems reformed after 1995 to focus on strategic investments that leverage partners • NCE, CFI, CRC, Genome Canada completing the check-off system, but mostly not strategically focused • Merging units into AAFC; regional directors shift in NRC to platforms • Move to exploit joint federal-provincial constitutional rules with the Growing Forward programs and 60/40 division of responsibilities • Multi-national companies largely defect from innovation system, leaving Canada as an adaptor and adopter rather than innovator	• Stagnation in regulatory systems slowed new product introduction to a crawl • Infill traits approved, but rest of portfolio (second- and third-generation GM crops, GM animals, GM microbes) all stalled in system • Sask.-led efforts on wheat; producer-led efforts on canola and pulses; Alta.-led efforts on cattle; Ontario-led efforts on horticulture

1993–2015: Anti-globalization, new models for social engagement, and GM wheat	• Post-1995 referendum efforts to reconstruct ideas/concepts of federalism, e.g., federal refocusing on unique federal authorities: North, trade, defence, stricter adherence to federal-provincial division of powers and shift to greater role for provinces • End of national consensus of agriculture as a corporatist enclave • CUSTA, NAFTA, Cairns Group, and WTO sharpened focus • Reframing of world as the "risk society" and the search for an engaged, reflexive society • Anti-globalization ideas emerged from 1999 WTO meeting in Seattle • Rise of the Internet and social media as medium/forum for debate; concomitant emergence of globalized effort: global messaging with local advocacy	• Globalized power emerges in agriculture and food • Farming becomes massively heterogeneous with more trade and foreign investment • Triggers collapse/winding down of old agri-food systems: pools merge, then privatize; most of the co-op centrals retrench; CFA loses control; end of CWB • Other ministries gain power and authority: IC (consumer, anti-trust), HC, EC • Multi-national companies organize but offer narrow message; advocacy groups emerge and largely frame public image of debate; retailers caught in middle • Dramatic changes in retailing: rise of big box and category-busting retailers at same time as organic diets emerge	• GM wheat developed but stalled and cancelled • House of Commons finds it difficult to manage continuing debate: no GM debate • End of expert advisory groups (CBAC, NRTEE) and clamp down on public role for federal scientists • Periodic industry-led experience and interest-based advice sought (Jenkins, Wilson, STIC) • Federal government and NGOs groping for a means to engage that satisfies both; dialogue of the deaf stalls policy file • Global/local controversies emerge and dominate debate: climate change; organic vs. GM futures; global food security; labelling and regulatory integrity focused dissent • STI message blunted and overshadowed • 5-year plans provide the focus; little room for public dialogue; institutionalized federalism

Note: see the chapter text for expansion of abbreviations in this table.

Anti-globalization ideas and supportive international ideas grew with the rise of the Internet and social media as legitimate arenas for debate.

In the context of agri-food S&T and innovation policy options, although the problems facing agri-food innovators often appeared local, for the most part, the most appropriate framings and solutions were globally generated or focused. The biotechnology revolution, in particular, introduced a transformative technology into the economy and society. While specific applications of the technology generally seemed benign, the cumulative effects promised to be profound. Given the concomitant emergence of the precautionary principle in the "risk society," demands rose for new means and methods of democratic engagement and dialogue. Agriculture and food is one of the few spaces where the demands for such socially responsible policy resonated with the broader citizenry.

Economic and Social Power

In the economic and social power dynamic, science, especially biotechnology, has become global. Significant effort is needed to be able to access the knowledge and tools, partly because of the diffusion of the capacity across industry, governments, and universities and partly because of the proprietary nature of many of the critical tools. The science is global, but the capacity to access and effectively use it is often agglomerated in key multinational networks, often anchored by large corporations (e.g., Monsanto) involving both subsidiaries and contractual partners. In some ways a more important development is that the scale and scope of biotechnology has led to the emergence of regional bioscience clusters, each involving an ecosystem of public labs, universities, entrepreneurs, firms, commodity groups, industrial associations, sophisticated buyers, dense labour markets, and effective linkages upstream and downstream of the innovative effort and government policy-makers. Canada has a number of sites – Edmonton, Saskatoon, Winnipeg, Guelph, and Montreal – that at times exhibit the dynamism one expects from such venues.

The agri-food system in the past generation has been admitted and subsumed into the increasingly globalized world. Trade barriers have been successively lowered, and agriculture's privileged status in the GATT system was given up in the WTO agreement in 1995. As a result production and markets have begun to be rationalized, with both producers and regions increasingly specializing and competing in global markets. Many less developed countries have used scientific and technological research (e.g., advanced breeding techniques and biotechnology) to rapidly adapt and adopt leading-edge technologies in order to transform their market position. India, Brazil, China, and South Africa, in particular, have gone from marginal players in world markets to holding leading market shares in key agri-food commodities. In many cases they have gone from importers to developers and exporters of technology.

Canada has only partly adapted to these pressures and reconfigurations of power. As a founding member of the Cairns Group of leading agri-food exporters, and as a principal in almost all of the leading trade negotiations, Canada has pushed forward the competitiveness agenda, but with the caveat that the supply-managed commodities primarily produced in Ontario and Quebec are to a great extent exempt. As a result, Western Canadian agriculture has almost fully adopted the innovation agenda, while Central Canadian producers and industry have not. This divergence blunts Canada's policy stance globally and divides the Canadian agri-food industry into two camps, making it harder to have industry- or sector-wide dialogue and to find universal policies and strategies.

Time, Temporal Realities, and Conflicts

Since the introduction of biotechnology and GM organisms almost thirty years ago, conflict has been almost continual about when or whether and over what time frames to introduce such products into use. The appropriate strategy for managing government investments in biotechnology remains undetermined and delayed, and the appropriate mechanisms for timely and effective public input are contested. Industry has gone with a minimalist short-term strategy of just dealing with those matters that directly relate to its part of the market, be it seeds, marketing, branded foods, or retailing. Similarly, farmers have ceased to project a single voice – instead, a plurality of views is now available, with some for the most part ignoring the S&T and innovation policy agenda, some actively embracing new technologies, and others spurning biotechnology in favour of differentiating products through their provenance (e.g., organic, non-GM, ethically produced, and locally produced). Sometimes the same producer might have apparently conflicting interests depending on the crop or product involved.

Against this backdrop, NGOs and individuals have mobilized a strong, loud, and sustained critique of industrialized and globalized agriculture, especially that part of the industry that has adopted genetically engineered plants or animals. Corn, canola, and soybeans continue to be cultivated and consumed in Canada, but no major new agri-food innovations have been introduced since the heyday of the mid-1990s. The cacophony of voices in this area is at times deafening; however, it has only stalled and not reversed the technology.

In an effort to leave no tool untried, social interests have tried partnerships with some selected actors in the differentiated agri-food system, including with the Canadian Wheat Board to oppose Roundup Ready wheat and with retailers to produce voluntarily labelled GM-free foods. They have also engaged with back-bench and opposition MPs and provincial legislators in an attempt to advance mandatory labelling and changes to the regulatory regime and used litigation in the courts to slow, stall, or punish adoption of the new technology.

With no end in sight for this stand-off between adopters and opponents, a majority of the global agri-food exporters continue to aggressively develop, adapt, adopt, and use new technology in key product categories. At some point in the not too distance future some tipping point may be reached, with either the economic imperative overriding the demands for better decisions or farmers and industry being forced to find alternatives to using biotechnology to sustain their competitiveness.

CONCLUSIONS

Ultimately, the three policy and governance histories in the agricultural, food, biosciences, and biotechnology domain reflect the normal conflict among three dominant goals for public policy: effectiveness, efficiency, and equity. New technologies may be effective and generate efficiencies, but various social interests and governments often want more. They are concerned with the balance (as are the authors) in the overall innovation economy-society nexus. Transformative technologies, by their very nature, make it difficult if not impossible to realize all three goals simultaneously either at an overall technology assessment level or at the product and process approval levels.

Agricultural is in many ways a crucible for a number of key issues triggered by innovation that divide society. Transformative change in the sector due to new technological development undercuts the autonomy of many producers, as the technologies only work with integrated, intensive farming systems. The presence of new products of biotechnology also contests both national and consumer sovereignty, as the global economic rules have been set to facilitate (some say promote) the proactive development and diffusion of productivity-enhancing innovations. Even if one can erect barriers to isolate a product category, sector, or region from the new technology, the competitive pressures ultimately undercut any short-term benefits of exclusion.

Technological change is seldom equitable – by definition new innovations create winners and losers. At a deeper level, the advent of biotechnology in the global agri-food system has caused governments, producers, and consumers to question our relationship with nature itself: does humankind hold dominion over nature, or should we position ourselves within and be bound by the limits of the natural world? It is unreasonable to expect Canada, a small part of a global system, to solve these problems alone, but we could and probably should be involved in the global effort to frame and address them.

The challenge is that while the agri-food R&D system is a venue of choice for contesting these issues, it is far from the perfect venue. The agri-food S&T and innovation system offers a unique opportunity to hold these debates as the context both means something to people and makes the trade-offs very explicit. The difficulty is that the impacts of the choices we make can be and are global, particularly on consumers and small producers in food-insecure regions of the world – who will ultimately pay the lion's share of the cost of any mistakes we might make.

10

The Genomics, Life Sciences, and Technology Domain

We turn now to the mapping and analysis of the genomics, life sciences, and technology domain in Canada's STI policy and governance system. As was the case in chapter 9, this domain has emerged internationally as a field of quickly developing S&T since the late 1980s. As in the agriculture-food STI story, the pace of current trends is anchored in a deeper history of how science and technology have raised questions about the ethics of new ways to intervene in life courses. Yet as a measure of the distance over which these discussions have travelled and changed, one only has to consider that within many people's lifetimes our society has learned of the structure of DNA, developed oral contraceptives, discovered the importance of stem cells, and worked out not just in vitro fertilization but also the possibility of "three parent" babies that use enucleated eggs as recipients of fertilized nucleases to avoid genetic diseases inherited with maternal mitochondrial DNA.

Life science and biotechnology related to humans has boundary links and overlaps with the life sciences and biotechnology more generally. The scope is actually vast and includes plants, animals, human beings, micro-organisms, and viruses, and hence linkages across the board from botany and zoology to genetics (US National Human Genome Research Institute 2015a). Technologies, including a vast array of diagnostic and testing approaches throughout the domain also cross over between microbial, plant, and animal domains as the potential to study, control, and manipulate life at fundamental levels pays little or no heed to biological disciplines. As new discoveries emerge they open up new adjacent potential possibilities. Equally, many of the non-scientific issues, including research ethics and the acceptable impacts on humans link to generalized concerns about dominion over nature and conceptions of power, including the power of the self and control of one's own body and personal health in a world of actual and potential targeted personalized medicine (Doern and Prince 2012; Doern and Phillips 2012).

Genomics is a term that has become a catch-all for the study of genes, genetics, and genomics. Genomics proper refers to the study of an organism's total complement of DNA – its genome – the foundation for the structure, development, and regulation of all life. Genes are principally involved in the coding for specific proteins, but increasingly it is becoming the case that genes, generically defined as regions or sequences of DNA, perform important regulatory functions within cells. The study of the structure and function of genes is genetics, a field that predates the emergence of genomics by several decades. The field of genomics, with its many non-human legacies, nevertheless is captured in the public mind by the inception of the international Human Genome Project (HGP) in the early 1990s.

Genomics, both human and as it applies to other domains, is among the most global because it is very difficult for any individual country to be wholly self-sufficient in the science and technology used to develop new products and processes. Equally, even if a country is less committed to research and development, it will be committed to being able to exploit relevant technologies needed to undertake regulatory assessments. Global exchange and use of genomics knowledge is continually challenged since the growth of much of this knowledge has arisen in the period, as chapter 8 discussed, when science and technology became increasingly protected and exploited under proprietary regimes (protected by patents, trademarks, trade secrets, and various commercial and contractual mechanisms). This system privileges some actors and forms of evidence at the expense of others – at least partly because most regulators prefer to work with owners of new products so that they can more easily assign fiduciary and residual obligations related to unintended and unanticipated consequences. This becomes more of a challenge once a product enters the market – consumers in Canada and globally have high expectations that they will be able to make their own choices about accessing and using foods, drugs, and other products (Phillips and McNeil 2001).

The genomics, life sciences, and technology domain is even more challenging as Internet-savvy Canadians and their families often have knowledge about, and access to, products including diagnostic tests well before they are approved in Canada, especially from the United States. This has triggered what has been called the "wow" to "whoa" phenomenon, as new discoveries and products with impacts on human health and life science are announced with excitement – almost immediately followed in the same rhetorical breath with human and social fears about such products and processes (Harris 2010; Wade 2010). Citizens, as individual patients and families, have mixed and diverse social preferences about how much they really want to know about findings from clinical genomic sequencing, seen in personal and intergenerational terms (Regier et al. 2015). For example, the more that stem cell research is pursued the more that the need for and advocacy of complex "collaborative regulation of innovation" is imperative (Devaney 2014).

The governance structure for the domain involves research funding and support by granting bodies and federal departments and also regulation by several agencies, including those that judge the science, safety, and marketing of drugs (both targeted and orphan and diagnostic tests and those who will pay for treatment under health care at the provincial level of government). The regulatory and guidance system also involves the decisions, norms, and ethics of health and social professions (Doern and Prince 2012, chapter 3). There are two main streams of government investment in research: (1) Genome Canada and granting council investments in basic genomics research discovery and development and (2) government intramural investment in the science needed to underpin public policy and regulation. Provinces, foundations, and especially private capital come on-line only as the research advanced and commercial opportunities became clear.

At the federal level, the intramural Genomics R&D Initiative (GRDI 2012; NRC 2010) finances and coordinates seven federal departments in the fields of genomics research. Over one billion dollars has been invested since 1999 in five, three-year funding cycles. This has generated a number of projects and discoveries that have importance to researchers and client genomic networks. But the links to regulatory issues are complex and often hard to track in any detail. The general GRDI (2012, 1) website says that "independent evaluations have found that GRDI is successfully supporting the core public policy, regulatory, and operational mandates of government." Regulatory science is part of the GRDI mandate, although an earlier evaluation of the program did not offer any specific insight into how this work has supported Canadian regulatory capacity (NRC 2010).

Although genomics and genetics have been around for just a few decades, their impact is widespread and diverse. Each of the federal agencies and related research organizations undertakes de facto premarket regulatory assessments, in that each has its own norms, rules, and processes for making choices that effectively pre-screen genomics applications before they reach the main product and process regulators (Doern and Phillips 2012). Almost all organizations now adhere to some articulated set of ethical norms that are then embodied in institutional rules, corporate pledges, corporate social responsibility (CSR) processes, or in mission and mandate statements.

Most institutions also operate a set of practices for codifying, disseminating, and asserting ownership for any inventions or discoveries and for biosafety (e.g., embedded in Good Laboratory Practices or other professional or industrial standards). As shown in chapter 6, the early phases of the regulatory process involve federal research granting bodies such as the Canadian Institutes of Health Research (CIHR), the Natural Sciences and Engineering Research Council of Canada (NSERC), the Social Sciences and Humanities Research Council of Canada (SSHRC), the Networks of Centres of Excellence (NCE), Genome Canada, and the Canada Foundation for Innovation (CFI) as they impose and enforce processes and norms through their grants, including the requirement for "leveraged money" and the changing and more complex structure and role of peer review and merit review.

As long as the efforts and the outcomes of research remain contained in laboratories or greenhouses, formal regulatory oversight does not really begin. If and when a genomics application is proposed for use beyond containment, then governments in Canada and around the world begin to respond. Canada uniquely uses the *novelty* of the product to trigger assessments – in conformity with World Trade Organization (WTO) norms. Thus, the regulator examines product attributes rather than the production and processing methods (PPM) used to produce the product. Novelty can arise either by simply transferring organisms to the ecosystem from other areas or by mutagenic or transgenic techniques. Other national systems tend to trigger assessments based on the use of transgenic methods, but they still focus their efforts on assessing the risk of using or consuming the resulting products. The practical effect is that the Canadian system catches and assesses more products (e.g., mutagenic crop varieties) than other competing systems. The actual assessment in Canada (and most other countries) is then based on the internationally accepted risk analysis frameworks, which broadly encompass three distinct phases – risk assessment, risk management, and risk communications – with different actors assigned different tasks (Doern and Reed 2000).

There are other response and regulatory examples as well such as the way that the health system has had to respond to a wide array of genetic tests, from quite advanced oncogenetic therapies all the way to on-line sales of genetic tests. Thus, Canada has had, like any other jurisdiction, to figure out where old regulatory frameworks and decision-making processes still apply and where new regulations and practices are required (see more below).

At the *federal level* the regulators with partial direct or indirect genomic-related mandates, laws, regulations, and guidance roles include the following: Health Canada, and several of its directorates such as the Health Products and Food Branch Inspectorate and Therapeutic Products Directorate; the Canadian Food Inspection Agency (CFIA); Environment Canada (EC), for those products that do not have Acts listed under the Canadian Environmental Protection Act; Fisheries and Oceans Canada; the Public Health Agency of Canada (PHAC); Industry Canada, for competition policy; the Canadian Intellectual Property Office (CIPO); the Patented Medicine Prices Review Board; the Canadian Agency for Drugs and Technologies in Health; and the Office of the Privacy Commissioner of Canada. All of these agencies operate in analogous and complementary ways, with delegated groups undertaking risk analysis that is then adjudicated by a risk management group and communicated more broadly to the public and interested parties. Each system attempts to reflect the "common regulatory principles" articulated in the 2007 Regulatory Cooperation Framework between Canada, the United States, and Mexico.

Genomics-based products are generally evaluated only once in Canada, at the federal level. Nevertheless, *provinces* at times can be key actors. As we see further below, many provinces have carved out niches in the genomics research area targeted on their specific economic priorities (often related to health, forestry, or the fisheries). British Columbia and Quebec, through Genome BC and Génome Québec, respectively, in particular, have invested in provincially targeted research competitions. At least as important are universities, which undertake much of the foundational research on genomics. As provincial entities governed on an arm's-length basis, universities have their own internal regulatory and policy provisions and processes regarding research ethics, patenting, commercialization, and public-good science (Doern and Stoney 2009).

At the other end of the research chain, provincial agencies are often the most important market for technologies and services. Health departments and provincial drug formularies, as the ultimate payers on behalf of taxpayers, make decisions about whether to purchase or fund within provincial medicare programs new genomics-based products or services, including drugs, devices, and tests. Products may have been granted patent rights and Health Canada approval, but this does not mean that they all will be funded or used in the provincial systems (Canadian Agency for Drugs and Technologies in Health 2015). Numerous provincially based groups of medical professionals, patients, carers, and disease advocates aggressively engage with provincial authorities, seeking to bring forth new and improved prospects. The combination of variable uptake of technology and uneven lobbying by interest groups opens up potential for conflict and differential access to drugs under government health care insurance.

International regulatory and policy bodies exert considerable direct and indirect influence on Canada, both through harmonizing the evidence and processes for regulating genomics research and applications and through various kinds of exhortative demonstration effects. In addition to the important scientific, commercial, and regulatory links Canada has built with its major trading partners – especially the United States, the members of the European Union, and other OECD member states – Canada belongs to a range of international organizations that work to normalize the models, methods, and metrics of regulatory practice. Regulators communicate almost daily with colleagues in competent regulatory agencies in other states to identify the appropriate ways to undertake the vital tasks of hazard identification, hazard characterization, exposure assessment, and risk characterization. Reforms to the US Food and Drug Administration have also been important to Canada's potential regulatory reform agenda (FDA 2013; Carpenter 2010). The Obama Administration's more aggressive stance in monitoring genomics-related research ethics is also noteworthy (Meslin 2010). And in the larger political-regulatory context, the role of religion in embryonic stem cell research in the United States has served as a cautionary tale in Canada's somewhat more secular political culture (Devaney 2014; Knowles 2010; Morris 2007).

FOUR DOMAIN POLICY AND GOVERNANCE HISTORIES

To develop a deeper sense of how and why this Canadian domain has been forged, we look at four partly and necessarily overlapping policy and governance histories: reproductive technologies, Assisted Human Reproduction Canada (AHRC), and federalism imperatives; the International Human Genome Project and early response support for Canadian life science research; the creation and evolution of Genome Canada; and the emergence of personalized medicine both as a socio-economic discursive arena and a "product" realm.

Several such technological and science policy and governance trajectories are essential to appreciate what is increasingly shaping this domain (McCabe and Husereau 2014; Doern and Prince 2012; Migone and Howlett 2009). The reproductive technologies trajectory reveals the importance of federalist constitutional imperatives and inertia in Canada but also the defining role of women in its politics and in the progression of views of the self and one's own life and person. The HGP trajectory helps reveal defining international S&T results and pressures and Canadian efforts to engage through increased research funding support by several granting bodies, which were themselves already going through reform in peer review and funding, as discussed in chapter 6. The Genome Canada story brings evidence of a more focused approach but also reveals in different ways the regulatory and ethical issues about the conduct of research. The personalized medicine trajectory partly involves a different kind of policy discourse but also complex grey zones regarding how to move to a new form and era of health care partly characterized by the federal government as personalized health.

Late 1980s to 2015: Reproductive Technologies, Assisted Human Reproduction Canada, and Federalism Imperatives

Reproductive technologies were brought to the national agenda mainly by women's groups and coalitions and were the focus of the work from 1989 to 1993 of the Royal Commission on New Reproductive Technologies (RCNRT 1993; Miller Chenier 1994). There was a decade-long delay in implementation, as legislative and regulatory approaches started nominally in 1996 finally resulted in legislation in 2004 and subsequent regulation in 2007. But then a Supreme Court of Canada constitutional ruling in 2010 forced a rethinking of federal versus provincial jurisdictional powers and legitimacy in this complex field, which by then was being joined by statutory mention of the integrity of the human genome in individuals.

The Commission's final report only referred briefly to biotechnologies and biosciences in a chapter on commercial interests and identified at that time very few Canadian companies engaged in research directly relevant to reproductive technologies. Most research was taking place in universities using funding from government or private foundations (RCNRT 1993, chapter 24). Nonetheless, the women-led reproductive technologies debate noticeably included early aspects of genetic testing and presaged many of the larger dimensions of the current genomics and life sciences and technology domain.

Although the Commission was aware of the just started International Human Genome Project and Canada's interest in it, the goal of determining the structure and location of human genes was not viewed by the Commission as part of its mandate with respect to genetic research and technology as related to human reproduction. But the Commission was already engaged with the same broad array of groups and players with a direct interest in assisted human reproduction, genetic research and its applications, and aspects of genomics. These included organizations representing women, children, individuals and families, persons with disabilities, faith and religion, industry/business, insurance and law firms, medicine and clinical laboratories, scientific research (natural and social sciences), universities, and governments (Doern and Prince 2012).

Women played a particularly crucial role in advocating the need for the Royal Commission on New Reproductive Technologies and in its subsequent research, findings, and recommendations. In 1987 feminist organizations, researchers, and health groups, who argued that discussions on reproductive technologies were adjudicated primarily by the (mostly male) medical profession with little attention to the broader implications of these technologies on women, initiated the Canadian Coalition for a Royal Commission on New Reproductive Technologies. Pressure from both the medical/scientific community and women's organizations found space inside and outside the royal commission, and in the commission's final report (Miller Chenier 1994).

The system to regulate activities and products regarding assisted human reproduction is, to a large degree, an unfinished work in governance and policy implementation and remains a contentious and contested realm in both Canadian society and federalism. Assisted Human Reproduction Canada, the federal regulatory body created in 2007, had a mandate founded in its enabling legislation, the 2004 Assisted Human Reproduction Act. The principles underpinning the Act are diverse and relate to criminal law provisions to protect women and children, but also measures to encourage and support women and families, human individuality and diversity, and the integrity of the human genome (AHRA, s. 4).

Decklha (2009, 23) shows how these varied pressures made the development of the Act a "discursive site" in that the rules came into effect in fits and starts: the law emerged late in the Chrétien era from work by the House of Commons Health Committee in 2001, was not adopted until 2004 (Hébert et al. 2004), and implementation has muddled along since. Its work captured the politically negotiated and multilayered nature of its actual and potential governance (Caulfield and Bubela 2007; Morris 2007; Scala 2007).

In particular, the Assisted Human Reproduction Act centres on four broad objectives: (1) protecting and promoting the health of donors, patients, and offspring; (2) protecting and promoting the human dignity and human rights of Canadians who use or are born of assisted human reproduction technologies; (3) fostering ethical principles in research and clinical practices; and (4) allowing scientific advances that benefit Canadians (AHRA, s. 24).

The legislation was enacted in March 2004, while various sections of the statute came into force in April 2004, further sections in January 2006, and another section in December 2007. Assisted Human Reproduction Canada, the resulting Health Canada regulatory agency, was not established until 2006 and its first president was only appointed in 2007. AHRC reports to Parliament through the minister of health and the minister and Health Canada officials are responsible for the development of strategic policy in respect to the legislation and its regulations.

In practical terms the mandate of Assisted Human Reproduction Canada emphasizes knowledge functions such as advice to the minister, communications with stakeholders, and information to the public. The AHRC developed a memorandum of understanding with Health Canada that specified that the Health Products and Food Branch Inspectorate of that department will undertake inspection and compliance activities as required, until such time as the relevant sections of the Assisted Human Reproduction Act that deal with inspection and enforcement came into force.

Constitutional concerns were present from the outset but became subject in 2010 to a definitive Supreme Court of Canada ruling centred on an appeal of a Quebec court ruling. Specifically with respect to the challenged sections of the Assisted Human ReproductionAct, the constitutional issue concerned whether the prohibitions, regulations, and penal sanctions in the legislation are suitably related to the criminal law powers of Parliament or fall under areas of provincial powers. In a complex and split 5–4 decision, the Supreme Court of Canada held that the federal government's appeal of the Quebec court's decision be allowed *in part* (Supreme Court of Canada 2011). But some sections were held to exceed the legislative authority of the Parliament of Canada, representing an attempt to regulate hospitals, clinics, medical research, property and civil rights, and matters of a local and private nature, all of which belong to the jurisdiction of the provinces.

A core issue dividing the Supreme Court concerned the purpose and nature of the legislation: is the Assisted Human Reproduction Act of 2004 primarily a law to prohibit certain inappropriate activities or does it include objectives and policy instruments aimed at promoting and controlling certain beneficial practices of assisted reproduction? Two very different aspects of genetic manipulation have been combined in a single piece of legislation. The social and ethical concerns underlying these two aspects appear to be distinct, and in some cases, even divergent. While the prohibited activities are deemed to be reprehensible, the controlled activities are considered to be legitimate. Parliament has therefore "made a specious attempt to exercise its criminal law power by merely juxtaposing provisions falling within provincial jurisdiction with others that in fact relate to the criminal power" (Supreme Court of Canada 2011, para. 278). In the end, a majority of the justices saw that the Act had sought to address both these aspects; hence the constitutional predicament.

Looking for activities and bodies to cut in its 2012 Budget, the Harper government used the Supreme Court decision as its main argument for cutting and abolishing Assisted Human Reproduction Canada as the main regulatory body. Any remaining parts of the Act under federal jurisdiction are handed by other parts of Health Canada. The Assisted Human Reproduction Act also spawned an agency for its implementation, based in Vancouver, which has spent considerable sums of money but was also targeted for failing to have delivered on the important parts of the Act.

Thus, the reproductive technologies trajectory reveals the importance of the federalist reality in Canada but also the defining role of women in its politics and in the progression of views of the self and one's own health, life, and person. Genomics emerged as a driving factor as the early work of the Royal Commission morphed into law and policy and was then confronted by a Supreme Court decision. That decision went to the heart of the preventative approach regarding abuses, some centred on criminal law, and the supportive approaches, which required careful health and risk management. Many of these governance imperatives are still central in the larger human genome domain story, as we see further below.

Early 1990s to 2003: The International Human Genome Project and Canadian Research Support for the Changing Life Sciences

Genomics differed from classical genetics research from the outset due to its large scale, broad scope, and heavy reliance on computer-based bio-informatics. The core team involved in mapping the human genome (International Human Genome Sequencing Consortium 2001, 860–1) saw and linked it to the earlier pivotal historic phases of related scientific progress in biology:

> The rediscovery of Mendel's laws of heredity in the opening weeks of the 20th century sparked a scientific quest to understand the nature and content of genetic information that has propelled biology for the last hundred years. The scientific progress made falls naturally into four main phases, corresponding roughly to the four quarters of the century. The first established the cellular basis of heredity: the chromosomes. The second defined the molecular basis of heredity: the DNA double helix. The third unlocked the informational basis of heredity, with the discovery of biological mechanism by which cells read the information contained in genes and with the invention of the recombinant DNA technologies of cloning, and fourth, sequencing by which scientists can do the same.

Although emerging from the discipline of biology, the leading scientists knew that genomics would burst the boundaries of biology due to its profound spillovers in many or indeed all life sciences. Many commentators and organizations saw it as the core of a "new economy" technology (Shreeve 2004; Wellcome Trust 1999; Thurow 1997).

The Human Genome Project was a global publicly funded project begun in the early 1990s to map the human genome or the genetic "book of humanity." The HGP produced a high-quality version of the human genome sequence, made freely available in public databases, "to generate a resource that could be used for a broad range of biomedical sciences." It provided "more powerful tools to study the role that multiple genetic factors acting together with the environment play in much more complex diseases … such as cancer, diabetes and cardiovascular disease" (US National Human Genome Research Institute, 2015a, 2). The scale of funding on the HGP and on other genomics research has been impressive. The United States and the United Kingdom took the lead. The US National Institutes of Health (NIH) helped to fund three Genome Centres; overall American public funding easily exceeded one billion dollars. In the United Kingdom, the Wellcome Trust put in the lion's share of support to the HGP. Canada officially joined the HGP in the late 1990s. Germany, France, and Japan also committed significantly to the project.

One of the key early political factors in genomic research globally (also in Canada) is that, unlike with biotechnology regarding foods, there were few overt political negatives in public discourse. The venture to sequence the human genome was immediately supported by governments as a means to secure new products and diagnostic approaches to cure or better manage human diseases (Doern and Prince 2012). This did not mean, as discussed in the introduction to this chapter, that there would not be complex issues to face and manage. Indeed Canadian policy and health scholars have cast genomics in terms of both the abovementioned multiple trajectories and multiple forms of framing and reframing. Indeed, genomics and its policy and governance processes changed multiple levels of public and private power, including the power of the self (Doern and Prince 2012; Knoppers and Isasi 2004; Migone and Howlett 2009).

Human genome research also proceeded at breakneck speed in the private sector with the support of private equity alongside public investments. Understandably, this trend resulted in a growing concern about the ownership of genomic data. Some of these issues emerged visibly in 2000 when negotiations broke down between the HGP and Celera Genomics, a private US company, which was moving even faster than the public project to map the human genome by using a propriety sequencing technology. While other private biotechnology companies feared that Celera would lock up rights under patent protection, public interest scientists and advocates feared that it would stifle the free exchange of scientific data and information (Shreeve 2004).

These concerns were partly reflected in a joint statement by US President Bill Clinton and UK Prime Minister Tony Blair issued on 14 March 2000. The two leaders stressed the need for the raw and fundamental human genome data to be kept in the public domain so that scientists could utilize it freely. The Clinton-Blair statement asserted that "unencumbered access to this information will promote discoveries that will reduce the burden of disease, improve health around the world and enhance the quality of life for all humankind" (quoted in the *Independent*, 15 Mar. 2000, 5). Reflecting the complex interplay of values, interests, and political power involved, the two leaders went on to say that "intellectual property protection for gene-based inventions will also play an important role in stimulating the development of important new health care products" (ibid.). Somewhat problematically, the statement referred to both free and patented use of genomics without specifying in what proportions.

Our next policy and governance history on Genome Canada and its research centres discusses Canada's central focus on genomics research support, although positive support for genomics research began earlier and then expanded to other federal research and granting bodies. The trajectories of support are complex, in part, because the life sciences did not begin with genomics, but also because, as shown in the analysis in chapter 6 of the research-granting and levered-money domain, the funding bodies were themselves changing in quite major ways that intersected with and complicated the arrival of the global genomics project. For example, the NRC's five biotechnology institutes by 1998 were already being impacted by the genomics momentum when NRC was promised $32 million by the Liberal government to create a genomics venture (Doern and Levesque 2002, 89). Earlier fields of specialization within NRC institutes already included human diagnostics and human therapeutics. The National Research Council thus started its Genomics and Health Initiative, which while not an institute per se, gave a more explicit focus to genomics. Centred in Ottawa, its work was intended to advance genome and health research in areas including the detection and treatment of infectious diseases, biorenewable oils, cancer therapeutics, the management of heart disease, and virtual reality systems for surgical oncology (NRC 2010).

The federal Networks of Centres of Excellence (NCE) programs were created in 1987 to foster research networks per se, and fairly quickly in the international genome project era two networks emerged: the Stem Cell Network (SCN) and the Centre of Excellence in Personalized Medicine (CEPMed; NCE 2014a). By 2015 twenty-five of the NCE's forty-nine funded networks were identified as being in the health and life sciences sector broadly defined, including the above-mentioned Stem Cell Network and such networks as Biotherapeutics for Cancer Treatment, the CellCAN Regenerative Medicine and Cell Therapy Network, and the Centre for the Commercialization of Regenerative Medicine (NCE 2015).

The Canada Institutes of Health Research (CIHR), formed in 2000, also moved into this space. Among its first thirteen virtual interdisciplinary institutes (mentioned in chapter 6) was one centred on genetics and another on cancer, both influenced and shaped by the genomics agenda. The most obvious institutional presence of genomics research in the CIHR is the Institute of Genetics (IG). Its mandate is to support research "on the human and model genomes and on all aspects of genetics, basic biochemistry and cell biology related to health and disease, including the translation of knowledge into health policy and practice, and the societal implications of genetic discoveries" (Institute of Genetics 2009). In keeping with this mandate, its research priority themes include from genes to genomic medicine; population genetics, genetic epidemiology, and complex diseases; and genetics and ethical, legal, and social issues. Each of the CIHR's institutes were to conduct research across the four pillars of health sciences inquiry noted in chapter 6, namely, biomedical, clinical, health systems and services, and population health.

The HGP policy and governance trajectory is pivotal largely because the project was avowedly international in nature and promised major benefits and progress in dealing with multiple diseases and multiple life science and technology fields. Both governments and the early private sector firms saw it mainly in positive public interest terms even though they possessed different conceptions of public goods versus private property. Early Canadian support for genomics research was offered from several granting councils and rapidly taken up by the university research community. But a more focused and comprehensive genome effort only emerged with the formation of Genome Canada in 2000 as a central part of a Liberals' S&T innovation economy effort that emerged when federal budgetary surpluses appeared after a decade of federal deficits and cuts to science.

2000–2015: The Creation of Genome Canada and Its Research Centres

Genome Canada provides a particularly interesting window into the federal government's efforts to increase public and private R&D expenditures, conduct research with commercial potential, close Canada's productivity gap, and bolster Canada's position as an innovation leader. In many ways, the emergence of Genome Canada is a direct result of Canada's small and delayed role in the HGP. The new organization was instigated by human health researchers and other health vested interests and was established as an arm's-length not-for-profit corporation in the February 2000 federal budget with an overarching mandate to ensure that Canada become a world leader in genomics research in targeted sectors, including agriculture, forestry, fisheries, health, the environment, and later, the accompanying social sciences.

Genome Canada was a creation of Industry Canada. In the period leading up to the announcement of its creation, various groups of academics and regional stakeholders advocated a range of options, including supplementing the existing granting councils; creating regionally situated, national centres of genomics expertise focused on specific domains (e.g., cancer, genetic diseases, agriculture, or fish); and creating a single national centre. In the end, the government chose to develop a new national funding agency outside of the main granting councils, with "regional" centres formed principally to develop regional projects and applications and to manage the administration of the programming. Three of the initial centres mapped on to provinces: Genome BC (GBC), the Ontario Genomics Institute (OGI), and Génome Québec (GQ). Two were "regional": Genome Prairie (GP), sponsored by and located in Alberta but also serving Manitoba and Saskatchewan, and Genome Atlantic (GA), located in Halifax and representing but not directly sponsored by any of the four Atlantic Provinces. In 2004 Alberta set up Genome Alberta (GA) and Genome Prairie moved to Saskatoon and to deliver services in Manitoba and Saskatchewan.

In many ways, Genome Canada resembles a special operating agency (SOA), a departmental form introduced by the Government of Canada in 1989 (Farazmad 1996). The SOAs were created, in general, to allow for autonomy of managerial practices. The goal was a focus on service delivery with a "more streamlined and less rule-bound approach" (ibid., 53). Genome Canada is managed this way, by a board of directors. The CEO reports to the board, which reports to the minister of industry through annual summaries of operations, maintaining the principles of ministerial responsibility. In a special operating agency, the CEO is not directly responsible to a minister or a board of directors, but rather through a deputy minister.

Initially, the SOA model was both convenient and facilitated the rapid development of a national enterprise. The federal government was able to advance a block of funds to the operation, effectively parking part of their budgetary surplus outside the strictures of accrual accounting and annual review. It also helped generate instant credibility for the new organization, as Genome Canada started with a succession of large capital infusions, which made it the largest funder in this area in Canada and a globally scaled institution. Moreover, Genome Canada was able to earn a return on the advanced funds, which helped them generate important operating capital in the early years. In April 2005 Genome Canada's initial funding agreement with the federal government came to an end. The Amended and Restated Funding Agreement on 30 March 2005 ended the practice of block grants and directed the corporation to submit a "strategic" budget proposal each year, so that the designated funds could be dispersed by Industry Canada based on annual need. Since then a hybrid system has evolved, with annual proposals being supplemented with a series of strategic "asks" for new investment capital.

Genome Canada and its six regional centres are maintained and staffed using federal conditional grants. From inception to 31 March 2014 Genome Canada has raised and disbursed $921.2 million for research projects, S&T Innovation Centres, and regional Genome Centre operations. The organization also spent in 2013–14 about $6.8 million annually on operations. The original plan was for all of the activities sponsored by Genome Canada to be cost-shared at least 50:50, and in most projects that ratio has been met or exceeded. The 2014 corporate annual statement reported that the $921 million of direct disbursals was estimated to have leveraged approximately $1.3 billion of other effort and funding.

Genome Canada is primarily focused on supporting large-scale research that could not otherwise be pursued. In that sense it has incentivized teams of researchers to bid on multiyear, interdisciplinary research contracts. To get a sense of this, Zhang (2013) estimated using agency data that the average competitive grant for genomics-related research from Genome Canada was about $4.4 million (2001–12), compared with $418,000 from CIHR (1999–2014) and $67,000 from NSERC (1999–2012).

Over the life of the organization, Genome Canada has matured, expanded, and modified its processes and procedures. In all, it has administered seven major competitive research competitions and almost twenty smaller-scale, more focused research initiatives. The first three competitive calls were open to any application, while the last four have been targeted on specific realms: applied genomics for bioproducts and crops (ABC), forestry and the environment, personalized health, and feeding the world. The seven competitive calls together have funded more than 130 teams of researchers and directly invested more than $620 million in genomics science. While the projects do not parse cleanly by sector or region, the approximate distribution of funds for Genome Canada's first decade was as follows: 65 percent for health; 17 percent for technology development; 16 percent for agriculture, forestry, fishing, and the environment; and 3 percent for socio-economic studies, administered by the following centres (shares of Genome Canada funds allocated): 35 percent by OGI, 26 percent by GQ, 23 percent by GBC, 12 percent in the prairie centres, and 4 percent by GA (KPMG 2009).

The somewhat disproportionate allocation of funding – with Quebec and British Columbia hosting more than their per capita share and the Ontario Genome Institute and the prairies hosting a lower share – is partly due to the matching requirements of competitions and the fact that host

Table 10.1 Genome Canada science initiatives, 2000–2015

Year of competition	*I 2000–01*	*II 2001–02*	*III 2004–05*	*ABC 2008–09*	*LSARP I 2010–11*	*LSARP II 2012–13*	*LSARP III 2014–15*	*Other 2003–15*
Focus of competition	All	All	All	Bioproducts & crops	Forest & All	Health	Feeding the world	19 Narrow calls
Approved projects (no.)	17	34	33	12	16	17	10–15 est.	~90
Success rate (% of full proposals)	28	53	36	25	NA	NA	NA	NA
GC funds allocated	$81M	$146M	$205M	$53M	$60M	$45.1M	$30M	> $300M

Source: Authors' calculations using data drawn from Genome Canada website.
Note: see the chapter text for expansion of abbreviations in this table.

provinces have aligned their policies differently. The BC government has frequently directed capital infusions to Genome BC to both match federal dollars and to administer their own competitions, and Génome Québec has a standing offer to co-fund any successful project sponsored and delivered in Quebec, while investigators working with Genome Alberta, Genome Prairie, the Ontario Genome Institute, and Genome Atlantic all need to make case-by-case applications for co-funding.

In addition to the general project competitions, Genome Canada has initiated about twenty other funding initiatives that have been more narrowly focused, ranging from new technology, bid data, specific international consortia, specific partners (Spain and Italy), specific problems (*E. coli*, listeria, cancer), entrepreneurship, and commercialization. Many of these initiatives were opportunistic, as Genome Canada was able to buy Canada or Canadian teams into larger efforts or to do one-on-one or small partnerships with key agencies or stakeholders willing to invest in the area. In many ways Genome Canada has moved away from being a stand-alone actor to being an integrator and project developer or manager, along the way exerting leadership in the genomics space both in the federal system and more broadly.

Initially the organization sought to maintain independence from most of the other actors in the genomics area, but as its first tranche of block grants were expended, it found value in partnering. As far back as 2004 Genome Canada collaborated with the Department of Fisheries and Oceans to co-fund the Genoma Espana initiative: Genome Canada ran the competition and DFO provided some of the capital and initial direction (Bearing Point 2004). This model has become much more common since then, with the last two large-scale applied research programs (LSARPS) co-funded, respectively, with CIHR and the Western Grains Research Foundation. All in all, Genome Canada has partnered with more than half a dozen co-funders, including the CIHR, NSERC, CFI, and various Alberta government agencies.

Genome Canada has introduced two main innovations into science programming in Canada. First, it has introduced the notion of professionally managed, large-scale science. Others pioneered many of the practices it uses, but Genome Canada was the first to put the pieces together and has gone the farthest. While the NCE program first introduced the concept of large-scale science teams and actually provided resources for project management, many of those ventures until recently were more virtual than directed, and they often took the form of managed competitions for smaller awards within the NCE themselves. Similarly, many grant programs now require investigators to set out milestones and to sign contractual agreements, but reporting has been modest and implementation has largely been left as voluntary. Granting programs around the world increasingly require investigators to consider how they will translate or commercialize their results, but this is more a reporting than performance goal. Genome Canada put all these ideas together in a proactive management system.

All of the projects Genome Canada funds are set up through research contracts, rather than grants in aid of research, with mandatory professional management, reporting, technology transfer programs, and evaluation. Recognizing the disconnect between its management and commercial objectives and the cultures in most Canadian universities, Genome Canada provides specific in-project funding for governance and administration and works with each project lead to hire a dedicated professional project manager, who is then paid for by the project but in the employ of the lead Genome Centre. Each project then has to manage its affairs proactively, with oversight and ultimate control residing with project management teams, overseen by independent external science advisory committees. For longer projects (more than two years), there have always been formal, mid-term peer reviews of the science, the management, and the personnel. All projects have to keep commercialization and knowledge translation in focus: projects are initially assessed and chosen based partly on their expected socio-economic benefits and the end-of-project evaluations look closely at the outputs and outcomes.

Genome Canada also has advanced the concept of engaging social scientists and those doing research that involves humans in an effort to create a reflexive science. Although this approach was pioneered in the Human Genome Project, where about 3 percent of the gross outlays were directed to examine the ethical, legal, and social implications (ELSI) of science, Genome Canada expanded it to address all of the realms, rebranding the concept as genomics and its ethical, environmental, economic, legal, and social aspects (GE3LS). Genome Canada has also tinkered with the model to change the nature of the engagement between the scientific and socio-ethical teams. Similar to the HGP ELSI efforts, in the first two open competitions (I and II) Genome Canada encouraged large-scale, stand-alone teams of social scientists and those doing research involving humans to carve out an area of investigation (e.g., health, agriculture, or fish) to focus their efforts on.

In the early years there was a tendency to focus much of the research on the overall appropriateness of the applications and to drill down to investigate claims of inequity. Beginning with Competition III, Genome Canada began to encourage and support GE3LS investigators to embed themselves into the science projects, in an effort to broaden the range of subjects investigated (e.g., the nature of the research enterprise and the potential for translating any outputs into use) and to facilitate knowledge translation and diffusion with and among the researchers. Up until that point, only a portion of the funded projects engaged with GE3LS scholars. With the ABC (applied genomics for bioproducts and crops) competition in 2008–09, Genome Canada mandated that henceforth all science projects would need to have an integrated GE3LS program. Stand-alone GE3LS projects were allowed, but they were not encouraged or privileged in subsequent competitions. A national debate continues over the appropriate balance between stand-alone, large-scale GE3LS projects and small-scale, embedded GE3LS scholarship – there is no agreement on which generates better scholarship or impacts. While that debate continues, it is undeniable that over the first fifteen years of project support, Genome Canada funded more than 150 GE3LS scholars from across Canada and encouraged those scholars to link internationally to the global ELSI community (VALGEN 2010).

The model has unambiguously delivered a scholarly and practitioner community empowered by the latest scholarly methods and an understanding and awareness of the challenges of genomics technology; however, there is some concern that the scholarly output from the effort has not had the impact on commercialization and policy they had hoped for. Many of the GE3LS efforts in recent competitions have been narrowly cast to examine barriers to commercialization, in some cases simply providing in a consultancy style freedom to operate studies, and market evaluations and regulatory and commercial strategies. Although of some value, those activities are probably more appropriately undertaken by consultants on a fee-for-service basis than by tenured scholars.

Overall, there is a sense that the genomics venture in Canada, focused on Genome Canada, has accelerated Canada's capacity in the field but has yet to lead to the promised socio-economic impacts. Although it missed contributing significantly to the HGP, Canada is now ranked variously as one of the top five producers of basic science, GE3LS social sciences, and patented technologies and products, but few if any of those have obviously generated economic or commercial gains. The long and uncertain development pathways make it difficult to sustain enthusiasm for such large and targeted investments.

Mid-1990s to 2015: The Emergence of Personalized Medicine as a Socio-Economic Genomics Discursive Arena and "Product" Realm

One particularly complex and broad domain policy and governance issue is personalized medicine, which is fundamentally built upon the genomics revolution (Personalized Medicine Coalition 2014; Boyer 2010; *Economist* 2010). We have already referred to it above regarding the NCE's early creation of its Centre of Excellence in Personalized Medicine. Personalized medicine was defined in early work by the Health Canada Personalized Medicine Working Group, which includes the CIHR, as (CIHR 2011, 1):

> the tailoring of preventative, diagnostic or therapeutic interventions to the characteristics of an individual or population. It does not mean the creation of health interventions targeted directly to an individual, but rather that the scientific advancements that underpin personalized medicine provide the ability to classify individuals into sub-populations based on their susceptibility to a disease, or response to a specific treatment. This can allow for prevention/intervention strategies and earlier and/or targeted interventions to improve health outcomes. Personalized medicine aims to transform the delivery of healthcare to patients such that it will evolve from a "one size fits all" system towards a common tailored system. Delivering more predictive, preventative, and precision care … will also enhance awareness about lifestyle and preventative lifestyle changes based on individual risk factors … [and] is expected to lead to better health

outcomes, improved treatments, and a reduction in unnecessary treatment with its associated costs and adverse events.

A CIHR (2013, 1) report referred to its 2010–13 personalized medicine "signature initiative" that centred on a "national framework" with five components: (1) building evidence, (2) addressing regulatory and policy challenges, (3) national infrastructure and services, (4) data handling, and (5) outreach and education. Regarding regulatory and policy challenges, it highlighted the initial formation of the above-mentioned Personalized Medicine Working Group within the federal government, its early planning meeting in 2011, the convening early in 2012 of a multi-stakeholder workshop, and finally, a CIHR–Health Canada workshop in March 2013 on orphan drugs (ibid., 4).

These reports and developments at the Canadian Institutes of Health Research and Health Canada (and the investments by Genome Canada in genomics directed at health) represented a major federal investment in research funding that was ultimately underpinning the personalized medicine transformation. Canada was also closely tied in with and aware of US developments, which were arguably even more advanced, particularly with regard to regulation and guidance, and geared much more to the private sector–dominated system of health care in the United States (FDA 2013; Personalized Medicine Coalition 2014). The FDA was especially focusing on its role in paving the way for a new era of medical product development.

Meanwhile, the Harper government, in March 2013, gave its then science minister, Gary Goodyear, the lead role in the announcement of $150 million of research grants in a CIHR and Genome Canada partnership labelled as "new personalized health projects across Canada" (Genome Canada 2013, 1). The minister emphasized that the government "recognizes that genomics is a science at the core of the global bio-economy and offers a myriad of social and economic benefits" (ibid.). Also emphasized was that these projects showed how "our government is supporting Canadian families" (ibid.). The announcement, not surprisingly, did not dwell on what personalized medicine or personalized health might mean but rather that the research would focus on disease and life science fields such as "epilepsy, autism, HIV/AIDS, cancer, cardiovascular disease, rare neurological diseases, and stroke, among others" (ibid.).

An early 2014 presentation in Washington by Genome Canada President Pierre Meulien (2014) focused on the challenges of advancing "personalized health" via genomics research. Regarding Canada he argued that "our ability to move the latest technology into the health-care system is traditionally low and the way technology is assessed across the country is heterogeneous" and that "new technologies are often seen as just an added cost and the economic analyses performed are not convincing enough for the payers" (3). Though Meulien provides evidence of a good genomics research support record in Canada, he posed several questions regarding how we translate research "when we need to consider a lot of complex issues" (5). These include (ibid.):

- How good is the technology? (clinical validation)
- In a fast moving field, when do we decide that "now is the time for transfer"?
- Is it easy to adapt existing clinical laboratory structures?
- Who will make these decisions? (and based on what criteria?) (technology assessment based on sound economics and clinical benefit?)
- Who will pay?

Complexity abounds as well in what specific interests, players, and health professions are or will be involved overall or for particular diseases and individuals and their families and with regard to both pre-market assessment and post-market monitoring. Both change and inertia bid for strategic and tactical attention on the way to the "who will pay" question and the myriad answers to it.

McCabe and Husereau (2014) provide considerable insight into one set of factors and puzzles about this journey by exploring issues about how one gets from "science to value" in the relationships between personalized medicine and health care policy. They explore the steps that are needed to "facilitate the translation and adoption of high value personalized medicine in the Canadian health care system" (1). They stress that "all stakeholders will need to fully understand and identify the societal value associated with research and commercialization … while ensuring that they do not drive out existing high value health care activities" (ibid.). Overall they see a need for various kinds of benefit-cost analysis with their recommendations aimed at (1) insisting that health system payers clearly define what constitutes value; (2) exploring options to more clearly align evidentiary requirements and processes between regulators and health technology assessment bodies that support payers; and (3) increasing strategic focus in applied and technology-oriented basic research" (8–10).

The policy and governance dilemma relates more precisely to the *orphan drugs* issue as a key aspect of the CIHR's personalized medicine's "signature initiative." Orphan drugs are those that would not otherwise have reached the market because of market imperatives but that were crucially needed and useful to smaller niche markets of patients. The United States and the European Union developed special tax-based incentives to enable some products to get to market (Maeder 2003), but Canada has no such system, seemingly because it just did not register as a higher priority tax measure.

There will, however, be many more potential orphans in the new high volume and complex product market at hand in this domain. More specifically, the notion of tailored products and small markets refers to a shift to many more products – drugs, medical devices, and combined drugs, and devices – derived from genome mapping and using DNA characteristics. Drug products historically have been aimed typically at huge blockbuster drug markets, but genome-centred products are actually different (Goldacre 2012; *Economist* 2007b). The extent to which this happens depends on the availability of risk capital for firms, especially for the numerous small firms developing these products.

Interestingly, Health Canada in its now planned *Orphan Drug Framework for Canada* defines orphan drugs as "those drugs used to treat rare diseases" (Health Canada 2012, 1). This is a somewhat misleading definition in the genome product context, in part, because it gives no related sense of what "rare" means and the volumes involved. Health Canada is developing what it describes as a "modern framework for the designation, authorization and monitoring of orphan drugs," a key focus of which "will be on international information-sharing and collaboration ... enabling Canadian scientists and regulators to participate with trusted global counterparts will make better use of scarce resources and benefit Canadian patients" (ibid.). The new framework will still be based on clinical trials, supported by greater international information sharing, and following authorization drugs will continue to be closely monitored "for effectiveness and safety while in use" (ibid.). No related tax-based incentives, however, are on offer to smaller firms thus far.

After consultation, the Harper government announced in October 2012 what it described as the first ever "Canadian framework to increase access to new treatments and information and Orphanet-Canada Online Portal" (Health Canada 2012, 1). The framework was described in terms of patient access but also as a system "designed to spur research and innovation in Canada" and to enable patient participation. Access is facilitated by funding that will now enable Canadians to use "Orphanet, the world's online reference portal for information on rare diseases" (ibid.). If genomics and personalized medicine proves to add value to the medical system, it also promises to cause significant changes in the current three-stage clinical trials process, and the related drug approval system (see chapter 5) may be fundamentally incompatible with the personalization of drugs and dosage in the context of a person's genomics and lifestyle.

Ultimately in this domain and in a personalized medicine context, one is speaking of a regulatory and a health *product* that inevitably takes on meanings that lose many of their traditional comfortable moorings compared with the other regulatory health and food product realms. Normally, a product can be assessed at the pre-market and post-market phases of assessment and regulation, and even be "withdrawn" from the market or the health and food systems. But in the genomic and personalized medicine realm, the defective "product" may be a child or offspring for whom the pre-life or early stages are crucial and need to be effective in profoundly human terms (Doern and Prince 2012).

There are also concerns about Canadian and American private companies simply selling their products without regulatory approval on-line or in drug stores. These products raise issues about where the post-market starts and which regulators loom largest and with what kinds of monitoring and review powers. There are also issues regarding policy and rules on bioethics at various stages in research, product development, and review. In both these areas, developments in the United States are especially germane, in part, because they will have cross-border impacts and demonstrable effects on Canada. They will be sold in Canada or at least to Canadians, and thus there are regulatory gaps potentially on both sides of the border depending on when and how Health Canada moves into this aspect of regulation and in what specific ways.

Last, but definitely not least, personalized medicine and its products face the ultimate issue of who will pay. The Canadian Agency for Drugs and Technologies in Health (CADTH) is the non-profit organization funded by the federal, provincial, and territorial governments that develops and provides evidence-based clinical and pharmico-economic reviews to assess a drug's cost-effectiveness. Drug approval may occur but public payers may not cover each drug approved because analysis by CADTH may show it to be not as cost-effective relative to an existing drug. CADTH is centrally involved in the Common Drug Review (CDR), which involves a single process to assess new drugs for potential coverage by participating federal, provincial, and territorial drug benefit plans (Doern and Prince 2012, 166–71). Provincial and territorial health ministries are the crucial players because they in the end make the decisions (regulatory and expenditure) as to whether a new drug will be funded and listed for reimbursement under medicare services delivered at the point of patient care.

CADTH's (2015) strategic plan for 2015–18 sets out how it intends to inform choices "in a new era of health care in Canada" by "delivering value: inform health policy and clinical practice by getting the right information to the right decision-makers at the right time; expanding reach: build receptivity for health evidence; and setting the standard: champion meaningful evidence and leading methods" (ibid., 1). Regarding receptivity for health evidence, CADTH had earlier reviewed its "patient input process" via a study by the consultancy firm SECOR (2012) that probed patient access in Canada in comparison with other selected countries. The individual patient focus was an increasing demand in the politics of personalized medicine but had not been assessed in any significant way by CADTH.

Ultimately, of course, the battle over who pays, and how much, enters the political-economic realm of federal-provincial health care funding overall and how agenda's might be developed and under what kind of discourse and areas of emphasis. For example, the provinces are increasingly hit by growing health care costs and budgets as a percentage of their total spending (around 50 percent) and which are seen as crowding out vital education and other spending.

The federal government in the Harper era has not been willing to discuss any major shifts or increases in its share of health care spending, given its view of "open federalism," and given its other economic and resource priorities, not to mention the deficits it has had to manage and survive (McGrane 2013). Other demands for spending have recently focused on how and when the Harper government or any other future government will plan for and execute funding for "expensive orphan drugs" (Picard 2015). Some pressures for reform focus on Canada's need for a national pharmacare plan as the next logical progressive phase of Canada's health care system (Gagnon 2013). But pharmacare is expensive and may or may not cover the new varied products of personalized medicine that are queuing up in the genomics and health sciences and technology domain.

THE THREE ELEMENTS

We now track with the aid of Table 10.2 the three elements being utilized to understand change and inertia across the four domain policy and governance histories.

Policy Ideas, Discourse, and Agendas

The ideas across the policy and governance histories are similar in that they are anchored in concerns about life and disease prevention and treatment and about the genuine prospect of making breakthroughs and gains in human health outcomes due to new technologies and science. But the specific ideas and the related discourse deployed in each of the histories examined are different as they entered evolving national S&T and innovation agendas.

Reproductive technologies and assisted human reproduction was the core discourse deployed in the first policy history anchored by advocacy by women seeking protection for themselves and children and also for families in the forging and nurturing of life. Ideas and discourse also centred on fostering stronger ethical principles in research and clinical practices. This issue might not have got onto the national agenda were it not for pressure for a royal commission and then coverage and follow-up reporting and discussion of its work and recommendations.

The history of the International Human Genome Project was anchored on the idea of the human genome itself and its massive international scientific body of work, centred in the United States and the United Kingdom. The HGP was from the outset presented as a positive human and social development and elicited in the main a positive sense of the pursuit of the public interest. The extent to which its benefits would be pursued as a public good as opposed to commercial property remains unclear, and despite efforts to quantify the impact of the HGP into a positive account of the benefits of innovation (Battelle 2011) there are doubts that even conservative multiplier effects caused by large public investments trigger innovation and the benefits thereof. There was widespread awareness in diverse life science communities of the project as it was being conducted and in its aftermath since 2003.

The HGP triggered and focused Canadian energies on genomics. Having missed being a major player in the sequencing of the human genome, Canada was determined to be a leader and winner in the race to sequence species and to use the new technologies to generate socio-economic benefits and commercial value. Genomics fit all of the criteria for a transformative, general purpose technology, in that most of our existing economy is based on exploiting our natural environment. Given the nature of genomics, however, the prevailing view was that we needed a major stimulus to reach our potential – Genome Canada was that push.

The emergence of personalized medicine as an overarching idea and discourse was placed on the S&T policy and health agenda as a way of

Table 10.2 Policy and governance histories in the genomics, life sciences, and technology domain: Three analytical elements

Policy and governance history	*Policy ideas, discourse, and agendas*	*Economic and social power*	*Time, temporal realities, and conflicts*
Late 1980s–2015: Reproductive technologies, Assisted Human Reproduction Canada, and federalist inertia and governance precursors	• Triggered by opportunities for and concerns about reproductive technologies • Protection for children, women, and families • Human individuality and diversity and the integrity of the human genome • Fostering ethical principles in research and clinical practices	• Women-led debate and lobby for royal commission • Feared power of the medical profession by women • AHRA as discursive statutory site for diverse interests • Emergence of the power of the self • House of Commons Health Committee negotiations • Supreme Court as decisive player on statute's limits regarding federal jurisdiction using its criminal law powers	• Long, slow process for Royal Commission work and then for development of the AHRA legislation and creation of regulatory body, AHRC • Sudden cancellation of AHRC in 2012 after Supreme Court decision
Early 1990s– 2003: International Human Genome Project and related Canadian initial research support for the changing life sciences	• HGP as major international S&T effort and achievement building on earlier key phases of biology development • Massive boundary spill-overs in most life sciences • Cast and seen as positive public interest effort regarding human diseases with few of the negatives that emerged initially with biotechnology and GM foods • Celera and US private corporate S&T development and entrepreneurial skill but centred on IP rights of an initially unknown nature and scope	• Power of S&T community internationally • Private power tied to property rights, especially in US private health care–dominated system • Canadian granting councils offer positive genomic research support, initially in late 1990s and then in parallel with Genome Canada after return of healthy fiscal surpluses in 2000 • Universities support early genome initiatives as well	• 10+ years to achieve mapping of the human genome • Early indications of both promising products and diagnostics, but also shortfalls and inertia in getting these adopted in clear-cut ways, given their underlying slow moving adoption at individual patient and family levels

2000–15: Creation and evolution of Genome Canada	• Genomics as a transformative technology for whole economy • Big Science as the solution • Professionally managed science • Socially responsible or reflexive science • Appropriate balance between the sciences and GE3LS	• Asymmetric federalism with some provinces engaging fully and others holding back • Efforts to leverage private capital delivering only limited impacts • SOAs as way to circumvent normal oversight and accelerate development • Interagency collaboration as a basis for growth	• Long and uncertain development times leading to conflict over investments and efforts
Mid-1990s to 2015: Emergence of personalized medicine as a socio-economic genomics discursive arena and "product" realm	• Personalized medicine as overarching idea and discourse (also cast as personalized health) • Tailoring for individual patients classified into subpopulations • No longer a "one size fits all" health care system • Better health outcomes • Personal health product research funding across Canada • "Supporting Canadian families" • Importance of identifying social value associated with research and commercialization • Orphan drugs vs. regular drugs • Patient access • Changing de facto notion of what is a "product" in human terms	• Power of the self as patient knowledge advances as social feature • Impetus is social from early NCE and CIHR support, later picked up by Health Canada via Personalized Medicine Working Group • Considerable awareness of key US developments in FDA • Canadians access on line to US products • Harper government seen in popular social terms as well as genomics as core of global bio-economy • Progress in government about the science support but not about the "who pays" issue in Canadian health care fiscal federalism	• Somewhat slower emergence of personalized medicine in central arenas of federal government • Need still to overcome past slow processes of adopting latest health technologies • Inertia in habit nationally of seeing new technologies as just a cost • Slow continuing delays in central government responses to the "who pays" question • Considerable uncertainty as to what new or changed "products" produce what kinds of costs and benefits and what still other products are already in the queue in high volumes

Note: see the chapter text for expansion of abbreviations in this table.

advocating ambitious reform and indeed of ending the "one size fits all" health care system. In its name, adapted as personalized health, the federal government was happy politically to see the development of personalized medicine as a way to support families and to show its role in developing Canada's bio-economy.

Economic and Social Power

In many ways the genomics story reflects the emergence of a new power system, with new actors, new processes, and new institutions largely driving, broadening, and democratizing the discourse and projecting power in new ways. This is not a story of elites defining the outcomes. Instead, the power dynamics in the story of reproductive technologies were undoubtedly mainly forged by women, who pushed for the royal commission study, in part, because they feared the power of the mainly male-dominated medical profession. The women's lobby also saw the issue as mainly a social one and indeed in different ways as involving the power of the self. When legislation on assisted human reproduction later was achieved, provincial power centred in Quebec emerged and led to the Supreme Court's judgment that weakened the Assisted Human Reproduction Act and ended the short existence of Assisted Human Reproduction Canada as a lead federal regulatory agency but still leaving the AHRA with some federal competences whose implementation by Health Canada is still very unclear.

Regarding the Human Genome Project, power initially rested at the international level. It was seen as a socio-economic research achievement but immediately revealed issues about relative power between different health care systems and economies, with the United States being the definitive example of a dominant private health care system and with strong firms interested in commercializing health technologies and products. In Canada, with its public health care system seen more as a sector than an industry, its already changing granting councils such as the CIHR and NCE were very supportive of new grants for genomic and life science research which was flourishing as a new century dawned and an earlier era of federal deficits and S&T budget cuts ended.

Power was widely diffused in the science system in the late 1990s, after a decade of cuts. Although the scientific community was solidly behind efforts to accelerate genomics research, there was no common agreement on how to do that. The confluence of the Liberal government under Chrétien, the rising influence of Industry Canada, and the need to find useful ways to park some of the growing fiscal surpluses led to the development of Genome Canada as a special operating agency, managed largely beyond the authority of executive government and independent of the traditional science power system (embodied in the NRC, the Councils, and intermural research branches in line departments). Nevertheless, in response to the regional strengths in Canada, a hybrid hub-and-spoke institutional design emerged, partly in reflection of the federalist nature of Canada and partly in defiance of the important role of provinces and not regions in the science and policy space.

In the personalized medicine history, Health Canada built on the earlier work of the main granting councils, and the foundation and work of Genome Canada to forge a more organized and visible Personalized Medicine Working Group. The influence of the United States via the FDA's regulatory and policy work was pivotal in Canada as was the growing availability to Canadians of products not yet necessarily approved in Canada. The larger pivotal power structure of the crucial federal-provincial financial system has still not made clear what will be paid for and by whom.

Time, Temporal Realities, and Conflicts

The temporal dynamics were fully on display across the policy and governance histories examined. As with all transformative technologies, the hype often is not met by visible action and impacts. There was a decade-plus slow process in the dynamics of getting a royal commission on reproductive technologies established, its actual work, and the eventual passage of legislation in 2007. But then came the short sharp impacts of the Supreme Court decision and the resulting decision by the Harper government to cancel Assisted Human Reproduction Canada.

These temporal features were also present in the Human Genome Project story overall. It took more than ten years to complete the mapping. There were then early indications of promising products and diagnostics but also shortfalls and inertia in getting these adopted in clear-cut ways at the level of individual patients and families. The Genome Canada story shows more action, with more than twenty-five programs and $2.2 billion of incremental investment, but the promised economic outcomes remain to be realized.

The personalized medicine story showed its somewhat slower emergence as a guiding concept. Compared with the United States, Canada faced lingering obstacles to adopting the latest technologies and institutional habits among payers in seeing new technologies as just a cost. There were also institutional barriers and uncertainties as to what new or changed "products" produce what kinds of costs and benefits and, all the while, the sense that still other products are already in the genomics and life science and disease queue.

CONCLUSIONS

The analysis in this chapter of S&T and innovation policy, governance, and democracy in the genomics, life sciences, and technology domain reveals the extent to which it emerges as an internationally driven science and technology assessment dynamic. From the outset, we have demonstrated how there are product and regulatory overlaps with the biotechnology domain. The analysis has shown that there were socio-economic political positives from the outset in large part because the politics was led domestically by women and then by research and values tied to disease prevention and diagnosis across both established and new broadening conceptions of the life sciences.

Notions of the power of the self and individual patient and patient access and more broadly personalized medicine have also been central. Indeed, these combined features anchor it fully into this book's interpretation of the economic and social innovation nexus. The focus initially is on social and health outcomes, but issues also emerge about economic concerns regarding property rights to new products and diagnostics and also to who pays.

The chapter has also drawn attention to the complex regulatory system on both the science granting side and on the later pre-market and post-market product side. The total governance system, and also its democratic features, involve a truly complex and diverse array of federal, provincial, and international bodies and numerous health, disease, and research professions and knowledge providers, including individual patients, mobilized in many ways by Internet-centred self-diagnosis information, hype, and puzzles as to what to buy and who to see.

Canada's health care system also looms large not only in its potential for positive reform of interest to patients and families, but also with regard to is ability to quickly assess new products and technologies regarding actual use across Canada and how to conceptualize what a "product" is in human terms. It is by no means clear as to who will pay for health care reforms and disease and diagnostic patient care priorities, including regular drugs, orphan drugs, and products that are neither but are growing nationally and internationally in high volumes and uncertain impacts.

We have also seen important features of the Internet in this chapter. We will look in more depth in chapter 11 at our final STI domain cast as the Internet, communications, and social media domain. It is also our third and final domain driven by international technology imperatives that Canada has had no choice but to respond to and seek to gain socio-economic gains from.

11

The Internet, Communications, and Social Media Domain

The Internet, communications, and social media domain is the final S&T and innovation policy domain we examine. As with chapters 9 and 10, this domain is fundamentally centred on a transformational international technology. With regard to the Internet, Canada reacted and then reconfigured its economy and society in a massive way from the early 1990s onwards to take advantage of its transformative benefits.

The *Internet* is often equated with the World Wide Web (WWW) or the Web, but they are different in that the former is the infrastructure while "web pages are just one of the many kinds of traffic that run on the infrastructure" (Naughton 2012, 40). The Internet has without a doubt changed the economy, social interaction, and the nature of democracy both for good and for ill (Dahlgren 2013; Borins 2010; Coleman and Blumler 2009).

This domain raises complex and often intractable issues about who controls or regulates the infrastructure. One view is that the Internet is the last great frontier that is not, or at least should not, be governed by or for any specific interests. In the context of S&T and innovation, this vision offers the prospect for "permissionless innovation," the basic freedom to experiment with new technologies and business models wherever they come from, whoever you are, and wherever and whenever you want (Thierer 2014; Naughton 2012). Goldsmith and Wu (2008), in contrast, write of the illusions of a borderless world and argue that the destiny of the Internet in future decades will reflect the interests of powerful nations and the conflicts within and between them. Indeed, on the basis of the power of the Internet and its key firms and states, others argue for a broader more democratic socio-economic basis of policy, control, and reform (Keen 2015; Fox 2015).

Communications in the domain refers to the terrain occupied by regulated broadcasting, telecommunications, television, cable, radio, and also mass print media. It is increasingly the world of computers, initially large mainframe computers (Dyson 2012; Allan 2001; Ceruzzi 1998), then research computers linked as the initial rationale for the Internet, then personal computers, and now a dizzying array of digital and wireless devices, including smart phones, iPads, and smaller devices that individuals use on an hourly or daily basis socially and economically. In its traditional pre-Internet forms, basic telecommunications were viewed as a "communication" industry of producers aiming product and programming at consumers, and they were regulated by agencies such as the Canadian Radio-television and Telecommunications Commission (CRTC; Doern, Prince, and Schultz 2014, 150–67; Miller 2012; Schultz 2003). However, in the Internet age, consumers and citizens are now as much creators as consumers of information products, benefiting from apparently free communications (van Dijck 2013) and the permissionless innovation ethos of the Internet platform.

We do not examine in any detail in this chapter the traditional telecommunications sector or the evolution of computers. Taking them as given, we focus on social media and the emergence of social production that is now generating as well an "Internet of things." We do, however, discuss the role of the CRTC and pressures on it to regulate the Internet in the context of the convergence of established norms and waves of new communication technologies and content; the related issued of Internet neutrality is also addressed (Hunter, Iacobucci, and Trebilcock 2010).

Social media, the final part of the domain designation, is defined as "Internet sites where people interact freely, sharing and discussing information about each other and their lives, using a multimedia mix of personal words, pictures, videos and audio. At these sites, individuals and groups create and exchange content and engage in person to person conversations. They appear in many forms including blogs and microblogs, forums and message boards, social networks, wikis, virtual worlds, social bookmarking, tagging and news, writing communities, digital storytelling and scrapbooking, and data, content, image and video sharing, podcast portals, and collective intelligence" (Curtis 2013, 1).

Social media overall are best known now in relation to giant firms such as Google, Facebook, and Twitter, which individually and collectively revolutionized the Internet and the Web; the domain also includes groups such as Wikipedia, You Tube, dating sites, comparison-shopping sites, servers (such as Amazon), and literally millions of sites globally, nationally and at the community level that have produced the network society (Castells 2011; Barratt and Shade 2007). The three social media giants and the general business models they have exploited are less than two decades old. Social media are now the main sites for so-called Big Data mining that is producing new correlational patterns and new theories of socio-economic reality and performance (see chapter 3).

The Internet, communications, and social media domain is having a wide range of S&T and innovation effects. We have already seen this in chapters 1 and 3 in terms of the changing nature of peer review and the number, variety, and reach of academic journals, most of which now have an on-line presence. This has also generated the world of e-books and digital publishing, dovetailing with the rise of firms like Amazon, which has a major effect on valuing, positioning, and distributing the world's written knowledge. The domain's centres of power and innovation bring both threats and opportunities to news media. Traditional print media such as newspapers have lost revenue to new modes of communication and been forced to adapt or perish (Lloyd and Toogood 2014; Marland, Glasson, and Small 2015). In some cases, this has generated whole new business models: in the United Kingdom it has transformed the BBC, one of the world's leading taxpayer-funded media institutions, from a broadcasting giant into one of the world's best on-line newspapers and largest media empires. BBC journalists no longer just talk, they also write on-line columns and blogs.

In Canadian institutional and governance terms, this domain is centred in Industry Canada and the CRTC. Industry Canada has expended significant effort to foster the Information Highway, broadband access, the e-economy, and the digital economy and society via spending programs and consultative processes. The CRTC, as mentioned above, is the main regulator of broadcasting and telecommunications (discussed further below). The federal government presence easily extends to a whole-of-government promotion of e-government and now digital government. Other regulatory agencies are periodically important: the Office of the Privacy Commissioner of Canada (OPC) defines rules for federal and commercial use of private information (Doern, Prince, and Schultz 2014), while policing agencies such as the RCMP investigate and pursue criminal offences committed on the Internet (RCMP 2014).

For Industry Canada the Internet is both an opportunity and a challenge. The department (and its predecessor organizations and ministries) had in the main been focused in the past on industries and sectors, on various forms of sectoral industrial policy, and on regional policy.

Beginning in the late 1980s with new trade agreements, Industry Canada's abilities to subsidize industries were sharply reduced and it had to shift to framework-oriented and cross-economy initiatives in response to many kinds of technological changes and other global imperatives (Doern 1996). Promotion of the Internet was one of these changes, but its speed and uncertainty meant policy could never catch up to international technological and market developments. Formal technology assessment processes were not even considered. Playing catch-up was the new normal, and it was hard to even tell when or if the catching had been achieved and, if so, in the name of what values, or what claimed better future.

The CRTC is a key player with greater regulatory power and more telecommunications knowledge than Industry Canada. Nevertheless, Industry Canada contributed to a number of vital regulatory matters, including on competition, consumers, and intellectual property. The department also had regulatory aspirations and under the Harper government consolidated some of them (see below). As the Internet emerged, the CRTC's instinct initially was not to regulate the Internet given its own history of dealing with earlier fast-moving technological changes; this conformed to the prevailing norm of deregulation or self-regulation that pervaded the federal system (Doern, Prince, and Schultz 2014; Schultz 2000; Doern 1997). The CRTC has, however, reviewed its stance on several occasions, as discussed later.

FOUR DOMAIN POLICY AND GOVERNANCE HISTORIES

We explore the domain empirically through four S&T and innovation policy and governance histories in a broadly chronological order, albeit noting that some developments show up in more than one policy story. These histories highlight the fast-moving and often unpredictable trajectory of the Internet and its uses.

The first policy history relates to early forms of recognition and reaction to the Internet as a transformational technology beginning in the 1970s; this period ended by the early 1990s. Next we examine Industry Canada's role in fostering the Canadian Information Highway strategy in 1994–97 and then the broadband access strategy in 2001–02, which included related programs such as the Community Access Program (CAP). We then trace the emergence and role of social media in the 1998–2015 period, focusing on the origins and evolution of Google, Facebook, and Twitter globally and in Canada. We also examine the impact of the Internet on economic developments centred partly on cluster formation and policy, but also on Canada's already quite strong information and communication technology (ICT) industry.

Finally, we examine in the 2008–15 period new policy and governance pressures, shocks, challenges, and tentative responses and debates arising internationally and within Canada from the following: the Edward Snowden whistle-blowing exposure of US spying practices on the Internet; the "right to be forgotten" court decision in Europe; the issue of fast track access to the Internet for some firms in the United States and in Canada and implications for Internet neutrality and innovation; and the release of the Harper government's *Digital Canada 150* strategy. These issues and developments compellingly raise features, battles, overlaps, and contradictions between the global and national regulation of the Internet and social media in relation to concepts of Internet neutrality and individual privacy and protection, but also global competition policy.

1970s to Early 1990s: Recognizing and Reacting to the Internet as a Transformative International Technology

Our first policy and governance history centres on how the Internet emerged and the varied initial responses internationally and in Canada. The first evidence of a focus on this domain can be found in the early 1970s reports by the Science Council of Canada (SCC 1971, 1973) that relate to computers and the computer industry. Several background reports led to a major study by the SCC (1982) on the information society. That report foresaw and examined many of the emerging and still crucial information society issues, including those on technology impacts on labour, health services, regulation, and privacy. The Science Council presciently recommended that "federal and provincial ministers of justice provide strong legal remedies, both civil and criminal for misuse of personal information" (ibid., 62).

The Internet in Canada thus emerged initially in discussions by and with scientists and technical communities, with relatively subdued political interest. To see further its early appearance and manifestations we explore briefly the following: the invention of the Internet in the 1970s, the invention of the World Wide Web in 1992, the formation of CA*net in the late 1980s, the establishment in 1991 of the Canadian Network for the Advancement of Research, Industry and Education (CANARIE), the formation and development of the SchoolNet program, and the creation of Research in Motion (RIM), the company that delivered the iconic Blackberry.

The origins of the Internet can be traced to the period 1966–72 during which the US Department of Defense developed the ARPAnet network. It took more specific shape in 1973 under the "internetworking" project led by Vinton Cerf and Robert Kahn, both engineers who had worked on ARPAnet. As Naughton (2012, 45–6) points out, the design problem was "how do you create a network that will seamlessly link other networks, and be, in some sense, future proof." Drawing on their assessment of the weaknesses of telephone networks as being both owned and tightly regulated, Cerf and Kahn's view was that two principles were pivotal for the Internet: (1) no central control and (2) the network should not be optimized for any particular application, now more commonly known as Internet neutrality. Cerf and Kahn developed the protocols that allow data packet switching between different kinds of machines, the foundation of the system. The Internet was "switched on" in January 1983 when ARPAnet migrated to the new protocol.

Indeed, "four ground rules" were critical to Kahn's early thinking (Leiner et al. 2014, 3):

1 Each distinct network would have to stand on its own and no internal changes could be required to any such network to connect it to the Internet.
2 Communications would be on a best-effort basis. If a packet didn't make it to the final destination, it would shortly be retransmitted from the source.
3 Black boxes would be used to connect the networks; these would later be called gateways and routers. There would be no information retained by the gateways about the individual flows of packets passing through them, thereby keeping them simple and avoiding complicated adaptation and recovery from various failure modes.
4 There would be no global control at the operations level.

The "end to end" principle is inherent in these kinds of ground rules. Naughton (2012, 50) sums up the Cerf and Kahn view by saying that the Internet essentially does only one thing: "It took data packets in at one end and did its best to deliver them to their destinations. But it was entirely indifferent as to what was in the packets; it didn't care whether they were fragments of an email message, a digitalized voice conversation, a pornographic video, or a music track: if you had an idea for a new application that could be achieved using data packets, then the network would do it for you, with no questions asked."

The Internet (later in combination with the Web) is a communications technology that fosters innovation without the need for permission by authorities. Indeed, the Internet itself and particularly the Web were themselves created without any authority's explicit permission.

The development of the World Wide Web was led by and essentially the creation of Tim Berners-Lee, a young British computer scientist working at CERN, the Geneva-based high-energy particle physics laboratory (Naughton 2012). Because of CERN's array of permanent and visiting scientists speaking many languages, Berners-Lee concluded early on that the crucial imperative at CERN and on the larger Internet was the need for documentation systems by computer software that allowed documents to be stored and later retrieved (Berners-Lee 1999; Pew Research Center 2014). Such a system would have to be decentralized, span the entire Internet, and need to use *hypertext* documents that had internal links to other documents.

More precisely, to create the Web, Berners-Lee had to (1) invent a way that gave every single Web page a unique identifier and machine-readable address, the uniform resource locator (URL); (2) design a technical protocol for interconnection, called the Hypertext Transport Protocol (HTTP) or the now familiar http://; (3) create software so that people could *browse and edit* Web pages and a *server* program that would enable a networked computer to "serve up" Web pages on demand; and (4) build a standard language for marking up Web pages so that they would be rendered consistently by any browser, which is now called the HyperText Markup Language (HTML; Naughton 2012, 54–5). The Web was launched not by some CERN announcement but by simply releasing its key features on to the Internet on 1 August 1991. As Naughton concludes, "This transformation was triggered by a single individual, working mostly alone on an unofficial project within a large organization which was, for the most part, blissfully unaware of what he was up to. He was able to trigger the avalanche without having to ask anyone's permission. Tim Berners-Lee had a great idea that could be implemented using data packets. He released those packets onto the Net and the network did what it was designed to do" (59). In 1994 Berners-Lee founded the World Wide Web Consortium as "a place for stakeholders to reach consensus around the specifications and guidelines to ensure that the Web works for everyone and that it evolves in a responsible manner" (WWW Foundation 2014).

It is obvious now that "the Web has changed the world ... it has become the most powerful communication medium the world has ever known ... it has changed the way we teach and learn, buy and sell, inform and are informed, agree and disagree, share and collaborate" (ibid., 3). We discuss its impacts in the policy and governance history of the social media that has since developed. Suffice it to say at this stage that the Web has enabled both permissionless innovations such as Wikipedia and WikiLeaks, for good or ill, and literally millions of other websites and communication channels allowing individuals, families, and groups to become producers of information rather than just consumers.

By design and intent, this structure is perpetually disruptive, as both benign interactions and planned subversive efforts can destabilize. Social movements and NGOs have interconnected in ways that challenge industry, governments, and societies everywhere to respond. Meanwhile, the absence of gatekeepers has enabled those with more negative goals to introduce a wave of destructive malware, including computer viruses and spam, and to engage in cybercrime, cyberwarfare, and cyberterrorism (Schneier 2015).

The CA*net and CANARIE story ultimately covers a period from the mid-1980s to the early 1990s. The Canadian Network for the Advancement of Research, Industry and Education emerged at an April 1991 conference sponsored by Industry Canada's predecessor department, Industry, Science and Technology Canada (ISTC) in response to its explicit S&T mandate to establish plans for creating a high-speed network. CANARIE had roots in CA*net, established in the late 1980s, which in turn grew out of NetNorth, established in 1984 as an initial network and set of protocols for the academic community (Shade 1994). CA*net, supported partly by NRC funding, became the backbone network through which the majority of Canadians could access the Internet. Its user community was mainly in universities but with "a secondary user base composed of community colleges, government and industry groups, commercial enterprises and growing number of Kindergarten to Grade 12 (K-12) schools and freenets, such as the ones sprouting up in Ottawa, Victoria and Vancouver" (ibid., 2).

By September 1993 the Internet "consisted of over 1.7 million computers and 15,000 nets (6000 + foreign) in over 130 countries ... and reache[d] between 15 and 30 million people" (Shade 1994, 1). Canadians who were not academics "encountered the Internet for the first time by a variety of means ... [but] the common and distinguishing feature of each of these methods was their limited availability, high true cost, and, in some instances, severe service shortages" (Tumin 2000, 6).

The perceived need for CANARIE and for high-speed connectivity was strongly influenced by US developments, including its then new National Research and Education Network (NREN), which coordinated the networking activities of key federal agencies such as the National Science Foundation (NSF). In Canada the federal government was the first and remains the main funder of CANARIE, although it has leveraged a number of public-private initiatives. CANARIE, operating as a federal non-profit corporation, is viewed as a critical infrastructure, and it has received significant financial infusions in the Mulroney, Chrétien-Martin, and Harper eras. CANARIE has "deployed one of the world's largest and fastest networks dedicated to cutting-edge research, education and innovation. Over 19,000 km of fibre-optic cable link Canada's most curious and creative minds and those of colleagues across the country and around the world. CANARIE additionally manages a range of funding programs that advance digital infrastructure in Canada" (CANARIE 2014, 1).

SchoolNet emerged in 1993, at the end of the Mulroney era, as Industry Canada's first national Internet-related initiative; it was led by Doug Hull, at the time Industry Canada's director general of science promotion and academic affairs (Tumin 2000). Initially, the effort began when the federal government transferred older, no longer needed government computers to Canada's K-12 schools. Connectivity emerged as an issue and SchoolNet was the response. It began with a pilot scheme linking schools in Ottawa to the Internet in October 1993 "using satellite links between a school in Newfoundland and one in Niagara" (ibid., 8).

Since education in Canada is clearly under provincial jurisdiction, the key to further progress was the development of a variety of federal-provincial and public-private partnerships (P3s), initially launched through a National SchoolNet Advisory Board. The private sector was also crucial; technology firms viewed SchoolNet as an avenue for building demand and creating new current and future users. By June 1994 SchoolNet had connected 3,000 schools, ten times its original target; a new target was 16,500 schools by 1999. A linked First Nations SchoolNet was also launched in 1996 (Indian Affairs and Northern Development Canada 2009).

SchoolNet began in the context of severe federal deficit-driven budget cuts, but its popularity among voters and ministers eventually yielded budgetary support. SchoolNet became the best known example of Industry Canada's soon-to-emerge Information Highway strategy and was the anchor for an extended Community Access Program in rural and urban communities (see more below).

Last but not least it is worth mentioning the creation of Research in Motion and its eventual development of the iconic electronic organizer, the Blackberry (Sweeny 2009; Gillette, Brady, and Winter 2013). In 1984 Mike Lazaridis, an engineering student at Canada's then-new University of Waterloo, joined with Douglas Fregin, an engineering student at the University of Windsor, to form RIM as an electronics and computer science consulting company. As in the Internet examples above, key individuals mattered. The early RIM leaders tinkered away on wireless technology, including point-of-sale terminals, wireless modems, and pagers, albeit with only modest commercial success. Jim Balsillie joined the company in 1992 and quickly became co-CEO with Lazaridis of RIM as it became a public company.

By the late 1990s Blackberry became a breakthrough technology that allowed workers to send and receive emails while away from the office. It was a simple technology and above all it was secure. From the outset RIM focused its sales and marketing on corporate customers rather than general consumers. For high-level and middle-level political and business professionals, then reliant on one-way communication pagers, the Blackberry became popular, useful, and even addictive (Sweeny 2009). The US government became Blackberry's single biggest user, while major corporations became avid subscribers.

By 2007 Blackberry grew from 2,000 to 12,000 employees and turned the city of Waterloo "into Canada's Silicon Valley" (Gillette et al. 2013, 6) with over 1,000 tech companies in the region. But 2007 was also the year in which the first Apple iPhone came on the market aimed at the consumer market. It presaged an eventual sharp decline for RIM. By 2011–12 RIM's profits and Blackberry sales globally plummeted, as the world of iPhones and iPads took on and outcompeted computers and email-related communication devices such as the Blackberry (Silcoff 2014; Yakabuski 2013). By 2014 the company had implemented a recovery plan and introduced new products that seemed to be stabilizing its market position, but technology analysts continued to debate its long-term future.

1994–2002: Developing a Canadian Strategy for the Information Highway and for High-Speed Broadband Access

The first comprehensive federal effort to respond to the Internet emerged in the 1994 Speech from the Throne, when the newly elected Chrétien government committed to develop and implement a Canadian strategy for the Information Highway, the term used to describe the Internet by then US Vice President Al Gore. The new government's first industry minister, John Manley, was put in charge of the initiative, and he began a three-year process that included major multidepartmental involvement within the government but, even more pivotal, an external Information

Highway Advisory Council (IHAC). The need for concerted thought and action was deemed necessary because other competitor countries were already ahead of Canada in the development of policies and institutional capacities (Turk 1998).

The internal processes involved key senior public servants but with active leadership by Manley, then a novice minister. He had to "weigh the associated political risks of bringing together competing interests to form an advisory council as opposed to using commissioned studies, small expert panels or other lower risk public policy mechanisms" (Turk 1998, 6). His decision to bring in David Johnston to lead IHAC supplied both expertise and socio-economic scope to the task ahead. Johnston had just stepped down as principal of McGill University after fifteen years in office, and he had been chairman of the Association of Colleges and Universities of Canada (AUCC) and chair of the relatively new National Round Table on the Environment and the Economy (NRTEE). Johnston pressed hard for a broad-based advisory council that would include representatives of the telecommunications and broadcasting industry (seen as at the apex of the Information Highway industry) as well as the Internet community, public interest and consumer groups, educators, librarians, labour unions, and the cultural community (Johnston 1996). Telemedicine and multimedia would also find a place on the council.

An initial government guidance discussion paper (Industry Canada 1994b) offered a set of objectives and principles for the council. The goals were to create jobs through innovation and investment in Canada, reinforce Canadian sovereignty and cultural identity, and ensure universal access at reasonable cost. The main operating principles were (ibid.):

- An interconnected and interoperable network of networks
- Collaborative private and public sector development
- Competition in facilities, products, and services
- Privacy protection and network security
- Lifelong learning as a key design element (added later).

The paper set out fifteen issues that were parsed out to the five working groups set up to conduct discussions. The topics included competitiveness, Canadian content and culture, learning and training, access, social impacts, privacy and security, copyright, intellectual property, and especially competition between telephone and cable companies.

There were, not surprisingly, many contentious battles among council members and within the various working groups. Ministers from relevant departments often attended council meetings or contacted Manley with their concerns. Despite these internal and external differences, some consensus was reached, or at least finessed through the language and discourse in the report, and the Information Highway Advisory Council reported its findings to the government and Canadians on 25 September 1995 (IHAC 1995).

The IHAC report had the same range and breadth of ideas, conclusions, and recommendations as the earlier discussion paper and a similar range of specific content. Criticism of the report and its processes centred on concerns about its content regarding sovereignty and cultural identity, which though part of past broadcasting and telecommunications policy, seemed out of step with the larger themes and uncertainties of the Internet and the importance of a competitive market-based approach. Indeed, somewhat complicating the discussion, the CRTC had been holding its own consultation, and it released a report shortly afterward that focused on the links and conflicts between competition and culture inherent in Canada's transition to an Information Highway age (CRTC 1994).

The Chrétien government's formal response came in May 1996. Presented as an action plan by Industry Canada (1996), it was cast in the language of an information society and a knowledge economy. The government response included discussion on building Canada's Information Highway, growing Canadian content, realizing the social and economic benefits for all Canadians, and getting government right. Among the specific measures were developing a "standards road map" to ensure compatibility with global systems, a promised Task Force on Digitalization, the introduction of privacy legislation to protect personal information, and a Student Connections Program to hire students to introduce and train over fifty thousand small businesses to use the Internet.

The IHAC report and government response both addressed the Community Access Program, an initiative that emerged out of the example and success of the Industry Canada SchoolNet program. The Community Access Program was an explicit response to concerns about equality of access, primarily in rural communities but also in some urban centres. The logic behind this priority ranking was that urban communities could more easily solve their own problems of access, but rural communities could not (Tumin 2000). The plan was that schools would become the CAP sites. However, as was the case with SchoolNet, this overlapped with provincial and local government jurisdiction, with the result that Industry Canada would need provincial and local allies and partners.

The initial lead partner was New Brunswick, a front-runner in the SchoolNet program. Premier Frank McKenna was an active supporter with strong cooperative instincts about overcoming the jurisdictional issues. A Canada–New Brunswick pilot project was quickly agreed to with five original sites for CAP identified. Problems arose over the rural-first policy at the beginning, but eventually urban communities were selected. Gagetown, New Brunswick, was the first CAP site to open. Doug Hull, who led the SchoolNet and CAP initiatives for Industry Canada, summed up what he heard locally from the Gagetown mayor, who said (quoted in Tumin 2000, 14), "When trade was by river, we used to be right on the main route. Then came the real highway, which bypassed this town, and we've been languishing for the last 100 years. Now with this program, the information highway is here and we don't mean to be left behind. We're tired of our kids leaving town. We want them to stay here. Until now there has been no prospect for development here, because the only highway has been far away. The Internet is changing all that." Word quickly spread about the New Brunswick pilot projects, and communities in several provinces lined up for CAP access. In December 1994 Industry Minister John Manley launched the national CAP program at Picton, Ontario, to emphasize its rural focus. In late 1997 the government issued a broad paper entitled *Connecting Canadians* that emphasized the importance of Internet connectivity for all Canadians (Industry Canada 1997). When federal budgets became more robust in 1998, an urban CAP program was formally launched.

It warrants emphasizing that these overall strategies and the hundreds of related "announcements" were extremely popular with ministers and officials in a period of mainly financial cuts and deficit reduction imperatives. The Information Highway process and programs became one of Industry Canada's most politically feasible, framework-oriented, cross-industry policy realms that was both popular and visible in a social context (Doern 1996). But Industry Canada also housed the telecommunications sectoral branch acquired after the demise of the former Department of Communications in 1993. This branch was anxious to show that it ought to be the lead source of policy advice and technical expertise within the government on telecommunications and Information Highway matters rather than the CRTC (Schultz 2003). The latter, of course, was the well-established and independent regulatory body.

The second major Internet-focused strategy came after the 2000 election when the Chrétien Liberals promised to solve the so-called digital divide by committing to making high-speed broadband access available to all Canadian communities by 2004. A National Broadband Task Force was established in 2001 as part of the Liberals' innovation economy approach. It had broad stakeholder representation and involvement but the power of the key communications firms in the cable and telecommunications sector was noticeable in its final recommendations (Schultz 2003, 194–5). The rhetorical flourish of the effort was captured in the report's title, *The New National Dream: Networking the Nation for Broadband Access* (National Broadband Task Force 2001). Schultz opined that one result of this "dancing around the digital divide" is that "few public reports have received the near universal condemnation that greeted the publication of the Task Force Report" (2003, 195).

The Task Force had "radically extended its mandate" beyond communities to businesses and residences, and its bias in favour of existing service providers seemed to give short shrift to the Task Force mandate to make recommendations that would be technologically neutral and would enhance competition (Schultz 2003, 196–7). The report's quick political demise was at least also partly due to the Liberal Party leadership battle to succeed Jean Chrétien, where Industry Minister Brian Tobin and Finance Minister Paul Martin were two of the leading candidates.

Following these Task Force–oriented formative initiatives, Industry Canada under the Liberals launched further Internet policy studies and strategies, but with the Information Highway moniker permanently supplanted first by "e-commerce" and then by "e-economy" discourse (OPC 2004). The department's Electronic Commerce Branch was renamed the Digital Policy Branch. One of the early initiatives in the Harper era was to convene in October 2007 the Canada Roundtable on the Future of the Internet Economy (Waverman 2007; Geist 2007; MacLean 2007). As we see below, the issue of Internet neutrality emerged, which generated new pressures for regulation.

1998–2015: Social Media and Economic and Social Innovation – Google, Facebook, and Twitter and Canadian ICT Impacts and Development

The definition of social media at the beginning of this chapter captures its breadth, its countless big and small sites, and affirms that its history predates the era of Google, Facebook, and Twitter. In many ways, the Internet and the Web were social platforms from the beginning, but quickly they became sites for, and drivers of, economic development globally and in Canada. We flesh out the Canadian ICT impacts and developments later in this section, but first we focus on the core notions of connectivity and connectedness that are central to this development. Indeed, the concept of networks took on a much more encompassing meaning as a socio-economic organizational platform in contrast to, and compared with, traditional hierarchies and markets (van Dijck 2013; Doern, Prince, and Schultz 2014; Benkler 2006). Businesses themselves, willingly or unwillingly, had to adapt to take advantage of these new fast-developing worlds of economic and social production and communication (Kietzmann et al. 2011; Edosomwan et al. 2011). At the same time, it is important to appreciate the history and evolution of Google, Facebook, and Twitter as dominant exemplars of social media and social and economic innovation. We look briefly at each in turn and then comment further on Canadian ICT, Internet, and social media in the context of economic and social development.

Google, the world's pre-eminent Internet search engine, began in 1996 as a doctoral research project by two Stanford University students, Larry Page and Sergey Brin, as they focused on the mathematical properties of the Web and the potential features of a searchable digital library (Auletta 2009; Stross 2008; Battelle 2005). Its domain website was formed in 1997, and Google was incorporated in 1998 with an index of about sixty million pages, a small beginning to what was to rapidly become a Big Data colossus. Its business model was to sell keyword advertising. Mid-way through the second decade of the twenty-first century, Google is by far the dominant search engine on the Web, with 68 percent of the searches done annually worldwide. It was also, as a result, beginning to attract considerable competitive attention from other communication giants, including Microsoft and Amazon (Streitfeld 2014).

When information was needed, the phrase "Google it" became arguably the best indicator of the company's central presence in everyday life, socially, academically for many students, and economically (Auletta 2009). This is equally true for scholarly research, where Google Scholar™ now is the go-to search engine for scholars looking to find published and working papers, partly because it is easily and freely accessible (other databases are behind pay-walls) and partly because it encompasses a larger share of the published and grey area literature, including both paper and on-line journals, books, and studies (other databases only sample the literature, focusing on the more established journals).

Google's profitability and global reach brought it into new realms of research, such as new driverless cars, corporate acquisitions of Internet social media firms such as YouTube, and, ultimately, philanthropy. It has also been targeted as a global company for not paying its fair share of taxes in some of the countries in which it is located and operating. Google Maps, one of its stable of products, were seen as both intriguing and inventive but also invasive, as local streets and individual homes are revealed close up in streetscape pictures freely available on the Web (Usborne 2014). Google, it was argued, "is now out to blur the lines between the digital world and the physical one … [including] trying to automate and simplify your daily interactions" (Akkad 2013, 1). In short, Google is a core contributor to the "Internet of Things" (Gershenfeld and Vasseur 2014). Its huge profits, technological research, staff, and knowledge base have turned Google into a huge R&D organization. At one level, this looks benign, but the algorithms underlying the company's search engine unconsciously accentuate our tendency to seek and find information that confirms our biases, which amplifies our differences and prejudices. The Internet for many people has replaced the conventional mass-media sources of information, which historically did much to create social norms and to mobilize citizens.

If Google is broadly economic and informational, Facebook is unambiguously a social media product. Created by Mark Zuckerberg and three fellow students at Harvard University, Facebook started in February 2004 as a social networking service (including the posting of photos of family members and friends). It began as a service limited to Harvard University students, extended to other Ivy League universities, and then to universities elsewhere in the United States and in Canada; ultimately it was opened up so anyone could use it (Kirkpatrick 2010; Phillips 2007). In 2014, 1.35 billion people used Facebook on a regular basis, almost one billion daily, and Facebook hosted over 100 billion photos. Members, families, and friends could, of course, still stay in touch via regular email, but Facebook expanded the personal basis of the human connection in many ways, including, as many now experience, on-line "face time."

Facebook attracted investment capital early, but it did not become profitable as quickly as had Google, in large part because advertising, though crucial to the Facebook business model, was and is not as easy to develop and sustain on this medium (Carlson 2011; Kirkpatrick 2010, chapter 13). It is now solidly profitable due to ads shown in the news feeds of Facebook users checking the service on their mobile phones. Mark Zuckerberg has said that the company will be spending "heavily for years on newer services like private messaging, virtual reality and Facebook search without any near-term prospects of making money from them" (quoted in Goel 2014). In some respects at least, this sounds like a strategy to make Facebook more Google-like.

As with Google, Facebook has new corporate research and market aspirations, such as using high-altitude, solar-powered drones as well as satellites and lasers to increase and deliver Internet access in the developing world (Vincent 2014). In Britain, Facebook and other social media firms have been invited by the Speaker of the House of Commons to examine how to enhance digital democracy, including on-line voting (Grice 2013). But it has not all been smooth sailing. As Facebook hit its global stride in the past five years, it was unexpectedly involved in a controversy related to job market scrutiny of potential employees on Facebook. Specialist firms searched out such information for client companies by looking at various social media sources, especially looking for embarrassing, explicit, or irresponsible photos and videos posted on Facebook and other sources; many job candidates found their on-line persona scuppered their employment prospects (Preston 2011). A report by Canada's Privacy Commissioner (OPC 2012) investigated concerns related to the "Facebook generation" of children and youth in a more basic and general way and cautioned: "All of that online communication creates a permanent record – and that could carry risks to their privacy and to their reputations. Not just today but perhaps even more in the future. Teenagers are expected to make mistakes – it's a natural part of growing up. The fact that electronic records of many of the mistakes of today's youth will persist for decades to come is cause for deep concern."

Twitter is the third and most recently established social media giant. This on-line service, established in March 2006, allows registered users to send and read very short messages, or "tweets," of 140-character (maximum) text messages. Twitter was created by Jack Dorsey, an undergraduate student at New York University, and three colleagues of an existing small company, Odeo (Bradshaw 2013; Carlson 2011). It was designed as a micro-blogging information service, with the name "twitter" chosen based on the dictionary definition of a short burst of "inconsequential information." Twitter grew rapidly and by 2014 had 284 million users who were sending over 500 million tweets daily, the majority from mobile devices. Users, as "followers," subscribe to other users' tweets (Johnson 2009). Unlike with the active Facebook community, many users are simply followers. Well over 40 percent of user accounts have not tweeted – these users simply watch (Murphy 2014). As a business, Twitter quickly attracted investors and capital, but it has proven difficult to generate hoped-for advertising revenue from the system (Kuchler 2014).

The notion that tweets consist of inconsequential information is in some sense true, though, in other senses, its patterns of communication have various kinds of impacts. Many governments and their businesses routinely communicate via Twitter, as do many interest groups and charities. In war zones and emergency situations, tweets can become emergency communications systems and can work to create political and social meaning (Castells 2012). Celebrity is now partly defined by a media star's number of Twitter followers. Some individuals seek publicity and even notoriety through their tweets. Others have been charged with criminal offences based on one or more tweets, and some have been sued for libel or damages based on the evidence supplied by tweets. For many non-users, the reaction to hearing about tweets is as often as not to wish that the "twitterati" would simply get a life. Nevertheless, Twitter appears to be here to stay, and for many it is as addictive in its own way as the Blackberry for its users and supporters.

Although the Internet and the big three social media giants are US-centred international technologies, Canadians have without doubt become active, willing, and diverse users of the Internet and social media. Recent surveys have shown Canada ranked either second or third behind the United States in Internet and related use of social media (Krashinsky 2014). But a recent study also points globally to what has been referred to as a growing lack of trust in the Internet (Dutton et al. 2013), and there have been growing calls for its reform (Keen 2015; Fox 2015).

Social media developments had immediate social and economic impacts in Canada both through Canadians' enthusiastic use of sites such as Google, Facebook, and Twitter and by the formation of tens of thousands of Canadian sites of social media communication, production, and involvement. The developments had immediate major economic and commercial impacts, including presenting overlapping boundary challenges between the established ICT industry centred on telecommunications firms and the closely linked and competing social media. Canada has, in many ways, a world-class ICT telecommunications industry and infrastructure including an export-oriented ICT and software industry centred on Ottawa, encompassing firms such as Nortel, Corel, Mitel, JDS Uniphase, and regional office operations of major global ICT companies. But Canada's telecommunications utility sector of telephone and cable companies, due to its oligopolistic and monopolistic features, imposed on Canadians the highest costs for mobile telephone use of any OECD country.

Recent assessments of Canada's system of industrial clusters showed that "service clusters such as business services, finance, ICT services, and creative and cultural industries tend to be located in the largest urban areas and are experiencing high levels of growth" (Spencer 2014, 2). A broadened categorization was needed because of boundary overlaps, although ICT cluster realms per se were identified in Toronto, Kitchener-Waterloo, and Vancouver. Other recent commentary has highlighted Toronto as an area of national and international focus of start-up companies involved in the emerging "Internet of things," centred on wearable technology but linked to the fashion industry (Cosco 2015). Canada and Canadians are also impacted by what some are calling a "post-TV" world (Strangelove 2015), with a fast-changing world of firms and competition offering direct on-line programing, services, and innovation, exemplified by Netflix versus traditional habits of TV watching. The CRTC, as we will learn below, has been challenged to figure out what to regulate and in whose interests (Anders 2014).

2008–2015: Internet Neutrality and Privacy – National and Global Governance Pressures, Shocks, Challenges, and Responses

Our final policy and governance history examines recent national and global governance challenges and responses to issues concerning Internet neutrality and privacy policy. We look in turn at the following: Canadian debates and pressures regarding regulation of the Internet as a market; the Internet versus spy agency impacts of Edward Snowden, the US National Security Agency (NSA) employee who released classified documents in 2013; the decision by the European Court on the "right to be forgotten" of users of social media sites such as Google; the decision by the US Federal Communications Commission (FCC) to open hearings on non-neutral access to the Internet and its decision to confer on the Internet the legal designation of a regulated public utility as in other telecommunications fields; and Canada's 2014 *Digital Canada 150* strategy, including its renewed concerns about federal privacy law and accountability.

Canadian debates and pressures regarding Internet regulation versus markets increased markedly in the context of the massive use and growth of social media. Political, economic, social, and policy pressures created a maelstrom of activity. CRTC and Industry Canada offered competing views about how to assess new technology developments, including involving arm's-length regulators, line departments, and ministers. The debate highlighted the conflict between the highly regulated, established telecommunications oligopolies and the usually small, new market entrants. This battle increasingly played out in the Harper era as firms contested what Internet neutrality means (Barratt and Shade 2007; Geist 2007).

Several examples illustrate the domestic debates and shifts at play. One that predates the period was the CRTC's quite sensible conclusion in 1994 not to regulate the then-new Internet. Significantly, it opted for a radical reorientation of the respective roles of regulation and competition, with the latter not simply one of the objectives but henceforth the primary lens through which the CRTC would fulfill its responsibilities. The CRTC (1994) justified its new approach on the grounds that the telecommunications environment was changing "in ways that outpace the ability of regulators to recognize and define, let alone control." As a result the Commission concluded that regulation should concentrate on those services where monopoly still prevailed and should strive to open all telecommunications to competition, including the local telephone market, the last remnant of monopoly (ibid.).

Concerns about the nature of telecommunications policy and the CRTC role increased, however, to the point where the Martin Liberal government, when pressed by Bell Canada, established a review panel. As with many of these ventures, the report only came out after the government changed. The report of the Telecommunications Policy Review Panel (2006) was sharply critical of the CRTC's "balancing approach," which was interpreted as pursuing the goal of promoting interests of new entrants over the incumbent firms. The Panel concluded (ibid.): "Application of the doctrine has resulted in a new, high level of regulatory intervention aimed at shaping the structure of markets, rather than allowing market forces to determine the success or failure of different service providers. The relative degree of intervention by the CRTC on behalf of the new entrants has been very substantial and has led to the imposition of extensive constraints by the CRTC on the activities of the major suppliers of many telecommunications services (the incumbent firms)." The Panel not only recommended that this approach be terminated but that "market forces should be relied upon to the maximum extent feasible as the means of achieving Canada's telecommunications policy objectives" and that the role of regulation should consequently be significantly reduced (ibid.).

The new Conservative minister of industry, Maxime Bernier, seized upon the report by the Telecommunications Policy Review Panel and was able to reverse the CRTC's decisions and to fundamentally recast federal telecommunications policy (Barrett and Shade 2007; Schultz 2008). The federal government issued a policy directive to the CRTC that did not simply reorder the existing statutory objectives but recast them in a major way. Drawing on the Review Panel's recommendations, the CRTC was ordered in its decision making "to rely on market forces to the maximum extent possible as the means of achieving the telecommunications policy objectives" and concomitantly to reduce regulation as much as possible so as not to interfere with the operation of market forces (quoted in Schultz 2008, 159).

Although the directive and its use to justify overturning a specific CRTC decision was criticized as unprecedented "political" decision making, Bernier and the government were unwilling to back down. The government had established, for the first time since the creation of the CRTC and after a number of unsuccessful attempts by governments of different political stripes to do so, that political authorities would henceforth control policy-making in telecommunications and that the CRTC's role would be circumscribed. In other ways, however, Janisch (2012) argues the relationship between Cabinet and the CRTC on policy matters was becoming less clear, particularly in the context of regulating the Internet. One particular conflict involved the 2011 CRTC decision to impose a *usage-based pricing* scheme for Internet access (CRTC 2011a). It was quickly rescinded by the CRTC under intense ministerial and government pressure (CRTC 2011b), which Janisch (2012, 47) describes as "a complete disregard for the legal regime."

The issue here centred both on what is meant by Internet neutrality and market forces. For the core oligopolistic telecommunications firms, Internet neutrality and market forces meant competition among themselves. For the Harper government Internet neutrality and market forces increasingly meant encouraging new entrants into the market, including via the aggressive use of new spectrum auctions. Auctions previously had been technical issues but now were political events, with licensing decisions coming more through Cabinet direction than arm's-length CRTC decisions (Schultz 2011).

For other players and commentators, there were still other interests at play and other policy boundaries to be redrawn or crossed. Geist (2007), for example, argues that the network neutrality issue that was emerging in the United States (see more below) was also now present in Canada and that, indeed, "Canada is arguably the leading source for illustrations of how Internet service providers (ISPs) may prioritize or block content and applications when it was in their interests to do so" (3). Still others point to the ways in which the Internet is not the only technology involved. The development of so-called deep packet inspection (DPI) technologies, which allow ISPs "to monitor the content of data packets in real time and make decisions about how to handle them" has profound implications for Internet governance (Bendrath and Mueller 2011,

1142). DPI implementation in Canada and the United States has led to "public protests, litigation and major regulatory proceedings ... in which network neutrality norms were used to challenge DPI deployments" (Mueller and Asghari 2012, 462). At another level of analysis entirely, Hunter, Iacobucci, and Trebilcock (2010) examined Canadian content policies vis-à-vis the Internet in an era they refer to as "an abundance of technologies," where technological convergence makes it difficult to predict and regulate. Caution is their watchword; they posit the need for greater focus on relevant and workable governance forms and policy instruments.

Events, however, are moving at a speed that is scarcely manageable. Recent CRTC processes and decisions exhibit this quite starkly. At its 2014–15 "let's talk TV" public hearings, Netflix, the US streaming and on-line production company, attended one of the hearings. The CRTC tried to get Netflix to provide data on its Canadian users, but the company flatly refused on the grounds that the CRTC had no jurisdiction. Undoubtedly, Netflix is popular with Canadians. As of October 2014, 29 percent of anglophone respondents to a user survey reported they subscribe to Netflix, up 40 percent in just two years. Unable to beat Netflix, Canadian utilities such as Bell announced plans to introduce their own streaming services – Bell introduced Crave TV, but the kicker was a user had to be a cable subscriber to access the content (Berry 2014).

Early in 2015, following public hearings, the CRTC reaffirmed its position that it would continue to regulate Canadian television, including on-line services, in an effort to foster true innovation (CRTC 2015; Csanady 2015). The CRTC committed to promoting Internet neutrality, measured by how content is received; in short, it did not want new services from the cable and telecommunications companies where Canadians had to subscribe to get them (Dobby 2015).

Edward Snowden's whistle-blowing in 2013 of details of the scale of global surveillance by the US National Security Agency has raised any number of issues about privacy and Internet neutrality. Snowden was enabled by his own access to complex computer and Internet data as a NSA contract employee. His revelations about the extent of US surveillance raised questions about the activities of spy agencies in general, including those in the United Kingdom and Canada. Countries such as Brazil and Germany responded by proposing "to encourage regional online traffic to be routed locally rather than through the US ... [as] the first steps in a fundamental shift in the way the Internet works" (Taylor, Hopkins, and Kiss 2013, 3).

One strong, almost universal response from the Internet servers, content providers, and technology companies is that NSA surveillance is an assault on the trust of their own users (Miller 2013). As a result many have begun to encrypt email search queries and other information flows. In essence, the response is to make spy agency snooping infinitely more difficult (Hodson 2013; *Economist* 2013c).

The principle of the "right to be forgotten" is loosely related to the other controversies. It was a central feature of a 2012 European Commission plan announced as a part of its effort to modernize its 1995 Data Protection Directive. The European legislation regarding the 2012 plan has yet to be approved and would need the approval of all twenty-eight EU member countries; however, it is being pushed along by a May 2014 ruling by the European Union Court of Justice that Google "must amend some search results at the request of ordinary people in a test of the so-called 'right to be forgotten'" (BBC News 2014). A Spanish man had brought the case to the court to have a long-past auction notice for his repossessed home removed from Google, alleging that it was a violation of his privacy. The initial Google response was to reiterate that the search engine does not control data – it simply offers links to information freely available on the Internet. The Court of Justice ruled that the rights of the individual are paramount "although there was a public interest defence when it comes to people in public life" (BBC News 2014).

The aftermath of the court decision produced a vast array of responses, including some that linked the EU court decision with the visceral European reactions to the Snowden revelations (one salient discourse was based on the fact that the US had spied on German Chancellor Angela Merkel's telephone calls). The response from Google was initially to remove some of the data as requested (since requests were coming in immediately), but then also revoking some of these decisions when adverse reactions occurred on the grounds of freedom speech and public interest. The UK's information commissioner reacted to Google's "avoidance of responsibility" by comparing it to the polluter-pay principle. Google began a series of meetings with EU member country data protection bodies in 2014 and the final outcome remains unclear (Lee 2014).

Many reactions focused on the extent to which the European court "gave no guidance on how their ruling was to be carried out ... and gave no indication that they recognized the Pandora's box their ruling might open" (Scott 2014a, 1). Others drew attention to the popularity and dominance of Google and other Internet giants, stressing that "85 percent of market share for search in [Europe's] five largest economies ...

compared to Google's share of the American market [which] stands roughly at 65 percent" (Scott 2014b, 1).

The modes of discourse regarding possible directions and solutions were diverse. Google CEO Larry Page promised cooperation and consultation but stressed that the court ruling would damage the next generation of Internet start-up companies in Europe (Waters 2014). In essence, he paired the Internet and Google as complementary sources of competition. Germany and France, drawing upon traditional competition law and policy, argued that Google's excessive market share in the European Union restrained competition and as a result should be broken up (Fontanella-Khan, Barker, and Vasagar, 2014). Early in 2015 the European Competition Commission announced that it would enquire into Google's search engine services and strategies based on complaints that they favoured Google's own shopping services (Waters 2015). At some deep level, all the combatants had some considerable belief in Internet neutrality and thus de facto qualified notions of privacy rights. But there were also concerns about competitive advantage.

There have been to date no formal responses by Canadian authorities such as the CRTC to the "right to be forgotten" principle or to how it might be adopted or implemented formally or informally in a North American context.

In April 2014 the US Federal Communications Commission began a discussion process to consider whether there should be *fast lanes* for Web traffic, in effect, ending the concept of net neutrality. There had been considerable pressure since 2010 and even earlier from firms arguing that they should be able to pay service providers for "special, faster lanes to send video and other content to their customers" (Wyatt 2014a). Firms seeking such special access included Disney, Google, and Netflix. Earlier, a US Federal Court of Appeal had struck down for a second time FCC rules intended to guarantee a free and open Internet. Thus, the FCC needed to devise new rules regarding net neutrality that would meet court tests. But it was fully aware of serious consumer opposition as well as opposition from new start-up businesses that wanted an open Internet, the kind that Google and other giants had when they were starting.

The Federal Communications Commission by a vote of 3–2 supported a proposal for an open public debate on "different approaches to accomplish the same goal, an open Internet" (Wyatt 2014b). Left vague, however, is the "connection between an Internet service provider, which sells a connection to consumers, and the operators of backbone transport networks that connect various parts of the Internet's central plumbing

... [which] means that as long as an Internet service provider ... does not slow the service that a consumer buys, the provider can give faster service to a company that pays to get its content to consumers unimpeded" (ibid.). As noted earlier, this was already a feature offered by some Canadian Internet service providers.

Tim Wu, a Canadian-born lawyer and strong advocate of net neutrality, a professor, and a practising lawyer in the United States, sees net neutrality as an idea that is overwhelmingly strong and supported and indeed stronger than any particular rules that may be in place or planned (Sommer 2014). But there are more occasions now, as we have already seen, where what net neutrality means in practice is being debated and contested.

Essentially, at stake is whether a new approach could involve the designation of the Internet as a public utility, or a common carrier, analogous to other telecommunications, broadcasting, and transportation utilities, such as rail lines. By early 2015 the US Federal Communications Commission was giving strong public indications that it would opt for a strong "open Internet order" to ensure that no content is blocked and that the Internet is not divided into pay-to-play fast lanes (Lohr 2015). In March 2015 the FCC announced that it would regulate broadband Internet providers as a public utility, a decision that "sets the stage for extended legal fights" (Ruiz 2015, 1). While Canada has been watching the US debate, similar pressures have been slower to develop in Canada (Miller 2012) but which now, as indicated above, have led to a stronger CRTC position.

The Harper government's plan, *Digital Canada 150* (Industry Canada 2014a), is the most recent, relatively comprehensive update and extension of Internet-related policy geared to the content and discourse of the digital age. Industry Minister James Moore was the main advocate and architect of this plan within the Cabinet with the "150" in the title linked to Canada's 150th anniversary as a country in 2017 (Geist 2014a). The strategy is anchored around five pillars: connecting Canadians, protecting Canadians, economic opportunities, digital government, and Canadian content. The discussion of each pillar included some new initiatives, but the plan devoted much of its space to listing and mentioning almost every federal policy since 2006 that could be said with hindsight to be digitally related (ibid.). Nonetheless, the strategy document is instructive, offering a vision of what Digital Canada aspires to look like and how its regulation and governance is changing and may yet need to change. Several items are highlighted in Table 11.1.

More than most past Internet policies and strategies, *Digital Canada 150* speaks openly about regulation. Shortly afterwards, Industry Canada announced new legislation: Bill S-4 (the Digital Privacy Act) and Bill C-13, on cyberbullying. Geist's (2014b) initial assessment of the bills is that they are positive and needed in some key respects, but they leave huge gaps that may encourage firms to engage in new anti-privacy practices. Geist argues that within Bill S-4 there is "a provision that threatens to massively expand warrantless disclosure of personal information," while Bill C-13 "creates an incentive for companies to voluntarily disclose to law enforcement by granting them full immunity from any civil or criminal liability for doing so" (ibid.). Although this is already done thousands of times each year, the concern among privacy advocates is that the immunity provisions will make this practice ubiquitous.

Privacy regulation, it is worth reiterating, predates the Internet and the Web. The Office of the Privacy Commissioner of Canada was created in 1983 under the Privacy Act (OPC 2012). The Commissioner acts as an independent parliamentary officer to advocate for the privacy rights of Canadians related to the federal public sector and the private sector operating within the federal jurisdiction. The diverse and changing views about privacy in a liberal political society, combined with rising concern about a growing surveillance society, make it difficult if not potentially impossible to define and to deal with abuses effectively, efficiently, and equitably (Schneier 2015; Etzioni 2012; Center for Digital Democracy 2012).

In response to on-line behavioural advertising, the Privacy Commissioner expressed concerns about the massive growth in "the practice of tracking a consumer's on-line activities in order to deliver advertising geared to that consumer's inferred interests" (OPC 2012). The Commissioner pointed to the regulatory challenges and gaps, arguing that "what it means in practice is that Internet ad networks follow you around online, watching what you do so that they can serve you targeted ads… [and] we have specifically pointed out that organizations engaged in online behavioural advertising should avoid tracking children – or tracking on websites aimed at children – since meaningful consent may be difficult to obtain" (ibid., 3).

A related crucial privacy regulation debate in 2012 followed federal Public Safety Minister Vic Toews' introduction and defence of the Harper government's lawful access legislation, Bill C-10, which was given the title of Protecting Children from Online Predators Act. The Bill would allow police without a warrant from the courts to access basic personal information about Internet users as a way to track and catch persons who prey on children. The minister initially bombastically labelled any critics of his Bill as being on the side of the child predators, a bellicose discourse

Table 11.1 Sample statements and features of the *Digital Canada 150* plan

CONNECTING CANADIANS

- An effective digital policy is one that connects Canadians through high-speed Internet access and the latest wireless technologies.
- [The government] will enhance broadband service to a target speed of 5 mega bits per second for up to an additional 280,000 Canadian households, which expands coverage to 98 per cent.
- [The government will] work with CRTC to unbundle television channels and ensure that cable and satellite providers offer Canadians the option to pick and choose the combination of television channels they want.
- [The government will] make broadband and connectivity projects eligible for federal support under the Infrastructure component of Building Canada Funds.

PROTECTING CANADIANS

- Canadians will be protected from on-line threats and misuse of digital technology.
- [The government] will strengthen the Personal Information Protection and Electronic Documents Act to better protect the on-line privacy of all Canadians.
- [The government] will introduce comprehensive cyberbullying legislation to protect Canadians from invasion of privacy, intimidation, and personal abuse including the non-consensual distribution of intimate images.

ECONOMIC OPPORTUNITIES

- Canadians will have the skills and opportunities necessary to succeed in an interconnected global economy.
- Business Development Bank of Canada will allocate additional $200 million to support small and medium-sized businesses with digital technology adoption; and additional $300 million in venture capital for companies in the information and communication technologies sector.
- Support for the Computers for Schools Program will continue.
- Institute for Quantum Computing (at University of Waterloo) will carry out and commercialize leading-edge research in quantum technologies.

DIGITAL GOVERNMENT

- Government of Canada will demonstrate leadership in the use of digital technologies and open data.
- [The government] will develop Open Science to facilitate open access to the publications and related data resulting from federally funded research.
- Creation of an Open Data Institute will encourage the use of open data to raise productivity and create new products and services to benefit Canadians.
- [The government] will continue to support and stimulate the app economy and create home-grown open data developer ecosystem in Canada.

CANADIAN CONTENT

- Providing easy on-line access to Canadian content will allow us to celebrate our history, arts, and culture and share it with the world.
- Our partnership with Historica Canada will create two new Heritage Minutes – short films on key events in Canadian history – every year from now until 2017.
- [The government will] continue to support the Canada Book Fund and the Canada Music Fund with permanent funding to create new digital content and enhance the visibility of Canadian content on digital platforms.
- [The government] will provide continued support for Virtual Museum of Canada and the Online Works of Reference in the new Canadian Museum of History to share Canada's stories and treasures on line.

Source: Compiled from *Digital Canada 150* (Industry Canada 2014a), 7–22.

that generated a fierce counterattack from a number of civil liberties advocates, including normally Conservative media such as the *National Post* (Selley 2012). Some commentators judged the counterattack was itself over the top because some of the contested provisions in the legislation had been in a similar 2004 Martin Liberal government bill that drew little attention but was later withdrawn because of similar attacks (Coyne 2012). The Privacy Commissioner also waded into this issue, writing Toews in October 2011 that "my provincial and territorial privacy colleagues … have called upon the federal government in 2009 and 2011 to take a cautious approach to legislative proposals to create an expanded surveillance regime that would have serious repercussions for privacy rights" (OPC 2011, 1).

The dispute about the appropriate balance between privacy rights and public security is unlikely to go away. As the Internet becomes more ubiquitous, the purely private space is shrinking, forcing governments everywhere to consider where to draw the dividing line and how to govern incursions across that line by both individuals and organizations of all stripes.

THE THREE ELEMENTS

We now track with the aid of Table 11.2 the three elements being utilized to understand change and inertia across the four domain policy and governance histories presented in this chapter.

Policy Ideas, Discourse, and Agendas

The Internet is as much a social movement as it is a technology. A number of subversive and disruptive transformative ideas, discourse, and agendas have been driven by the Internet, communications, and social media domain since Day 1. As we have seen, the ideas of no central control and Internet neutrality were embedded in the creation of the Internet as a network of networks. For the World Wide Web the exchange, storage, and retrieval of data packets was necessarily decentralized, spanning the entire Internet; the invention of hypertext protocols allowed universal interoperability. Permissionless innovation was both present in Internet design but also the goal of many in developing the system. Gradually, the concept joined the lexicon of discourse. High-speed connectivity, schoolnets, and community access featured in both discourse and early initiatives, as both means and ends.

Table 11.2 Policy and governance histories in the Internet, communications, and social media domain: Three analytical elements

Policy and governance history	*Policy ideas, discourse, and agendas*	*Economic and social power*	*Time, temporal realities, and conflicts*
1970s–Early 1990s: Recognizing and reacting to the Internet as a transformative international technology	• No central control and Internet neutrality • "End to end" principle • Permissionless innovation • World Wide Web • High-speed connectivity • SchoolNet • Community Access Program • Internet malware	• US-based and driven technology imperative • Role of individuals as permissionless innovators, such as Cerf and Kahn on the Internet and Berners-Lee on the World Wide Web • Influence of lab sites such as CERN • Power and influence of creators of websites such as Wikipedia and WikiLeaks • Influence and harm of varied malware creators/practitioners • Industry Canada (and former ISTC) with S&T mandates create and support CANARIE as research network • National SchoolNet Advisory Board • Early work of RIM and Lazaridis, Fregin, and later Balsillie, leading to Blackberry	• Presented as long-term technology infrastructure that should be technically managed, but not controlled • Informational innovations, good and bad and in-between, should be assessed if necessary, later • Slow or slowish national responses in Canada and elsewhere • Need to construct and invent new Internet-informed policy discourse and also periodic updates of the discourse
1994–2002: Developing a Canadian strategy for the Information Highway and for broadband access	• Information Highway • Broadband access and digital divide • Interconnected and interoperable network of networks • Create jobs • Reinforce Canadian sovereignty and cultural identity • Universal access at reasonable cost • Privacy protection and network security • Equality of access for rural and urban communities • Connecting Canadians • e-Commerce • e-Economy	• Active role and support of Liberal Industry Minister John Manley • Information Highway Advisory Council led by David Johnston • Battles/differences among IHAC members, and industry and social sectors and relevant federal ministers • Influence and energy of communities anxious to be connected (including schools, mayors, and provincial premiers) • Telecom firms as players, but initially on the defensive • Federal support with modest funding, given federal fiscal deficit • Broadband access review and strategy of 2000–02 sharply criticized, in part by co-optation of its agenda by telecom oligopoly	• Need to develop strategy but vaguely re the medium term • Telecom firms being more on the defensive compared with 1960s–1980s • e-Commerce and e-economy eras foreseen, but not yet e-society • But some notions of information society projected in 1982 Science Council report

1998–2015: Social media and economic and social innovation – Google, Facebook, and Twitter, and Canadian ICT impacts and development	• Social media • Social production and communication • Searchable digital library • Big Data • Google Maps • Internet of things • Digital democracy • Inconsequential information • Blogging and micro-blogging • Face Time • "Twitterati" • Cyberbullying	• US-centred and US-driven era of creation of Google, • Facebook, and Twitter as sites of social production and economic innovation • Individual ideas and innovation by Page and Brin, Zuckerberg, and Dorsey • Strong support by Canadians of social media • Combined strength of earlier ICT industry, but increasing development of broader service sectors with IT content in larger cities in Canada • Good and adverse impacts on democracy globally and in war zones and emergencies • Growing lack of trust in the Internet	• Massive speed and volume of data leading to Big Data designation and short- and long-term challenges and concerns • New fast but also long-term presence of social production • Intergenerational impacts on children • Google and other Web giants as growing R&D firms looking at launching next generation of products and processes
2008–15: Internet neutrality and privacy – national and global governance pressures, shocks, challenges, and responses	• CRTC shift to greater competition and market-driven ideas in areas where no oligopolistic firms • Competition seen by telecom oligopolies as competition among themselves • Competition by Harper government seen as coming from new entrants • Internet neutrality challenged in Canada as to its practical meaning re fast lanes for Web traffic • Surveillance by US and other spy agencies • Privacy and freedom of speech and communication • EU "right to be forgotten" principle • Protecting Canadians • Canadian digital content • Cyberbullying	• Shift of power from CRTC to Industry Canada under the Harper government • Power of Canadian ISPs to create own multitrack service speeds and concepts despite net neutrality • Snowden as whistle-blower on US NSA spying • Spy agency uses and abuses of Internet and social media sites • European court decision linked to effort to reform EU privacy law • More overt clash between US and EU • Web giant efforts to protect themselves from spy agencies • Key US firms press for paid fast track access to Internet • Industry Canada push for *Digital Canada 150* strategy, includes privacy law reforms and need for systemwide digital government	• Current and potential long-term threats to Internet neutrality • Fear of but also more support for complex Internet regulation and impacts on privacy and security • Fast pace of development of new social media–centred firms, sites, and producers of social content

Metaphorical references to the Information Highway were common in Canada and elsewhere as overriding descriptors of early formal agendas and policy strategies; so, too, was broadband access. More technical discourse on the Internet and the Web as an interoperable network of networks was common in public debate. Privacy protection and network security were seen as key public values and potential areas where legislation would or might be needed. Equality of access for rural and urban users and communities was a social norm from the earliest stages. When its full potential was visualized economically, ideas and discourse such as e-commerce and e-economy were deployed.

As the domain morphed into its more explicit social media growth phase, ideas such as social production emerged, and descriptors such a Big Data were more common. Diverse kinds of discourse emerged concerning blogging and Twitter messaging and growing harms such as cyberbullying. More recently, in the flush of Internet and social media realities, ideas and concerns about privacy, Internet spying, surveillance, and the European Union's "right to be forgotten" concept join the lexicon in juxtaposition with the long-standing right to freedom of speech and the communication of ideas. Issues and ideas about excessive Internet power also emerge, defined in the context of Google's ubiquitous role in defining what we know and in the debate about net neutrality. Some called for utility-style regulatory approaches to bring more "control" of the Internet. The notion of a digital economy and society and government has become the language of federal Internet strategies.

Economic and Social Power

The structure and nature of economic and social power in this domain originates above all in the United States, the source of most of the technology and commerce over our fifty-year period as a whole. Canada accepted the technology and the need to develop with it, albeit after a slow start.

A key feature of economic and social power is the role of key individuals in visualizing and designing the Internet and the Web. New products, services, technologies, and firms were developed out of profound individual energy and invention. We see this in the history of the Internet, the Web, key social media firms, and also with regard to the early dynamics of RIM and the Blackberry in Canada.

Within the Canadian government, Industry Canada at both the ministerial and senior public service levels and the CRTC, the independent telecommunications regulator, worked closely with the emerging Internet companies and the main duopoly clientele of telephone and cable utilities and the Canadian broadcasting industry. Industry Canada had significant influence, establishing CANARIE and SchoolNet and designing the initial Information Highway and broadband access strategies that mobilized but also revealed diverse views among social and economic industrial interests, including dominant telecommunications firms. Schools, mayors, and their communities were key in making the early Community Access and connectivity programs work.

The American dominance in the social media field has been a major challenge. Pre-Internet broadcasting and telecommunications were unambiguously one-way distributions of content, often focused to realize a cultural policy objective; the Internet and social media era has transformed the world into a complex network of two-way, global interactive exchanges, which have changed power and its related democratic structure both profoundly and permanently.

While there has been little formal control of the Internet, in the ways that Internet and Web creators envisaged it, there have always been concerns about Internet neutrality, privacy, and surveillance of both appropriate and illicit uses of the Internet by individuals, firms, criminals, and spy agencies. This may ultimately change the power, governance, and democratic architecture of the Internet as the global giants such as Google respond to pressures to do more to govern themselves. The diverging approaches in the United States and Europe may simply spur this on.

Nevertheless, the Internet is not above the law or public policy. In many ways it has evolved in directions defined by law or encouraged by public policy. Ultimately, the Internet as an economic and social space has always been subject to the application of criminal and civil laws regardless of whether offences are revealed through and on the Internet or any other technology.

Time, Temporal Realities, and Conflicts

Both the Internet and the Web were forged as consummate long-term technological infrastructure that should be technically managed to keep it as an interoperable network of networks not controlled by any one entity. In short it should stretch and function across generations as a kind of guarantee of perpetual innovation and evolution. Informational innovations, good, bad, and in-between, were to be assessed later, with what "later" meant left to future generations to determine. Initially, national policy responses to the Internet were slow and tentative. Canada and other countries had to construct and invent new short- and medium-term Internet-informed policy discourse. This took time. In the mid-1990s, as we have seen, there were medium-term notions of a coming and imminent e-business presence, and an e-economy, but strong awareness of e-society only emerged later.

As social media developed, and in particular as Google emerged, a pivotal temporal feature has been the massive scale and speed of change. This has us on the threshold of a Big Data era. This deluge of data creates both short- and long-term challenges, some of which we have discussed in earlier chapters and domain analyses. Numerous forms of burgeoning social production are creating intergenerational impacts due to the divergent information literacy and capacities of the young and the old.

Some of the recent shocks to the Internet, communications, and social media domain, such as increased surveillance by spy agencies and domestic social intrusions such as cyberbullying, generated short-term crises that focused attention. Exactly how states and societies should respond in the medium and long term remains vexing, given the considerable uncertainty as to what solutions, if any, might work in complex networked democracies and global systems.

CONCLUSIONS

The Internet, communications, and social media domain is, without doubt, the most transformational domain in S&T and innovation policy and governance affecting the average Canadian and the Canadian economy. This largely US-based technological imperative has offered Canada (and most countries) little opportunity to assess it systematically. It was a technological imperative and juggernaut from the outset.

The four policy and governance histories reveal important phases and trajectories over our fifty-year period. From the 1970s to early 1990s governments in Canada and everywhere were challenged to figure out if and how to respond to the accelerating developments in the domain. The early work of the Science Council of Canada was important but is little recognized these days. The key here, internationally, was the pivotal role of individuals and the way they forged with little or no government engagement the key principles of no central control and net neutrality. The second history was more Canadian-focused, examining the initial development in the mid-1990s of an Information Highway strategy and then the strategy on broadband access, both anchored in Industry Canada as the lead S&T ministerial policy department. Throughout this period, the CRTC was a strong competitive presence vis-à-vis Industry Canada, functioning as an independent regulator that could use public decision processes and its technical expertise in the telecommunications industry to frame the debate.

The domain story then necessarily and crucially shifted to the impact of social media and the emergence of Google, Facebook, and Twitter as exemplars of social production and socio-economic change, first in the United States and then in Canada and elsewhere. These technologies cross borders with impunity; none of the crossings required or involved Canadian permission. Our final policy and governance history, centred in recent years, examined new national and global governance pressures, shocks, challenges, debates, and tentative responses in Canada to issues raised by Internet neutrality and adverse social impacts. The Snowden whistle-blowing affair, the "right to be forgotten" court decision in Europe, and the issue of fast track access to the Internet for some firms in the United States all raise serious and not necessarily resolvable issues about the modern meaning of Internet neutrality in the face of new definitions and forms of privacy and security globally. Nationally, Canada's *Digital Canada 150* strategy is our specific most recent attempt to respond.

Communications is the middle name of this domain because of the pivotal prior and current changing role of the Canadian telecommunications industry, both as it has been historically forged and as it has been affected by technological change (including but not only involving the Internet). We have emphasized the ways in which the domain's social production impacts have changed life.

The innovation economy and society nexus themes and evidence are arguably most in evidence in this domain. More than the other domains examined, the innovation society writ large is a pivotal driver and indeed is often an overall macro-descriptor and discourse. The Internet and the Web as technologies were crafted with broad social rationales (i.e., no control and network neutrality) and early policy initiatives in Canada (e.g., the Information Highway, SchoolNet, and the Community Access Program) conformed to that basic thrust. The emergence of Google, Facebook, and Twitter, but also millions of other sites and businesses, engaged and expanded social production in a massive way. The communications industry oligopoly, which had Canadian content and cultural values central to its regulation, continue to produce and direct product to paying consumers, but now in a world of multi-directional economic and social production that is well beyond its control. Policies about regulating and governing the Internet, social media, and communications are increasingly entwined in social concerns about cyberbullying, privacy, and social revulsion against the excesses of state spying nationally and internationally. Nevertheless, across the entire development trajectory, this domain was also undoubtedly defining the new economy and its changing lexicon of e-business, e-commerce, networked science, and the Big Data economy.

The concept of permissionless innovation as a feature of the Internet and social media on the Web has emerged in this chapter, but we wait until our conclusions in chapter 12 to comment further on this notion of innovation. There we will add it to the discussion of theories of innovation examined earlier and the impacts of Big Data and the Internet on the nature of research, peer review, and publication through new on-line research journals. We will contrast the impact of the Internet juggernaut with the practice of technology assessment as revealed by the analysis of the biotechnology, genomics, and life sciences domains, offering insights to future challenges and responses.

PART THREE

Conclusions

12

Canadian STI Policy: The Innovation Economy and Society Nexus

Canadian science and technology and innovation (STI) policy has been examined in any number of ways and with different focal points. What the STI literature needed for academics and students, STI practitioners, and Canadians interested in this crucial policy field was a book that told the STI policy story in greater breadth and depth historically and in the current era. The first major contribution of this book is that it provides by far the most extensive and in-depth account and explanation of Canada's changing STI policy and related governance system. This has been achieved through an analysis of eight STI policy domains in Canada covering a fifty-year period across the four prime ministerial eras from Trudeau to Harper and through an examination of thirty-one policy histories. All of this integrated empirical coverage has been set in the context of important international forces and imperatives and intricate realms of public and private governance and political, economic, and social power and agenda setting.

Our second, closely related major contribution centres on the development in this book of an enhanced understanding of the *innovation economy and society nexus* in operation across this fifty-year story. We have examined within and across the eight domains whether an S&T-centred innovation economy and society has been established in Canada and whether the STI framework adequately captures the way in which STI policy is being put into practice in Canada over the past half-century. As we have stressed, an innovation economy and society is one that aspires to and achieves some kind of moving and interacting balance between STI directed at commercial, private, market objectives and STI deployed to achieve social objectives, including delivering public goods and supporting values related to redistribution, fairness, and community and citizen empowerment, often judged and based in part on the extent of democratic engagement in governance and decision making.

In the course of this needed and different complex analytical journey, we have provided new critical analyses about related features of the story such as the following: significantly changed concepts of peer review and the emergence of Big Data in the digital age and Internet information economy and society; different ways in which federal-provincial STI policies have impacted on both levels of government; and numerous realms of networked technology across Canada in universities, business, and government, including biotechnology, genomics, and the Internet, but also earlier technologies.

DOMAIN ANALYSIS

We offer our final summary conclusions and observations in two stages: an overall comparative look within and across the eight STI policy domains with the aid of the three analytical elements deployed; and then through an interwoven discussion of our main arguments, we offer our final considered answers to the three main research questions posed in the introduction of this book:

1 How, why, and to what extent has Canadian S&T and innovation policy, governance, and democracy changed in the past fifty years?
2 Where has S&T and innovation policy resided in Canadian federal, political, and policy agendas across the past fifty years?
3 How have concepts of science-based governance prevailed or been changed across the past several decades?

The innovation economy and society nexus helps answer and inquire into all three of these questions, and thus we also look across the eight domains in our final concluding view of the innovation economy and society nexus and why it is essential.

Comparing Eight Domains

Our mapping, analysis, and comparison of the eight S&T and innovation policy domains reveals a varied and complex picture and diverse ranges of S&T and innovation policy content, change, inertia, scope, and coverage. A comparison of the domains has been central to the empirical story of the chapters in Part II. We have seen that while each is different in important respects, they are also linked.

The *macro S&T and innovation policy domain* determines the federal government's systemwide and system-defining S&T policy statements, rules, and budgetary decisions regarding this field of policy. The emergence of offices or ministers of science are a part of this domain, but typically they have not had much room to manoeuvre in the face of extended prime ministerial and central agency power. As a domain, it includes inputs and constraints flowing from the following: macro S&T policy statements and strategies; budgets, budget processes, and annual S&T policy and decisions; regulatory policy, processes, and S&T and related science activities; policies and guidance regarding science advice and the role of scientists; and periodic public service renewal review processes. The domain analysis has shown that S&T and innovation policy overall has been consistently on the national agenda, but rarely at the top of the list of priorities.

The *government S&T departments and agencies domain* encompasses a large number of departments and agencies such as Health Canada, Environment Canada, and Industry Canada, and particular agencies within their purview, such as the National Research Council of Canada. Federal S&T and innovation policies are directed in various ways at these and many other departments and agencies. We have seen, however, that they are immediately and necessarily conjoined with, and interpreted through, the imperatives of departmental mandates, laws, regulations, their S&T strategies, and their established cultures. They are further influenced and reshaped in relation to the inherent number and content of the statutes that they are responsible for, on their own or in concert with other departments. Government departments and agencies are influenced by different volumes and types of projects, products, and services, which are a part of their programmatic and regulatory tasks and cycles of operation. Their remit is broad, spanning pre-market and post-market review and management, and through the coordination of their mandates within the broader socio-economic context.

The *granting, universities, and levered-money domain* is centred on the three current independent federal granting councils – the Natural Sciences and Engineering Research Council of Canada (NSERC), the Canadian Institutes of Health Research (CIHR), and the Social Sciences and Humanities Research Council of Canada (SSHRC) – that provide funds to approximately 140 eligible institutions, primarily university academics and researchers in hospitals. We have seen that these granting agencies are increasingly quasi-regulatory bodies in that they set the rules regarding research ethics, the structures of peer review and merit review, the mixes of disciplinary and multidisciplinary research, and the peer-review selection committees. They also define and create funding competitions that are increasingly thematic, priority areas as opposed to basic, curiosity-driven research. The growing levered-money aspects of the domain are also regulatory in nature in that they specify or require in different ways that, in order to get money, applicants have to bring both network partners to defined projects, requiring money from university, business, or provinces as partners. Overall, in governance and political terms, the domain involves concepts regarding what research independence means vis-à-vis executive government, business, and society, and when scientists are researchers but also advocates – and funding lobbyists themselves – either overtly or functioning behind the scenes. The domain dynamics suggest also that in some respects the federal government was partly ceding STI content and influence to the provinces, given how the provinces had their own single integrated granting bodies, and also how provincial network involvement emerged on CIHR funding on the CIHR Strategy for Patient Oriented Research. But otherwise the CIHR was mainly a federal dynamic rather than one led by the provinces per se. The Canada Foundation for Innovation (CFI) was partly a provincially driven initiative regarding research infrastructure, but it was key university presidents rather than the provinces that pushed for the CFI, and the big powerful universities such as the University of British Columbia, the University of Toronto, and McGill University were dominant recipients of CFI grants. We have seen as well in this domain analysis and elsewhere that the provinces, despite some recent growth in R&D spending, are still very much the secondary player in this crucial feature of the overall STI system.

The *industrial S&T and innovation domain* analysis has shown a complex trajectory of policy and change, often without S&T policy being explicitly mentioned in any central way. This was typically the case in the dirigiste and industrial strategy policy history both in the sense of the historic National Policy, but also when industrial policy and industry departments were created after the Second World War and into the late 1970s. Changed macro-economic and micro-economic policies in different ways worked to create a climate for innovation for Canadian industry. These policies generally were focused on micro-economic fundamentals, expanding trade, and exploiting the benefits of globalization. This included efforts to privatize and deregulate sectors to foster an ethos of competitiveness that would translate into a greater export-oriented Canadian manufacturing industry. The domain story is complemented by more explicit business S&T and innovation tax and tax credit incentives such as those featured in the analysis of the Scientific Research and Experimental Development Tax Credit. We have shown how a partially separate policy track defined by clusters and local or city-regional innovation systems has helped to frame and focus policy. These more recent domain features are joined by frequent and recurring accounts about how Canadian S&T and innovation policy for business has continuously underperformed in comparison with other countries. In short, it represents continuous policy failure in all four prime ministerial eras examined.

The *intellectual property, invention, and innovation domain* represents an important framework policy that has come to symbolize the emergence and importance of science, technology, and innovation in a world of increasingly knowledge-based economies. In contrast to the other framework policy areas, where benign neglect and collateral change is often the norm, intervention in the intellectual property world is purposeful in seeking to support and protect the rights of inventors and creators, especially via patents, and also in promoting the dissemination of such inventions so that still further innovation occurs. Federal laws and policies underpin this domain, but to a great extent policy regarding intellectual property is not controlled by the federal government. Rather, other national governments have used international trade negotiations and industries, and firms have used litigation to expand the scope of private property, both by the length of protection and by the nature of the claims. Instead of pushing for this, Canadian governments, federal and provincial, have spent most of their time seeking to preserve some balance between public and private interests and to decide how to optimize the socio-economic benefits of these complex systems.

The *agriculture, food, biosciences, and biotechnology domain* story is set in the fundamental reality that agriculture and food occupy a unique place in the hearts and minds of people. As the first and until relatively recently the most important social and economic activity for mankind, agriculture and food gets special place in the science, technology, and innovation space for Canada. Canada's global position in markets was anchored on large, exportable surpluses of grains that generated large trade surpluses. In that context, Canadian science, technology, industrial, and innovation policies were firmly anchored on the land. The three policy and governance histories examined showed in different ways how the agriculture, food, biosciences, and biotechnology domain is in many ways a crucible for a number key issues triggered by innovation that divide society. Transformative change in the sector due to new technological development undercuts the autonomy of many primary producers, as the technologies only work with integrated, intensive farming systems. The presence of new products of biotechnology also challenges both national and consumer sovereignty, as the global economic rules have been set to facilitate (some say promote) the proactive development and diffusion of productivity-enhancing innovations. Even if one can erect barriers to isolate a product category, sector, or region from the new technology, the competitive pressures ultimately undercut any short-term benefits of exclusion. This domain shows that technological change is seldom equitable – by definition new innovations create winners and losers. At a deeper level, the advent of biotechnology in the global agri-food system has caused governments, producers, and citizens to question our relationship with nature itself. The domain analysis illustrates an early era of federal S&T policy leadership through the canola story, but also its later decline in the name of concepts such as the new public management and the increased appreciation for networked and internationalized science and public-private partnerships.

The *genomics, life sciences, and technology domain* is related to, but goes well beyond, the issues and concepts found in the agricultural, food, biosciences, and biotechnology domain. Of all the sciences and technologies, genomics is arguably the most global and potentially most transformative in terms of the industrial economy and society. No country is self-sufficient in the science used to develop new products and processes or in the technologies needed to undertake assessments, and the products of these technologies require large markets to be commercially viable. This is a particular challenge as much of the knowledge embedded in these new products is protected and exploited under proprietary regimes (protected by patents, trademarks, trade secrets, and various commercial and contractual mechanisms), and nation states are focused on evaluating them almost exclusively in terms of their impact locally. This system privileges some actors and forms of evidence at the expense of others, which becomes a challenge once a product enters the market, as governments and consumers in Canada and globally have high expectations that they will be able to make their own choices about accessing and using the resulting foods, drugs, and other products. The main findings from our policy and governance histories are that while the technology enticed governments everywhere to engage, if for no other reason from fear of losing out, their existing policy and program management models were often not up to the task. As a result, in spite of massive effort, some policies, such as rules governing reproductive technologies and personalized medicine, simply struggled or failed to gain traction, although there has been perpetual tinkering with the tools of government to handle the new properties of this technology. The dynamics of federal-provincial power and interdependence are also present in this domain, in part, because ultimately who will pay, and how much, for myriad new testing and other products becoming available under personalized medicine concepts are decisions in the main of provincial health authorities.

The *Internet, communications, and social media domain* has emerged as the most transformational technological opportunity that has an effect on the everyday lives of Canadians. The scale, scope, and speed of the introduction of ICT-based new technologies, products, processes, and organization systems has overwhelmed the capacity of most governments, including Canada, to conduct formal technology assessment in any meaningful way. In contrast to the intellectual property and genomics domains, where large multinational firms were critical actors, the key here internationally was the pivotal role of individuals, as they acted alone or in small groups early in the development of ICTs to forge the key Internet principles of no central control and Internet neutrality. Canadian-focused responses centred on the initial development in the mid-1990s to 2002 period of an Information Highway strategy and then later a strategy on broadband access, both of which were anchored in Industry Canada as the lead S&T ministerial policy department. Throughout this period, the Canadian Radio-television and Telecommunications Commission was a strong competitive presence vis-à-vis Industry Canada, functioning both as an independent regulator with access to the public through its decision processes and as a source of expertise through its long-standing regulatory engagement with the oligopolistic telecommunications industry.

The social media aspects of this domain centre on key US firms such as Google, Facebook, and Twitter as exemplars of social production and socio-economic change, first in the United States and immediately in border-crossing floods of social impacts on innovation in Canada and elsewhere. None of these border-crossing and defining impacts required or depended on Canadian government permission. In recent years, however, new national and global governance pressures and shocks have generated debates in Canada about Internet neutrality in the face of massive social media growth and new concerns about adverse social impacts regarding privacy and national security. Our policy histories emphasized the ways in which the domain's social production impacts have changed the lives of Canadians and their families, so much so that they now directly compete as producers with the largely one-way communication processes embodied in the telecommunications industries regulated as public utilities.

The Three Elements in Analyzing the Domains

Analysis in the chapters in Part II was conducted in terms of our three domain elements: (1) policy ideas, discourse, and agendas; (2) economic and social power; and (3) time, temporal realities, and conflicts. Both inertia and change were revealed through these elements seen across the five decades of coverage and through the thirty-one illustrative policy and governance histories examined across the eight domains. But they also help us see the innovation economy and society nexus. Our contextual review of prime ministerial eras and agendas in chapter 2 and of Canada-US and international pressures and developments in chapter 3 showed related aspects of the conceptual and empirical story. These international ideas and institutions include science as a social system itself, peer review as a social construct for the science community, and the combined emergence and massive growth of both Big Data and grey literature. We also found that some policy, governance, and democratic dynamics show up and are best understood via reference to more than one element.

With regard to *policy ideas, discourse, and agendas*, our journey across the domains shows an array of ideas that, once articulated, rarely leave the policy scene. From the outset, "policy for science" and "science in policy" were core concepts expressed, complementing the underlying pursuit of improved industrial R&D. Macro S&T ideas also periodically embraced the principles of science advice. The National Research Council of Canada has always embodied the notion of science for nation building; in spite of the recent contested changes in its mandate and management, the idea still resonates. Ideas and discourse shifted in the 1990s to include notions of micro-economic endogenous growth, but also often considered in the context of local or regional innovation systems and clusters, which triggered discussion of often simplistic market-pull versus technology-push pathways.

Quite sweeping extensions of policy purpose involved moving away from just pre-market product safety to post-market monitoring in a full life-cycle context and also to broad notions of sustainable development and precaution. In the granting and funding process, new ideas moved the focus from curiosity-driven, uni-disciplinary research to thematic, priority-driven, interdisciplinary research. Peer-review goals and -processes were augmented by merit review, which involved a new set of metrics that rewarded priority-driven knowledge translation and commercialization efforts. New and expanded intellectual property rights and practices focused effort on a new set of opportunities. The concept and imperative of complexity entered the system. Virtual institutes and networks for funding and organization became the norm as agendas and discourse shifted. In the industrial S&T domain, policies in the dirigiste era mattered for technology, but they were more collateral impacts rather than intended, stated S&T ones. Later when clusters and local-city innovation systems emerged as the main policy discourse, innovation became viewed as more local and sectoral in nature. The same is true in the agriculture, food, biosciences, and biotechnology domain where federal and provincial research support, much of it in Western Canada, was pivotal. The domain was challenged, however, when the Canadian and global focus on biotechnology shifted to the question of genetically modified food products, where opinions about the benefits and risks associated with the technology varied widely. The emergence of genomics simply amplified the sense of urgency and salience of science. The Internet's creators helped generate ideas that challenged the traditions of government and the market, promoting and entrenching the concepts of no central control and Internet neutrality, anchored in a related ethos of permissionless innovation. Meanwhile, ideas and technologies promoting social media advanced the concept of social production and amplified concerns about the privacy of individuals and other Internet-centred social harms.

Our focus on the element of *economic and social power* reveals complex dynamics in the exercise of power within and beyond the state and in the context of the following competing and overlapping arenas and concepts of Canadian democracy: prime ministerial and Cabinet-parliamentary, federalism, interest group pluralism, civil society, and direct democracy centred on the burgeoning Internet-based social media and networks. In the macro S&T domain, economic actors and departments within Cabinet and priorities dominated, and there was a tendency for policy to be discussed with, and aimed at, S&T elites rather than broader interest groups, communities, or citizens. The weakness and high turnover of science ministers limited their ability to have much impact on the government agenda. Social power emerges, however, in macro-regulatory policy as it addresses regulatory safety and risk, macro-policy about universities and higher education, and ultimately, through individual and collective choices as consumers or social actors. Inside the federal system, health interests and consumer politics united to forge strict regulatory safety as the goal in the post-thalidomide era and its continued implications for government in the form of compensation claims being made by organized and vocal advocates. Environmental NGOs provided Environment Canada with a social base that remains relevant – the notion of sustainable development as the triple bottom line and responsible resource development both present in the Harper era but with a considerable emphasis on the latter.

In the granting domain there are also more precise structures of power centred around client and research community interests at the main granting councils and in Canadian federalism. We saw this emerge first in the health and medical research space through the development of the Medical Research Council and later, more broadly under the CIHR where the health profession, hospitals, and related networks asserted and gained influence. More strategic and applied research was a priority of both the state and many industrial proponents and a key impetus for the formation of the Networks of Centres of Excellence, and for networked granting programs both in the Councils and in Genome Canada. The direct lobbying power of university presidents was pivotal in the formation and structure of the Canada Foundation for Innovation and for new norms and processes of merit review as opposed to just peer review. These structures were also underpinned by the corresponding changes in intellectual property protection and dissemination policy and management, which were both promoted and exploited by industry and the academy. In the industrial S&T domain, power resided in the dirigiste and industrial policy era mostly with the state, but that was challenged and lessened as cities, clusters, and innovation networks emerged as alternate loci of activity and power. The reframing of the policy dialogue to focus on micro-economic and pro-market initiatives in support of growth spread widely, ultimately pushing out most other frameworks both nationally and internationally, except for the fiscal deficit crisis and recession from 2008 to 2013 when spending and stimulus programs centred on infrastructure re-emerged.

The agriculture, food, biosciences, and biotechnology domain saw a strong social base of power for food and for farms as a strategic export business, at least in the West. Although of a slightly different type, we also saw a strong social base of power in some of the politics of genetically modified foods and in broader anti-globalization politics, again both globally and nationally. Genomics also became one of the key battlegrounds (along with IP policy) where governments, elites, the market, and the social base are most engaged – so far this has not led to any major social or institutional breakthroughs, but it is still early days.

The Internet, communications, and social media domain involved a fast-changing and interacting set of socio-economic players, including the following: US technology power centres and firms, key individuals in the design of the Internet, the Web, and in dominant social media firms; established Canadian telecommunications interests that functioned in and protected their oligopolistic markets in the face of intense Internet competition; and the vast array and volume of social production that defines the social media age. Industry Canada and the CRTC competed for policy and regulatory authority but both in many ways were simply overwhelmed by the pace and scope of technological change.

The element of *time, temporal realities, and conflicts*, as stressed throughout, requires analytical detective work in interpreting its changing presence. Overall, S&T policy statements seek to capture the coming medium-term or even longer-term challenges as they are seen or guessed at and surmised. Support for industrial R&D and for closing Canada's R&D and/or innovation gap is stressed in virtually all S&T policy statements and in many Budget Speeches over the fifty-year period. But it remains unsolved. S&T provisions in Budget Speeches and Speeches from the Throne are often influenced by electoral calculus both in the pre-election and post-election contexts; minority or majority governments also change the dynamics. Policy crises and noticeable risk management failures, one at a time, or in clusters (as in the late 1990s) are themselves sudden and, in some basic sense, unexpected.

The current debate about the appropriate role for the National Research Council exemplifies the notion of supporting research and analysis only or mainly when proposals exhibit market-pull versus technology-push; this debate is replete with temporal judgments, vagueness, and assumptions overall and on a case-by-case basis. For Health Canada, temporal issues and factors were exhibited by the staying power of the primacy of the safety mandate in the post-thalidomide era. Different senses of time were inherently involved in the 2006 *Blueprint for Renewal* that advocated the much more elongated and needed time scales of progressive licensing and increased post-market monitoring (possibly triggering drug recalls) and re-licensing. The analysis of Environment Canada focused on the changing nature of its overall S&T strategies as they aspired to medium-term relevance and effectiveness but also were buffeted by budget cuts and by reorganizations of the S&T management structure.

The three foundational granting councils have not been immune to pressures that have emerged and had their effects over several decades. Growing funding needs and new pressures arose from the post-war baby-boomer demographics of the 1970s and 1980s. There were more universities established and more students conducting research under the supervision of a larger number of faculty members. Currently, a faculty retirement bulge from the same demographic trajectory means that new faculty had to be found and retained and given research support, this time in an even more competitive global university market. The Canada Research Chairs program, implemented by the Chrétien Liberals and sustained by the Harper Conservatives, was a partial effort to deal with recruitment, retention, and reversing the perceived "brain drain" to the United States. Even the newer granting structures have not been static. Genome Canada, for instance, has been under constant pressure to show socio-economic impact, which has forced it to try a range of new and different management and organizational approaches that they revisit and renew with virtually every competition.

The NCEs and CIHR institutes have attempted to respond to the need to address both short-term and longer-term needs. The challenge is that they cannot do it all. But they are continually asking how should new centres or institutes be chosen as medium-term creatures and objects of funding? Both these funders, as well as the Canada Foundation for Innovation, have mandates to support permanent highly qualified personnel (HQP), but in a period of changing human capital need for an innovation economy and society. Last but not least, the CFI, as a supporter of complex research infrastructure, is quintessentially grounded in a world of complex physical assets that need both reinvestment and renewal over longer time periods, with those physical assets linked closely to comparatively mobile human capital. None of these councils or institutions has figured out how to achieve these lofty goals individually, and cross-council interactions have been less in evidence than other jurisdictions, such as among the research councils in the United Kingdom.

Temporal realities and features of the industrial S&T and innovation domain again showed the steady and repeated focus on improving Canadian R&D and commercial performance and the equally steady both short-term and long-term failure to achieve it. While many of the policies were short-lived, some policies, such as the Canada-US Automotive Products Agreement, signed in 1965 and commonly known as the Auto Pact and procurement-led technology measures, lasted through most of the period. More recently, cluster initiatives were mostly structured as medium- and long-term ventures to build the conditions for success, in a world of often uncertain and non-linear outcomes. Early agriculture research policies also had considerable strength and staying power but later became cluster-like as they involved diverse partners. The genomics domain was built as a long-term, generational initiative, but it still has had a series of short-term crises or challenges to handle. One difficulty of the short-termism that is creeping into all the policy domains is that performance is increasingly difficult to assess: what can be done quickly and easily seldom has much lasting impact, while truly transformative investments have long, variable, and frequently unanticipated collateral impacts. In the life sciences more generally, whether it is drug discovery or agricultural biotechnology, investments typically take much more than a decade to yield a return.

In many ways the Internet and the Web exhibit both pressures. They were forged as consummate long-term technological infrastructure ideas and investments that should be technically managed to be an interoperable "network of networks"; however, the actual parts of the system have brief half-lives that require constant reinvestment. The absence of control makes this investment somewhat uncertain, as there is no obvious way to coordinate the respective bits. Nevertheless, the goal is to make a system that should stretch and function within and across generations as a kind of guarantee of perpetual information-centred innovation. At a minimum, Canada and other countries had to construct and invent new short- and medium-term Internet-informed policy discourse. In the mid-1990s, as we have seen, there were medium-term notions of a coming and imminent e-business presence and an e-economy, but, not until later, the e-society and the Internet of things.

As social media developed, the costs of computing declined, and the capacity to store large amounts of data increased, a pivotal temporal feature has been the massive speed and volume of data to emerge and hence the designation of an emerging Big Data era where the consolidation of a bewildering array and volume of information and knowledge is expected to generate unfathomable socio-economic value. Numerous forms of burgeoning social production are creating intergenerational impacts on children and the elderly, with differing information literacy and capacities.

OUR FIVE MAIN ARGUMENTS IN THE CONTEXT OF THE THREE RESEARCH QUESTIONS

In the context of this book's overall conceptual and empirical journey, we offer a concluding summary of our five main arguments interwoven with our answers to the three research questions posed by the authors in the introduction to this book: (1) How, why, and to what extent has Canadian S&T and innovation policy, governance, and democracy changed in the past forty to fifty years? (2) Where has S&T and innovation policy resided in Canadian federal political and policy agendas across the past fifty years? (3) How have concepts of science-based governance prevailed or been changed across the past several decades?

The Innovation Economy and Society Nexus

Our central overall argument is that scholars and policy practitioners need a more explicit understanding of the *innovation economy and society nexus*. This understanding is necessary to differentiate it from earlier economy-social policy relations centred on the more familiar post–Second World War to late 1960s development of capitalism, the Keynesian social welfare state, and more recently, the focus on micro-economic growth and the liberal-market state. This argument is central to answering the first research question, but also the second and third questions as well. It helps to show how, why, and to what extent Canadian S&T and innovation policy, governance, and democracy has changed.

The innovation economy and society is, in the simplest terms, a two-category analytical world, but the nexus of relationships between the two is complex and replete with boundary overlaps, complex transactional dynamics, and changing discourse that affect how it can be understood. The innovation descriptor is a policy term that emerged more formally in the late 1980s and dominates policy discourse in the present day. The roots of this concept can be found in earlier eras as well when policy discourse focused on the balance between social and economic progress. This is true, in part, because the entire field of S&T and innovation policy is often categorized in other sets of two-category worlds, including the following: science and technology, research and development, linear and non-linear or networked innovation, the social sciences and the natural sciences, and disciplinary and interdisciplinary research. In short, innovation at the extreme is about transformation rather than accommodation. Thus, Table 12.1 summarizes our concluding evidentiary sense of the innovation economy and society nexus present in each of our domains across the fifty-year story. It is worth reiterating, however, that economy-society comparisons and tensions in policy terms often centre on the predominant power of capitalism and the reality that in most instances social change and reform are often playing catch-up in the face of the greater relational and structural power of business in both liberal market economies and the global capitalist system (as discussed in the introduction and chapter 1).

Table 12.1 The innovation economy and society nexus in diverse S&T and innovation policy domains

S&T and innovation policy domains	*Innovation economy and society nexus: Examples, features, and complexities*
Macro S&T and innovation policy	• Overall the innovation economy is the larger theme in the S&T strategy statements even though S&T is overall the nomenclature of choice. • Social themes predominate in many Budget Speeches, especially when new knowledge is emphasized and linked to education and universities. • Under the Cabinet Directive on Regulatory Management regulatory policy, social issues, if not the innovation society, have some prominence as stated values, including those regarding health, safety, and the well-being of Canadians. • In the late 1990s social concerns were raised regarding health threats such as blood supplies and declining fish stocks, which led to calls for greater transparency and social access, to advocacy for better processes and principles regarding science advice, and for greater attention to ethics in science.
Government departments and agencies	• Government exhibits a more obvious presence of innovation society dynamics as a partial counterweight to innovation economy norms and pressures. • Health Canada policy history, where at present the stated mandate relates to making Canadians the healthiest people in the world, remains historically anchored on lessons from the 1960s thalidomide crisis; this is reflected but also extended in the current broad-based aspirations to create a full-scale pre- and post-market progressive licensing system for health and food products. • History shows an increasing recognition of health as an industry rather than just a policy sector, including the pharmaceutical industry, which has generated the need for speedier drug and product approvals. • Environment Canada policy history shows the socio-ecological breadth of its mandate and the presence of social science in its functioning, in addition to the natural and environmental sciences, but there was also push back in economic terms, in the Harper era manifested in the combination of safety, quality, and "long-run competitiveness" goals and aspirations. • NRC policy history shows that major reform efforts to realize better economic innovation and performance regarding industrial R&D always generated some social push back, including in the area of research in the life sciences; NRC's role in nation building remains salient.

Table 12.1 The innovation economy and society nexus in diverse S&T and innovation policy domains (*Continued*)

S&T and innovation policy domains	*Innovation economy and society nexus: Examples, features, and complexities*
Granting, universities, and levered money	• The granting sector brings out important examples of how innovation policy affects the structure and function of the granting system across various funding bodies; original MRC was driven by social content and pressures flowing from hospitals and health care, including research ethics, and these were extended massively under the CIHR and its virtual institutes; new forms of peer and merit review have significantly increased the factors and drivers involved. • SSHRC is the social sciences and social content incarnate and extends in its granting structure and networks to areas such as education and learning, the environment, communities, and women in the labour force. • NCE, the Canada Research Chairs, and Canada Excellence Research Chairs programs each in a different ways supports highly qualified personnel (HQP), but also creates and attracts talent; CFI with its infrastructure focus also funded diverse areas of new capital and infrastructure to foster social, environmental, and health benefits. • NSERC had the least obvious "social" content in its mandate and its funding except perhaps for the HQP goal which it expressed frequently. • The innovation economy is an increasing part of the discourse regarding all these entities, with notions of innovation expressed in various ways including "cultures of innovation," "commercialization," and also partnership-networked funding on a significantly increasing scale.
Industrial S&T and innovation	• Dirigiste industrial policy was initially seen as nation building via infant industries; regional policy was then, as now, cast as addressing society's employment needs and opportunities. • Clusters and local-regional innovation systems now embrace both the economic and social aspects of development, including the involvement of universities, community colleges, and in some cases, social economy actors. • Tacit knowledge, the lubricant that makes the system work, is fundamentally a socio-economic construct. • In spite of a long-standing focus on S&T in industrial manufacturing, involving repeated restatements of the goal and repeated program innovation, performance was never very strong when compared with other countries.
Intellectual property, invention, and innovation	• Such a nexus is central to the IP domain in that it is ultimately driven by a core protection vs. dissemination values trade-off. The natural temptation is to treat the protection idea and value as the economic dimension and the dissemination idea of invention and knowledge as the social dimension, but the protection side of the IP value trade-off also at its heart is designed in a social and even cultural sense to reward and encourage inventors and creators, many of whom are individuals working on their own or in small teams in various institutional and non-institutional community settings. The dissemination side of the IP value trade-off is also replete with economic innovation nexus attributes, given that complex chains of learning and distributed knowledge can produce economic gains of crucial kinds in a modern economy. • This domain has also been driven by big ideas such as the concepts of public and private goods, and by the disruptive concept of endogenous growth.

	• Change in policy and programming in this domain is more about the exercise of power by the US as the prevailing hegemon, as its industry, its social elites, and its mass movements to a great extent define the choices and determine the outcomes. • Time is a critical variable when it comes to intellectual property. Time is money. Impatience generates action; delays generate profound socio-economic losses.
Agriculture, food, biosciences, and biotechnology	• Agriculture is cast historically in social terms of needs for food and quality and safety of food and farm systems, but rapidly diverging socio-economic power structures between farm interests in Western Canada vs. Eastern Canada have changed the policy discourse. • Agriculture is increasingly about engineered productivity, which challenges conceptions of what is natural and safe; this is most evident in the debate about biotech GM crops and foods. • Agriculture has been faced with the emergence of the "risk society," which now dominates discourse globally and in Canada. • Anti-globalization has emerged in society, leading to new forms of democracy and debate about the merits of relevant vs. excessive democracy; the emergence and demise of engagement bodies such as CBAC and NRTEE partly exemplify the conflict.
Genomics, life sciences, and technology	• The scale, scope, and speed of the technology caused governments everywhere to jump on board, seeking to benefit from or at least not lose from the technological transformation. • The massive boundary spillovers created both threats (to our ability to regulate areas like reproductive technologies and personalized medicine) and opportunities (to gain economically and socially from the transformative effects of the technology). • Governments found themselves in a difficult position, with a range of new actors pressing for changes in ways they could not control: women's groups, university professors, multinational corporations, and consumers of genetic tests all challenged the governing system. • The long, variable, and uncertain timelines in this domain created major challenges for governments as the impacts were often so far removed from the decision point that they found it difficult to respond appropriately.
Internet, communications, and social media	• The innovation economy and society nexus themes and evidence are most in evidence in this domain; indeed, the information society writ large is a pivotal driver overall and is often its overall macro-descriptor and discourse. • The Internet and the Web as technologies were crafted with broad social rationales inherent in the principles of no control and network neutrality. Early policy initiatives such as the Information Highway and SchoolNet and later the Community Access Program and general access norms were social in their basic thrust. • Policies about regulating and governing the Internet, social media, and communications are increasingly entwined in social dialogue about threats such as cyberbullying, privacy, and social revulsion against the excesses of state spying nationally and internationally. • The emergence of Google, Facebook, and Twitter, but also millions of other sites and businesses, involve social production in a massive way. The communications industry and oligopoly, which in Canada always had Canadian content and cultural values as central to its regulation, now exists in a world where production and product is not just directed at paying consumers, but now has to function in a world of multi-directional social production that is well beyond their control. • Across the entire development trajectory, this domain was undoubtedly defining the new economy and its changing lexicon of e-business, e-commerce, e-government, and the information and Big Data economy.

But under the dynamics of technology- and innovation-linked forces for change, the innovation economy and society nexus is bound to be present but also different and require different kinds of awareness of boundaries, tensions, and benefits. The notions of what empirically is economic versus social is complex when research and innovation and technology relate simultaneously to realms such as health, safety, the environment, education, research ethics, and the impacts on populations, subsets of individuals, and communities that are more particular and perhaps harder to pinpoint and define.

The broad STI system story shows the scale and scope of change. In earlier periods, agriculture, farming, telecommunications, health, and the National Research Council, for example, were nation-building efforts; now they are instruments for a much wider range of goals and interests. There is, as a result, greater competition, not to mention complexity, boundary overlaps and tensions, among the domains overall.

The analysis overall has also shown how peer-review underpinnings of S&T have been changed and made more complex and managerial in nature. Peer review as the social and ethical norm for academic journals remains important but has changed, in part, because the Internet and the information revolution have allowed vast increases in the number of new journals with different disciplinary and interdisciplinary conceptions of who the peers are. As we have seen, the research-granting domain has been transformed, with many more peer-review committees and a layering of merit criteria and review processes that force research and infrastructure to comply with targeted, strategic goals. Curiosity-driven research is not dead, but it is no longer the only or even best path to academic career success.

Change in the system is also being propelled by the Internet and digital processes that in some ways are yielding a Big Data economy and society and in other ways are foreshadowing major changes that may not yet be fully formed. The combination of many new sources of grey literature and new input from social media (and its vast realms of social production by mobilized citizens who previously were mainly consumers of content produced by businesses and telecommunications industries) is poised to transform our future in ways unimaginable.

S&T and Innovation Policy as Typically a Low-Priority Agenda Field

Second, we argue that S&T and Innovation policy is typically a low priority field and activity in relation to larger and more mainstream and compelling macro-economic fiscal policy and micro-economic growth priorities and agendas. Equally, for the most part, it resides on the periphery of public opinion in Canada. The review of the agendas and Throne Speeches in chapter 2 and the Budget Speeches in chapter 4 illustrated this. Moreover, the main S&T policy statements and strategies surveyed in chapter 4 tended to be aimed at elite audiences. This low priority was also reflected in the fact that federal S&T ministers were shuffled in and out of office by successive prime ministers, on average on an annual basis, and the role of science adviser to the prime minister was abolished.

Agenda paradoxes, however, have also been in evidence in the eight domain stories overall. For example, as we have seen, in virtually every S&T statement in all prime ministerial eras, the need for Canada to have stronger industrial R&D support and commercialization performance was the top priority expressed. But, in each case, it was also consistently shown that performance did not improve. The agriculture and genomics domains offered tantalizing hope that new investments could move Canada up both the research rankings and lead to commercial and economic impact; however, the history in manufacturing and ICT suggests those lessons and successes may not be easily generalized.

One of the S&T and innovation agenda areas that did capture attention were the early and sustained Internet developments and policies. Governments and their changing industry departments launched the Information Highway and broadband access policies and strategies, in part because they had a quasi-populist appeal as focused policy goals in areas such as province-based schools, small communities, and cities. These have changed into a focus on digital rights and access, but the current Internet era has more negatives in the form of concerns about national security and forms of social abuse carried out via the Internet and social media. But there is no guarantee that these agendas can be convincingly or practically managed.

Natural Resources: Continuing Staples Trap or Base for Competitive Innovation?

Third, we argue that natural resources in Canada are still powerful in Canada's political economy. Some argue, as we have seen, that they thus constitute a form of embedded staples trap that harms the continuous development of both economic and social innovation; nevertheless, we think there is a more nuanced interpretation. First, we stressed in chapter 5 that Canada's natural resource development historically was anchored in S&T work and applications. This was true regarding the crucial role of the Geological Survey of Canada as Canada was mapped and settled, but it was also a crucial feature of oil and gas industry development, including the eventual development of the oil sands. Second, we stressed that industrial S&T domain analysis in chapter 7 did not examine Canada's natural resource industries – it actually ignored them as a source of international industrial competitiveness. But any answer to our second question about S&T and innovation agendas needs to convey a deeper historical sense about the natural resource tension at the heart of Canadian S&T and innovation policy. Canada developed its resource base using foreign capital, imported technologies, and accessing global markets, albeit with Canadian innovations to make them work in Canada's climate and economy. In this sense, as we have shown in chapter 9 and elsewhere, Canada was one of the first and most globalized economies. But once the first phase of development was over, and for an extended period of time, nation builders downplayed our historical resource-based roots and projected an image of Canada as a vibrant manufacturing and services-producing economy.

In spite of all the efforts to develop capacity that is disconnected from our resource roots, it remains true that much of Canada's industry that is thriving is also inextricably connected to the primary sector – providing inputs, processing outputs and financing, developing new production technology, and moving and marketing those products. There have been efforts to show the interconnectivity and mutual future prosperity from further engagement between some of the resource sectors – e.g., agriculture in the 1980s and oil and gas in the 2000s – and manufacturing and services sectors, but that message is still often subsumed into a partially but importantly misleading two-category world of old economy/new economy.

Chapter 9's account of the agriculture, food, biosciences, and biotechnology domain showed unambiguously that it has been S&T and innovation oriented from the outset. In the context of agri-food S&T and innovation policy options, the problems facing agri-food innovators often appeared local, but for the most part, the most appropriate framings and solutions were globally generated or focused. The biotechnology revolution, both in agriculture and in the larger genomics field, in particular, introduced a transformative technology into the economy and society. Specific applications of the technology generally seemed benign, and the cumulative effects promised to be profound. Given the concomitant emergence of the "risk society" and risk-benefit regulation, demands arose for new means and methods of democratic engagement and dialogue. The Canadian Biotechnology Advisory Council (CBAC), citizens' juries, and a plethora of new engagement models were tried, and mostly failed to meet the high and rising expectations of government and NGOs. Agriculture and food is one of the few spaces where the demands for more reflexive socially responsible policy resonated with the broader citizenry and has been attempted.

Science-Based Regulation as an Incomplete Descriptor or Theory

Our fourth argument relates directly to the question posed about how the concept of science-based governance has prevailed or been changed. We argue that science-based regulation has never been a complete descriptor or theory for de facto regulatory and risk-benefit governance. Science, technology, and knowledge in policy and governance have been, are now, and will likely increasingly be embedded in social context. At present, Canada is engaged in a policy discourse where these choices are to a disturbing extent submerged and subsumed under political and partisan communication and continuous attack-politics strategies in an increasingly socially mediated world.

Science-based regulation as a concept and as discourse arose mainly in the late 1980s and early 1990s in the wake of free trade agreements that normalized and codified regulatory norms and practices. These measures sought to ensure that any future efforts to create regulations would not be barriers to trade; rather, by making them science based or based on sound science, it was hoped they would erect dams against the feared flood of regulatory protectionism triggered by reduced capacity to use tariffs to control trade. But even early on, notions of "science" in policy and regulation always involved not just the natural sciences but also the social and managerial sciences, including and drawing on related kinds of evidence such as that supplied in regulatory benefit-cost analysis, red-tape analysis, and concepts of justice at the heart of administrative law.

Regulatory evidence also widened when regulatory systems were expanded to include not just delegated law but also guidelines and codes underpinned by various practices regarding the arts of judgment and discretion (as found in many assessment and governance systems). We have also seen in various ways how different kinds of knowledge were present in related science activities, embodied in the brains, experience, and social networks of front-line S&T personnel. This has also been extended as risk regulation expanded from safety as a predominant goal to more complex notions of risk-benefit and smart regulation. The more that regulation aspired to be not just the pre-market regulation of products and processes but also extended to longer-term, post-market, life-cycle monitoring, the wider the range of professional knowledge engagement. All of these seriously question the strict validity of "science-based" decision making in STI policy overall.

Assessment of Transformational Technology Processes across Three Domains

Fifth, and finally, we argue that Canada's formal transformative technology assessment processes and their viability and efficacy vary enormously regarding technologies such as biotechnology, genomics, and the Internet and social media. Part of the challenge is that what constitutes a "product" varies tremendously. This is pivotal in deriving answers to all three of the research questions posed for our analysis. Their cumulative and disbursed presence in recent decades has changed the nature of Canada's S&T and innovation policy, governance, and democratic system, in part, because transformative technology has, as always, been mainly internationally driven. As a result, however, these technologies are never assessed in an overall way but rather are addressed through their applications in an array of product or process assessments, approvals, rejections, or delays. In an important sense, patents, as the dominant form of intellectual property, are also driven at the "product" level, regarding individual patents with much harder struggles needed to assess larger groupings of patents, such as, for example, patenting higher life forms.

Our final three domain chapters in different ways showed how the products of transformative technologies were or were not assessed. Biotechnology did generate regulatory processes for different kinds of genetically modified foods but in quite controversial ways and always linked to the background politics of different levels of overall acceptability in the United States versus greater opposition in the European Union. Genomics and life-science products were underpinned in some respects by positive politics about the potential to cure or reduce the harms of particular diseases, but also by bouts of hype of the potential benefits of individual inventions. This in some ways morphed into much more difficult notions of persons, family members, and children being the "product." The Internet, on the other hand, was constructed on the notion that information packets should not be controlled at all. Anarchy, the invisible hand, or some social utopia was supposed to operate. This underpinned the concept of permissionless innovation, even though new social production exploded, generating bewilderingly high volumes and varieties of outputs, ranging from the positive to highly negative, often spilling well beyond the boundaries or capacities of any individual, firm, or government to control.

Thus, in any number of ways Canada's STI policy story, both through its successes and its failures, needs to be understood in a deeper historical context with the full array of S&T and innovation domains fully appreciated and fully considered. This is what our innovation economy and society nexus approach and our examination of thirty-one policy histories has sought to do. We anticipate an engaging and challenging array of new forms of democratic policy debate and analysis to ensue.

Postscript: The New Justin Trudeau Liberal Government

Completing this book as a new Liberal government takes office, it is hard to tell exactly which of the many promises and priorities raised in the preceding election and in the Liberal platform will be acted on first (or even at all) both in an overall sense and with regard to science, technology, and innovation (STI) policy. We can opine to some extent based on a review of the election debates and policy platforms and on the structure, style, personalities, and early prognostications of the new Trudeau Cabinet.

The seventy-eight-day election campaign, the longest in modern times, offered ample time to have a fulsome debate about STI policy in Canada, but for the most part the opportunity was forgone. Although there were lots of relatively small and targeted promises of change from all three parties, especially from the victorious Liberal Party, none of the debates and little of the campaigning or reporting focused explicitly on STI. At the five debates, three unfocused, one on security, and one on the economy, little of the formal questioning focused on related STI topics, and none of the take-away moments reflected any of the parties' views on STI.

The party platforms in some ways are more helpful in understanding the focus and emphasis the new government (and the opposition) might put on STI policy (Liberal Party of Canada 2015; Stoney and Doern 2015). During the campaign, and core to the electoral debate, was a focus on the macro-economic choices Canada can make. The biggest divergence in the parties was on the fiscal balance, with the Conservatives and NDP proposing to maintain balanced budgets (with the Tories offering legislation to entrench that promise), while the Liberals broke ranks, took leadership risks, and proposed to accelerate spending on infrastructure (up by 1 percent of GDP to $10 billion annually) and to run a short-term deficit to accommodate that growth. As part of that commitment, the Liberals proposed to allocate up to $6 billion of the infrastructure budget over the next four years (or about 15 percent) for green infrastructure, for a range of potentially innovative water and wastewater facilities, climate-resilient infrastructure, clean energy, clean-up of contaminated sites to facilitate new construction, and protection against changing weather patterns (Liberal Party of Canada 2015, 11–15). Taxes also generated significant debate. The Conservatives proposed to hold steady on tax policy, and the NDP and Liberals both proposed rebalancing the tax load by shifting taxes more to higher-income Canadians and reducing the burden on the middle class. Although the NDP favoured doing a similar rebalancing on the corporate side, the Liberals were more restrained, simply offered modest cuts to the small business tax rate (from 11 percent to 9 percent), offset by measures to avoid professionals from using this system to avoid income tax.

In terms of more focused STI policy, each party offered somewhat different options. The Conservatives mostly proposed to continue their support for the existing set of programs in areas such as the tri-councils, Canada Foundation for Innovation, and the National Research Council of Canada, while the NDP proposed a number of industrial-style innovation programs for major sectors, such as the auto sector. The Liberals offered a more aggressive "new Innovation Agenda" (Liberal Party of Canada 2015, 15): in total, they proposed more than $1.6 billion new or expanded spending on STI, including $200 million per year over the next three years to support technology incubators, research facilities, and financing for small businesses; $200 million each year to support innovation and the use of clean technologies in natural resource sectors, including forestry, fisheries, mining, energy, and agriculture; $100 million per year for the next three years for the Industrial Research Assistance Program; $100 million over four years for agricultural research; and $80 million over four years for the Canadian Food Inspection Agency (15–16). Other parts of the innovation package included setting aside a portion of federal procurement budgets for "promising new Canadian technologies and businesses, consistent with international trade agreements"; establishing Canada Research Chairs in sustainable technology; reinstating the Labour-Sponsored Venture Capital Corporation tax credit; and a general promise to base decisions related to GM crop production on sound science and transparency.

There is certainly some sense that the new administration will present a different style of governing. The unveiling and swearing in of the new

Cabinet was heralded as the "return to government by Cabinet," with individual ministers presented as sources of power and influence in their own right. Prime Minister Justin Trudeau made significant efforts to signal that he was ending the tight central control by the prime minister and the Prime Minister's Office exercised by Stephen Harper. This was reflected in the following days as more than a dozen individual ministers engaged in scrums with media to lay out early priorities for the government (some of which directly focused on STI policy and governance).

The Cabinet structure and membership suggests science and innovation could factor more in policy debates and possibly in policy actions. Although the superstructure of departments and agencies has not changed in any obvious way, the new government has signalled a change in emphasis, renaming Industry Canada as the Department of Innovation, Science, and Economic Development, while the minister of state for science continues, with the impression of a double focus on science in the new government (Spears 2015), even though there were also two ministers in earlier periods. Early opinion suggested the new science minister, Dr Kirsty Duncan, a medical geographer who served on the International Panel for Climate Change, would focus on "renewal" of the federal science system, which many assert declined under the Harper government. Some speculated that the Science portfolio would rebalance investment in research, one of the themes explored in this volume. Many expect the new government to focus more on "basic" research driven by curiosity, in contrast with applied science, research, and development. Meanwhile Environment Canada was renamed Environment and Climate Change, and Public Safety was enjoined with Emergency Preparedness. The orientation of the new Innovation Agenda, plus the addition of two ministers interested in infectious diseases (Kirsty Duncan and Jane Philpott, a medical doctor named as minister of health), may signal a subtle shift of the innovation agenda towards a focus more on sustainability and away from broader industrial innovation and also on "pan-Canadian collaboration on health innovation" (Liberal Party of Canada 2015, 9).

However, the absence of any discussion of technology in the policy platforms suggests that the long-standing gap between basic science and industrial application will continue. Technology issues began to arise in a subdued way when the content of the Trans-Pacific Partnership (TPP) trade pact became known as the new government took office. Key technology businessmen such as Jim Balsillie immediately voiced serious concerns about possible adverse impacts for Canada in some of its provisions related to increased intellectual property rules and standards insisted on by the United States and achieved in the TPP (Blatchford 2015).

The style of government and its relationship with both its scientists and the scientific community has undoubtedly shifted. The Liberal platform promised that a Trudeau government "will appoint a Chief Science Advisor who will ensure that government science is fully available to the public, that scientists are able to speak freely about their work, and that scientific analyses are considered when the government makes decisions" (Liberal Party of Canada 2015, 36). The early announcement that the mandatory long-form census is reinstated, a litmus test for many of the commitment of government for evidence, satisfied many. Thus, Trudeau and his science ministers were at great pains in early days to declare that the role of public scientists would be changed. For the time being, there are no apparent limits; undoubtedly, the new policy will evolve as the government begins to address the many controversies and conflicts where science helps define the problems and options but often offers little in the way of specific direction for action. A period of enthusiastic outreach and engagement is likely, but the latent conflict between science and practical politics will remain.

Ultimately, the portents for a new innovation agenda are promising, with the rhetoric supporting a new focus on innovation and sustainability, but old-style industrial policy has hardly vanished. Bombardier, one of the perennial national champions of high tech Canada, has suffered a series of delays recently in its new CSeries of narrow-body, twin-engine, medium-range jet airlines, causing serious erosion in its market value. As a result, the Government of Quebec on 29 October 2015 committed US$1 billion of new support for the company and, within thirty minutes of the swearing-in of the new government, challenged the new minister of Innovation, Science and Economic Development to match the Quebec contribution. How the federal government responds could signal whether we are truly on a new policy track or returning to a focus on large, strategic plays as occurred in the 2008–12 recession and in earlier periods.

References

GOVERNMENT OF CANADA AND ITS AGENCIES: DOCUMENTS AND REPORTS

Canada Research Chairs (CRC). 2014a. *About Us*. http://www.chairs-chaires.gc.ca/about_us-a_notre_sujet/index-eng.aspx

– 2014b. *Program Statistics*. http://www.chairs-chaires.gc.ca/about_us-a_notre_sujet/statistics-statistiques-eng.aspx

– 2014c. *Method of Allocating Chairs* http://www.chairs-chaires.gc.ca/program-programme/allocation-attribution-eng.aspx

Canadian Environmental Assessment Agency (CEAA). 2010. *Annual Report 2009–2010*. Ottawa: Author. http://www.hc-sc.gc.ca/hcs-sss/alt_formats/pdf/pubs/cha-ics/2010-cha-ics-ar-ra-eng.pdf

Canadian General Standards Board. 2004. *Voluntary Labelling and Advertising of Foods that Are and Are Not Products of Genetic Engineering*. CAN/CGSB-32.315-2004. (The National Standard of Canada). http://www.tpsgc-pwgsc.gc.ca/ongc-cgsb/programme-program/normes-standards/internet/032-0315/index-eng.html

Canadian Institutes of Health Research (CIHR). 1999. *A New Approach to Health Research for the 21st Century: The Canadian Institutes of Health Research*. Ottawa: Author.

– 2011. *CIHR Personalized Medicine Planning Meeting*. Ottawa: Author.

– 2013. *Personalized Health in Canada*. Ottawa: Author.

– 2014a. *CIHR Institutes*. http://www.cihr-irsc.gc.ca/e/9466.html

– 2014b. *2014–15 Report on Plans and Priorities*. http://www.cihr-irsc.gc.ca/e/47939.html

– 2014c. *2014–15 Peer Review Committees and Mandates – Open Operating Grant Program (OOGP)*. http://www.cihr-irsc.gc.ca/e/4657.html

– 2014d. *Message from Governing Council: Institutes Model Review.* http://www.cihr-irsc.gc.ca/e/48534.html

– 2015. *CIHR Peer Review Manual for Grant Applications.* http://www.cihr-irsc.gc.ca/e/4656.html

Canadian Institutes of Health Research, Natural Sciences and Engineering Research Council of Canada, and Social Sciences and Humanities Research Council of Canada. 2010. *Tri-Council Policy Statement: Ethical Conduct for Research Involving Humans.* (TCPS-2). Ottawa: Interagency Panel on Research Ethics.

Canadian Intellectual Property Office (CIPO). 2011a. *Business Plan 2011–2012.* Ottawa: Industry Canada.

– 2011b. *Legislation.* Ottawa: Industry Canada.

– 2015. *Modernizing the IP Community.* http://www.ic.gc.ca/eic/site/cipointernet-internetopic.nsf/eng/h_wr03829.html?Open&pv=1

Canadian Radio-television and Telecommunications Commission (CRTC). 1994. *Review of Regulatory Framework.* Telecom Decision CRTC 94-19, 16 Sept. http://www.crtc.gc.ca/eng/archive/1994/dt94-19.htm

– 2011a. "Usage Based Billing for Gateway Access and Third Party Internet Access Services." CRTC Telecom Decision 2011-44, 25 Jan. http://www.crtc.gc.ca/eng/archive/2011/2011-44.htm

– 2011b. "Billing Practices for Wholesale Residential High Speed Access Services." Telecom Regulatory Policy CRTC 2011-703, 15 Nov. http://www.crtc.gc.ca/eng/archive/2011/2011-703.htm

– 2015. *CRTC Continues to Set the Course for the Future of Television with Let's Talk TV Decisions.* 29 Jan. http://news.gc.ca/web/article-en.do?nid=926529

Canadian Trade Commissioner Service (CTCS). 2013. *Science, Technology and Innovation International.* http://www.tradecommissioner.gc.ca/eng/funding/ciip/ciip.jsp

Commission of Inquiry on the Blood System of Canada. (The Krever Inquiry). 1997. *Final Report.* Ottawa: Minister of Public Works and Government Services.

Competition Bureau. 2007. *Self-regulated Professions: Balancing Competition and Regulation.* http://www.competitionbureau.gc.ca/eic/site/cb-bc.nsf/eng/02523.html

Council of Science and Technology Advisors (CSTA). 1999. *Science Advice for Government Effectiveness.* Ottawa: Industry Canada.

– 2000. *Building Excellence in Science and Technology (BEST): The Federal Roles in Performing Science and Technology.* Ottawa: Industry Canada.

– 2001. *Reinforcing External Advice to Departments.* Ottawa: Author.

– 2002. *Employees Driving Government Excellence: Renewing S&T Human Resources in the Federal Public Service*. Ottawa: Author.
– 2003. *Science Communications and Opportunities for Public Engagement*. Ottawa: Author.
– 2005. *Linkages in the National Knowledge System*. Ottawa: Author.
Department of Finance. 1983. *Budget Papers: Research and Development Tax Policy*. Ottawa: Author.
Dominion Bureau of Statistics. Canada. 1970. *Federal Government Expenditures on Scientific Activities*. Ottawa: Author.
Environment Canada. 1990. *Canada's Green Plan for a Healthy Environment*. Ottawa: Author.
– 2000. "National Implementation Strategy on Climate Change." In *Canada's Third National Report on Climate Change: Actions to Meet Commitments under the United Nations Framework Convention on Climate Change*, 57–8. http://publications.gc.ca/collections/Collection/En21-125-2001E.pdf
– 2007. *Environment Canada's Science Plan*. http://www.ec.gc.ca/scitech/9FA49B9A-2A69-4BE9-AA4C-526C406AE3F7/EC_SciencePlanEn_2006.pdf
– 2013. *About Environment Canada*. http://www.ec.gc.ca/default.asp?lang=En&n=BD3CE17D-1
Expert Panel on Federal Support to Research and Development (EPFSR&D). 2011. *Innovation Canada: A Call to Action; Expert Panel Report* (The Jenkins Report). Ottawa: Public Works and Government Services Canada. http://rd-review.ca/eic/site/033.nsf/vwapj/R-D_InnovationCanada_Final-eng.pdf/$FILE/R-D_InnovationCanada_Final-eng.pdf
External Advisory Committee on Smart Regulation (EACSR). 2004. *Smart Regulation: A Regulatory Strategy for Canada*. A Report to the Government of Canada. http://publications.gc.ca/collections/Collection/CP22-78-2004E.pdf
Genomics R&D Initiative (GRDI). 2012. *About the Genomics R&D Initiative*. http://grdi-irdg.collaboration.gc.ca/eng/about/index.html
Health Canada. 2004. *Learning from SARS: Renewal of Public Health in Canada*. A report of the National Advisory Committee on SARS and Public Health October 2003.http://www.phac-aspc.gc.ca/publicat/sars-sras/naylor/index-eng.php
– 2006. *Blueprint for Renewal: Transforming Canada's Approach to Regulating Health Products and Food*. http://www.hc-sc.gc.ca/ahc-asc/branch-dirgen/hpfb-dgpsa/blueprint-plan/index-eng.php
– 2007. *Blueprint for Renewal II: Modernizing Canada's Regulatory System for Health Products and Food*. http://www.hc-sc.gc.ca/ahc-asc/branch-dirgen/hpfb-dgpsa/blueprint-plan/index-eng.php

– 2009a. *Post-Market Surveillance of Drug Products Derived from Biotechnology*. http://www.hc-sc.gc.ca/sr-sr/alt_formats/hpfb-dgpsa/pdf/pubs/post_mark-apres_mise-eng.pdf
– 2009b. *The Progressive Licensing Framework: Concept Paper for Discussion*. http://www.hc-sc.gc.ca/dhp-mps/homologation-licensing/docs/index-eng.php
– 2009c. *Drug Licensing Process*. http://www.hc-sc.gc.ca/dhp-mps/homologation-licensing/system/map-carte/index-eng.php
– 2012. "Harper Government Takes Action to Help Canadians with Rare Diseases." 3 Oct. http://news.gc.ca/web/article-en.do?nid=698449
– 2014a. *Branches Agencies*. http://www.hc-sc.gc.ca/ahc-asc/branch-dirgen/index-eng.php
– 2014b. *Acts and Regulations*. http://www.hc-sc.gc.ca/ahc-asc/legislation/acts-reg-lois/index-eng.php
– 2014c. *Health Portfolio*. http://healthycanadians.gc.ca/minister-ministre/portfolio-eng.php
– 2014d. *Health Products and Food Branch*. http://www.hc-sc.gc.ca/ahc-asc/branch-dirgen/hpfb-dgpsa/index-eng.php
– 2014e. *Science Policy Directorate*. http://www.hc-sc.gc.ca/ahc-asc/branch-dirgen/spb-dgps/spd-dgps/index-eng.php
– 2014f. *Health Products and Food Regulatory Modernization*. http://www.hc-sc.gc.ca/ahc-asc/activit/strateg/mod/index-eng.php

Indian Affairs and Northern Development Canada. 2009. *Evaluation of the First Nations SchoolNet Program: Final Report*. Ottawa: Author.

Industry Canada. 1994a. *Building a More Innovative Economy*. Ottawa: Author.
– 1994b. *Canadian Information Highway: Building Canada's Information and Communications Infrastructure*. Ottawa: Author.
– 1996. *Building the Information Society: Moving Canada into the 21st Century*.
– 1997. *Connecting Canadians*. Ottawa: Author.
– 2002. *Achieving Excellence: Investing in People, Knowledge and Opportunity*. Ottawa: Author.
– 2003. *Federal Science and Technology: The Pursuit of Knowledge*. Ottawa: Author.
– 2007. *Mobilizing Science and Technology to Canada's Advantage*. Ottawa: Author.
– 2011. *Intellectual Property Policy*. Ottawa: Author.
– 2014a. *Digital Canada 150*. Ottawa: Author.
– 2014b. *Seizing the Moment: Moving Forward in Science, Technology and Innovation*. Ottawa: Author.

– 2014c. *S&T and Innovation Strategy: Seizing the Moment*. Ottawa: Author.
Institute of Genetics. 2009. *About IG*. http://www.cihr-irsc.gc.ca/e/13102.html
Medical Research Council of Canada (MRC). 2000. *Where Health Research Meets the Future: The Final Report of the Interim Governing Council of the Canadian Institutes of Health Research*. Ottawa: Author.
Minister of Agriculture. 1925. *The Dominion Experimental Farms*. Ottawa: Printed by Authority of the Hon. W.R. Motherwell, Minister of Agriculture. http://agrienvarchive.ca/download/dominion_exp_farms_1925.pdf http://grdi-irdg.collaboration.gc.ca
Minister of Finance. 2015. *Strong Leadership: A Balanced Budget, Low-Tax Plan for Jobs, Growth and Security*. Tabled in the House of Commons by the Hon. Joe Oliver, Minister of Finance, 21 Apr. http://www.budget.gc.ca/2015/docs/plan/budget2015-eng.pdf
Ministry of State for Science and Technology (MOSST). 1975. *The Make-or-Buy Policy 1973–1975*. Ottawa: Author.
– 1980a. *Annual Report 1978–79*. Ottawa: Author.
– 1980b. *Federal Science Activities 1980–81*. Ottawa: Author.
– 1983a. *A National Science and Technology Policy for Canada*. Ottawa: Author.
– 1983b. *Towards 1990: Technology Development for Canada*. Ottawa: Author.
– 1984. *Report of the Task Force on Federal Policies and Programs for Technology Development. A Report to the Honourable Edward C. Lumley, Minister of State, Science and Technology*. (The Wright Report). Ottawa: Author.
– 1985. *A National Science and Technology Policy for Canada*. Ottawa: Author.
National Advisory Board on Science and Technology. 1990. *Revitalizing Science and Technology in the Government of Canada*. Ottawa: Author.
– 1995. *Healthy, Wealthy and Wise: A Framework for an Integrated Federal Science and Technology Strategy*. Ottawa: Author.
National Research Council of Canada (NRC). 1990. *The Competitive Edge: NRC Long-Range Plan 1990–1995*. Ottawa: Author.
– 1990–91. *Annual Report*. Ottawa: Author.
– 1992. *From Rapeseed to Canola: The Billion Dollar Success Story*. Saskatoon: NRC, Plant Biotechnology Institute.
– 1995. *Vision to 2001*. Ottawa: Author.
– 2010. *Evaluation of the Genomics R&D Initiative (GRDI)*. http://www.nrc-cnrc.gc.ca/eng/about/planning_reporting/evaluation/2011_2012/grdi.html
– 2013a. *NRC's Membership in International Scientific Unions*. http://www.nrc-cnrc.gc.ca/eng/about/global/international.html

– 2013b. *Strategy 2013–2018*. http://www.nrc-cnrc.gc.ca/obj/doc/reports-rapports/NRC_Strategy_2013_2018_e.pdf

– 2014. *About NRC*. http://www.nrc-cnrc.gc.ca/eng/about/index.html

National Round Table on the Environment and the Economy (NRTEE). 2012. *Canada's Opportunity: Adopting Life Cycle Approaches for Sustainable Development*. http://publications.gc.ca/collections/collection_2012/trnee-nrtee/En134-56-2012-eng.pdf

– 2013. *NRT Reports by Issue*. http://collectionscanada.gc.ca/webarchives2/20130322143450/http:/nrtee-trnee.ca/publications-2

Natural Sciences and Engineering Research Council of Canada (NSERC). 1979. *A Five-Year Plan for the Programs of the Natural Sciences and Engineering Research Council*. Ottawa: Author.

– 1999. *Ten Years to 2000: A Strategic Document*. Ottawa: Author.

– 2013a. *NSERC Peer Review Manual 2013–2014*. Ottawa: Author.

– 2013b. *Peer Review Process*. http://www.nserc-crsng.gc.ca/NSERC-CRSNG/Reviewers-Examinateurs/PeerReview-EvalPairs_eng.asp

– 2014. *NSERC's Statement of Values and Goals*. http://www.nserc-crsng.gc.ca/NSERC-CRSNG/values-valeurs_eng.asp

Networks of Centres of Excellence (NCE). 2013. *Building an Ecosystem of Collaboration*. http://www.nce-rce.gc.ca/Research-Recherche/2013/Index_eng.asp

– 2014a. *About the Networks of Centres of Excellence*. http://www.nce-rce.gc.ca/About-APropos/Index_eng.asp

– 2014b. *Building on 25 Years of R&D Excellence*. http://www.nce-rce.gc.ca/_docs/Anniversary-25-Anniversaire_eng.pdf

– 2015. *Networks and Centres by Sector*. http://www.nce-rce.gc.ca/NetworksCentres-CentresReseaux/BySector-ParSecteur_eng.asp

Office of the Privacy Commissioner of Canada (OPC). 2004. "E-Commerce to E-Economy: Strategies for the 21st Century." Background Paper for the Conference on the Challenge of Change: Building the 21st Century Economy, 27–8 Sept., Ottawa. https://www.priv.gc.ca/media/sp-d/2004/sp-d_040928_e.asp

– 2011. *Annual Report to Parliament*. Ottawa: Office of the Privacy Commissioner of Canada.

– 2012. *Annual Report to Parliament*. Ottawa: Office of the Privacy Commissioner of Canada.

Polar Knowledge Canada. 2015. Polar Knowledge Canada. http://www.canada.ca/en/polar-knowledge/

Policy Research Initiative. 2003. *Social Capital as a Public Policy Tool. Horizons: Policy Research Initiative* 6(3). http://www.horizons.gc.ca/sites/default/files/Publication-alt-format/2003-0183-eng.pdf

– 2006. *Social Capital as a Public Policy Tool: Objectives and Background.* Policy Research Initiative, Government of Canada.

Privy Council. 2014a. *Blueprint 2020 Summary Interim Progress Report.* Ottawa: Clerk of the Privy Council.

– 2014b. *What We Have Heard: Blueprint 2020 Summary Interim Progress Report.* http://www.clerk.gc.ca/eng/feature.asp?pageId=361

Privy Council Office. Prime Minister's Advisory Committee on the Public Service. 2013. *Seventh Report to the Prime Minister: Modernizing the Employment Model.* . http://www.clerk.gc.ca/local_grfx/docs/pmac-ccpm/7-2013-eng.pdf

Public Health Agency of Canada (PHAC). 2013a. *About the Agency.* http://www.phac-aspc.gc.ca/about_apropos/index-eng.php

– 2013b. *2013–14 Report on Plans and Priorities.* http://www.phac-aspc.gc.ca/rpp/2013-2014/assets/pdf/rpp-2013-2014-eng.pdf

Royal Canadian Mounted Police (RCMP). 2014. *Internet Security.* http://www.rcmp-grc.gc.ca/qc/pub/cybercrime-eng.htm

Royal Commission on Canada's Economic Prospects. (The Gordon Commission). 1957. *Final Report.* Ottawa: Queen's Printer.

Royal Commission on the Economic Union and Development Prospects for Canada. (The Macdonald Commission). 1985. *Final Report.* Ottawa: Minister of Supply and Services Canada.

Royal Commission on Government Organization (RCGO). (The Glassco Commission). 1963. *Scientific Research and Development*, Report 23 in vol. 4. Ottawa: Queen's Printer.

Royal Commission on New Reproductive Technologies (RCNRT). (The Baird Commission). 1993. *Proceed with Care: Final Report.* Ottawa: Minister of Government Services Canada.

Senate Special Committee on Science Policy (The Lamontagne Committee). 1970. *A Science Policy for Canada: Report of the Senate Special Committee on Science Policy*, vol. 1, *A Critical Review: Past and Present.* Ottawa: Queen's Printer.

Social Sciences and Humanities Research Council of Canada (SSHRC). 2004. *From Granting Council to Knowledge Council: Renewing the Social Sciences and Humanities in Canada.* Ottawa: Author.

– 2006. *The National Discussion on Protection of Human Subjects in Research.* Ottawa: Author.

– 2013. *Strengthening Canada's Cultures of Innovation: Strategic Plan 2013–2016.* http://www.sshrc-crsh.gc.ca/about-au_sujet/publications/strategic_plan_2013-16-plan_strategique_2013-2016_e.pdf

– 2014a. *Merit Review.* http://www.sshrc-crsh.gc.ca/funding-financement/merit_review-evaluation_du_merite/index-eng.aspx

– 2014b. *Facts and Figures*. http://www.sshrc-crsh.gc.ca/about-au_sujet/facts-faits/index-eng.aspx

Speech from the Throne (SFT). 1970. Parliament of Canada, 8 Oct. 28th Parliament, 3rd Session (1970.10.08 - 1972.02.16). http://www.parl.gc.ca/Parlinfo/Documents/ThroneSpeech/28-03-e.pdf

– 1973. Parliament of Canada, 4 Jan. 29th Parliament, 1st Session (1973.01.04 - 1974.02.26). http://www.parl.gc.ca/Parlinfo/Documents/ThroneSpeech/29-01-e.pdf

– 1974. Parliament of Canada, 27 Feb. 29th Parliament, 2nd Session (1974.02.27 - 1974.05.09). http://www.parl.gc.ca/Parlinfo/Documents/ThroneSpeech/29-02-e.pdf

– 1980. Parliament of Canada, 14 Apr. 32nd Parliament, 1st Session (1980.04.14 - 1983.11.30). http://www.parl.gc.ca/Parlinfo/Documents/ThroneSpeech/32-01-e.pdf

– 1983. Parliament of Canada, 7 Dec. 32nd Parliament, 2nd Session (1983.12.07 - 1984.07.09). http://www.parl.gc.ca/Parlinfo/Documents/ThroneSpeech/32-02-e.pdf

– 1984. Parliament of Canada, 5 Nov. 33rd Parliament, 1st Session (1984.11.05 - 1986.08.28). http://www.parl.gc.ca/Parlinfo/Documents/ThroneSpeech/33-01-e.pdf

– 1986. Parliament of Canada, 1 Oct. 33rd Parliament, 2nd Session (1986.09.30 - 1988.10.01). http://www.parl.gc.ca/Parlinfo/Documents/ThroneSpeech/33-02-e.pdf

– 1989. Parliament of Canada, 3 Apr. 34th Parliament, 2nd Session (1989.04.03 - 1991.05.12). http://www.parl.gc.ca/Parlinfo/Documents/ThroneSpeech/34-02-e.pdf

– 1991. Parliament of Canada, 13 May. 34th Parliament, 3rd Session (1991.05.13 - 1993.09.08). http://www.parl.gc.ca/Parlinfo/Documents/ThroneSpeech/34-3-e.html

– 1994. Parliament of Canada, 18 Jan. 35th Parliament, 1st Session (1994.01.17 - 1996.02.02). http://www.parl.gc.ca/Parlinfo/Documents/ThroneSpeech/35-1-e.html

– 1996. Parliament of Canada, 27 Feb. 35th Parliament, 2nd Session (1996.02.27 - 1997.04.27). http://www.parl.gc.ca/Parlinfo/Documents/ThroneSpeech/35-2-e.html

– 1997. Parliament of Canada, 23 Sept. 36th Parliament, 1st Session (1997.09.22 - 1999.09.18). http://www.parl.gc.ca/Parlinfo/Documents/ThroneSpeech/36-1-e.html

– 1999. Parliament of Canada, 12 Oct. 36th Parliament, 2nd Session (1999.10.12 - 2000.10.22). http://www.parl.gc.ca/Parlinfo/Documents/ThroneSpeech/36-2-e.html

– 2001. Parliament of Canada, 30 Jan., 37th Parliament, 1st Session (2001.01.29 - 2002.09.16). http://www.parl.gc.ca/Parlinfo/Documents/ThroneSpeech/37-1-e.html
– 2002. Parliament of Canada, 30 Sept. 37th Parliament, 2nd Session (2002.09.30 - 2003.11.12). http://www.parl.gc.ca/Parlinfo/Documents/ThroneSpeech/37-2-e.html
– 2004. Parliament of Canada, 2 Feb. 37th Parliament, 3rd Session (2004.02.02 - 2004.05.23). http://www.parl.gc.ca/Parlinfo/Documents/ThroneSpeech/37-3-e.html
– 2004. Parliament of Canada, 5 Oct. 38th Parliament, 1st Session 1st Session (2004.10.04 - 2005.11.29).
– 2006. Parliament of Canada, 3 Apr. 39th Parliament, 1st Session (2006.04.03 - 2007.09.14). http://www.parl.gc.ca/Parlinfo/Documents/ThroneSpeech/39-1-e.html
– 2007. Parliament of Canada, 16 Oct. 39th Parliament, 2nd Session (2007.10.16 - 2008.09.07). http://www.parl.gc.ca/Parlinfo/Documents/ThroneSpeech/39-2-e.html
– 2008. Parliament of Canada, 18 Nov. 40th Parliament, 1st Session (2008.11.18 - 2008.12.04). http://www.parl.gc.ca/Parlinfo/Documents/ThroneSpeech/40-1-e.html
– 2009. Parliament of Canada, 26 Jan. 40th Parliament, 2nd Session (2009.01.26 - 2009.12.30). http://www.parl.gc.ca/Parlinfo/Documents/ThroneSpeech/40-2-e.html
– 2010. Parliament of Canada, 3 Mar. 40th Parliament, 3rd Session (2010.03.03 - 2011.03.26). http://www.parl.gc.ca/Parlinfo/Documents/ThroneSpeech/40-3-e.html
– 2011. Parliament of Canada, 3 June. 41st Parliament, 1st Session (2011.06.02 - 2013.09.13. http://www.parl.gc.ca/Parlinfo/Documents/ThroneSpeech/41-1-e.html
– 2013. Parliament of Canada, 16 Oct. 41st Parliament, 2nd Session (2013.10.16 - 2015.08.02). http://www.parl.gc.ca/Parlinfo/Documents/ThroneSpeech/41-2-e.html

Statistics Canada. 2014. *Gross Domestic Expenditures on Research and Development in Canada (GERD), the Provinces and Territories National estimates 2004 to 2014 Provincial and Territorial estimates 2008 to 2012.* http://www.statcan.gc.ca/pub/88-221-x/88-221-x2014001-eng.htm

Task Force on Foreign Ownership and the Structure of Canadian Industry (The Watkins Report) Report. 1968. Ottawa: Author.

Telecommunications Policy Review Panel (TPRP). 2006. *Final Report.* Ottawa: Industry Canada. https://www.ic.gc.ca/eic/site/smt-gst.nsf/vwapj/tprp-final-report-2006.pdf/$FILE/tprp-final-report-2006.pdf

Treasury Board Secretariat (TBS). 1973. *Guidelines for the Implementation of the Make or Buy Policy Concerning Research and Development Requirements in the Natural Sciences*. Ottawa: Author.
– 1977. *Policy and Guidelines on Contracting Out the Government's Requirements in Science and Technology*. Ottawa: Author.
– 2005. *Canadian Institutes of Health Research 2005–2006 Report on Plans and Priorities*. Ottawa: Author.
– 2007. *The Cabinet Directive on Streamlining Regulation*. Ottawa: Author.
– 2011. *A Guide to the Regulatory Process*. Ottawa: Author.
– 2012. *The Triage Process*. Ottawa: Author.
– 2013. *Cabinet Directive on Regulatory Management*. Ottawa: Author.

LEGISLATION AND LEGAL CASES

Canada Assisted Human Reproduction Act (AHRA). S.C. 2004, c.2.
Harvard College v. *Canada Commissioner of Patents*. 2000. 7 C.P.R. (4th) 1, 189 D.L.R. (4th) 385 (F.C.A.).
Monsanto Canada Inc. v. *Schmeiser*. 2004. 1 S.C.R. 902, 2004 SCC 34. http://scc.lexum.org/en/2004/2004csc34/2004csc34.html
Supreme Court of Canada. 2011. *Reference Re: Assisted Human Reproduction Act (2010)*. 2010 SCC 61, [2010] 3 S.C.R. 457.

ALL OTHER SOURCES CITED

Akkad, Omar. 2013. "What's Next for Google?" *Globe and Mail*, 26 Oct.
Allan, Roy A. 2001. *A History of the Personal Computer: The People and the Technology*. Birmingham: Allan Publishing.
Anders, George. 2014. "Letting Go of an Obsession with Net Neutrality Could Free Technologists to Make Online Services Even Better." *MIT Technology Review* 117 (6): 29–34.
Andersson, Claes, Anton Tornberg, and Petter Tormberg. 2013. "Societal System – Complex or Worse?" http://www.insiteproject.org/wp-content/uploads/2013/05/RP_2013.pdf
Aronczyk, Melissa, and Graeme Auld. 2013. "Tar, Ethics and Other Tactical Repertoires: The Co-evolution of Movements for and against the Tar Sands." Paper presented at the International Studies Association Meeting, 3–6 Apr., San Francisco.
Association of Universities and Colleges of Canada (AUCC). 2014. *Submission to the Federal Science, Technology and Innovation Review*. 24 Jan. http://www.univcan.ca/media-room/publications/federal-science-technology-innovation-review/

Atkinson, Michael M. 2011. "Lindblom's Lament: Incrementalism and the Consistent Pull of the Status Quo." *Policy and Society* 30 (1): 9–18.

Atkinson, Michael, and William Coleman. 1989. *The State, Business, and Industrial Change*. Toronto: University of Toronto Press.

Atkinson, Michael M., Daniel Béland, Gregory P. Marchildon, Kathleen McNutt, Peter W.B. Phillips, and Ken Rasmussen. 2013. *Governance and Public Policy in Canada: A View from the Provinces*. Toronto: University of Toronto Press.

Atkinson-Grosjean, Janet. 2006. *Public Science, Private Interests: Culture and Commerce in Canada's Networks of Centres of Excellence*. Toronto: University of Toronto Press.

Aucoin, Peter. 1971. "The Role of Functional Advisory Councils." In G. Bruce Doern and Peter Aucoin (eds.), *The Structures of Policy-Making in Canada*, 154–78. Toronto: Macmillan.

– 1972. *Health Scientists and the Making of Health Science Policy in Canada*. Doctoral dissertation, Queen's University, Kingston, Ontario.

– 1997. *The New Public Management: Canada in Comparative Perspective*. Montreal and Kingston: McGill-Queen's University Press.

– 2008. "New Public Management and New Public Governance: Finding the Balance." In David Siegel and Ken Rasmussen (eds.), *Professionalism and Public Service: Essays in Honour of Kenneth Kernaghan*, 16–33. Toronto: University of Toronto Press.

Aucoin, Peter, and Richard French. 1974. *Knowledge, Power and Public Policy*. Ottawa: Science Council of Canada.

Aucoin, Peter, and Mark D. Jarvis. 2005. *Modernizing Government Accountability: A Framework for Reform*. Ottawa: Canada School of Public Service.

Aucoin, Peter, Mark D. Jarvis, and Lori Turnbull. 2011. *Democratizing the Constitution: Reforming Responsible Government*. Toronto: Edmond Montgomery.

Auld, Douglas. 1985. *Budget Reform: Should There Be a Capital Budget for the Public Sector?* Toronto: C.D. Howe Institute.

Auletta, Ken. 2009. *Googled: The End of the World as We Know It*. New York: Penguin.

Baker, Andrew. 2013. "The Gradual Transformation? The Incremental Dynamics of Macroprudential Regulation." *Regulation and Governance* 7 (4): 417–34.

Baldwin, Peter. 2014. *The Copyright Wars: Three Centuries of Trans-Atlantic Battle*. Princeton, NJ: Princeton University Press.

Barratt, Neil, and Leslie Regan Shade. 2007. "Net Neutrality: Telecom Policy and the Public Interest." *Canadian Journal of Communication* 32 (2): 295–305.

Barrett, Katherine, and Elizabeth Abergel. 2002. "Defining a Safe Genetically Modified Organism: Boundaries of Scientific Risk Assessment." *Science and Public Policy* 29 (1): 47–58.

Barry, Andrew. 2013. *Interdisciplinarity: Reconfigurations of the Social and Natural Sciences*. London: Routledge.

Battelle, John. 2005. *The Search*. London: Penguin.

Bauer, Martin W. 2015. *Atoms, Bytes and Genes*. London: Routledge.

Baumgartner, Frank R., and Brian D. Jones. 1993. *Agendas and Instability in American Politics*. Chicago: University of Chicago Press.

BBC News, 2014. "EU Backs 'Right to Be Forgotten' in Google Privacy Case." *BBC News*, 13 May.

Bearing Point (formerly KPMG Consulting Ltd). 2003. *Evaluation of the CFI Innovation Fund, University Research Development Fund, and College Research Development Fund: Final Report*. http://www.innovation.ca/sites/default/files/pdf/2003_bearing2_e.pdf

– 2004. *Interim Evaluation of Genome Canada*. http://www.genomecanada.ca/medias/PDF/EN/GCReport-Mar31041.pdf

Beck, Ulrich. 1992. *Risk Society: Towards a New Modernity*. Thousand Oaks, CA: Sage.

Behrens, Kristian. 2013. *Strength in Numbers? The Weak Effect of Manufacturing Clusters on Canadian Productivity*. Toronto: C.D. Howe Institute.

Bell, Stephen, and Andrew Hindmoor. 2009. *Rethinking Governance: The Centrality of the State in Modern Society*. Cambridge: Cambridge University Press.

Bendrath, Ralph, and Milton L. Mueller. 2011. "The End of the Net as We Know It? Deep Packet Inspection and Internet Governance." *New Media and Society* 13 (7): 1142–60.

Benkler, Yochai. 2006. *The Wealth of Networks*. New Haven, CT: Yale University Press.

Bennett, Alan B., and Sara Boettiger. 2009. "Case 5. The Public Intellectual Property Resource for Agriculture (PIPRA): A Standard License Public Clearinghouse for Agricultural IP." In Geertrui Van Overwalle (ed.), *Gene Patents and Collaborative Licensing Models: Patent Pools, Clearinghouses, Open Source Models and Liability Regimes*, 135–42. Cambridge: Cambridge University Press.

Bennett, Scott. 2012. "Federal Infrastructure Program Impacts: Perceptions at the Community Level." In G. Bruce Doern and Christopher Stoney (eds.), *How Ottawa Spends 2012–13: The Harper Majority, Budget Cuts and the New Opposition*, 190–206. Montreal and Kingston: McGill-Queen's University Press.

Berman, Elizabeth Popp. 2012. *Creating the Market University: How Academic Science Became an Economic Engine*. Princeton, NJ: Princeton University Press.

Berners-Lee, Tim. 1999. *Weaving the Web*. London: HarperCollins.

Bernstein, Jeffrey. 1986. *Research and Development, Tax Incentives, and the Structure of Production and Financing*. Toronto: University of Toronto Press.

Bernstein, Steven, and Benjamin Cashore. 2002. "Globalization, Internationalization, and Liberal Environmentalism: Exploring Non-Domestic Sources of Influence in Canadian Environmental Policy." In Debora L. VanNijnatten and Robert Boardman (eds.), *Canadian Environmental Policy: Context and Cases*, 2nd ed., 212–32. Toronto: Oxford University Press.

Berry, David. 2014. "Cable's Long Shadow: Streamlining Won't Save Current TV Model." *National Post*, 11 Dec.

Bessen, James. 2014. "The Evidence Is In: Patent Trolls Do Hurt Innovation." *Harvard Business Review* "HBR *Blob Network*." https://hbr.org/2014/07/the-evidence-is-in-patent-trolls-do-hurt-innovation

Bessen, James, and Michael James Meurer. 2008. *Patent Failure*. Princeton, NJ: Princeton University Press.

– 2014. "The Direct Costs from NPE Disputes." *Cornell Law Review* 99 (2): 12–34.

Bickerton, James, and Alain-G. Gagnon. 2009. *Canadian Politics*. 5th ed. Toronto: University of Toronto Press.

Biermann, Frank, and Steffan Bauer, eds. 2005. *A World Environmental Organization: Solutions or Threat for Effective International Environmental Governance?* London: Ashgate.

Bird, Malcolm. 2009. "Intellectual Property, Technology Offices, and Political Capital: Canadian Universities in the Innovation Era." In G. Bruce Doern and Christopher Stoney (eds.), *Research and Innovation Policy: Changing Federal Government–University Relations*, 172–91. Toronto: University of Toronto Press.

Black, Julia. 2002. "Regulatory Conversations." *Journal of Law and Society* 29 (1): 163–96. Special issue *New Directions in Regulatory Theory*, edited by Sol Picciotto and David Campbell.

Blais, Andre, ed. 1986. *Industrial Policy*. Toronto: University of Toronto Press.

Blatchford, Andy. 2015. "Jim Baisillie Fears TPP Could Cost Canada Hundreds of Billions of Dollars and Become 'Worst Ever' Policy Move." Financial Post, 8 Nov. 2015. http://business.fiancialpost.com/news/economy/jim-balsillie-fears

Boden, Rebecca, Deborah Cox, Maria Nedeva, and Katharine Barker. 2004. *Scrutinizing Science: The Changing UK Government of Science*. London: Palgrave Macmillan.

Bohannon, John. 2013. "Who's Afraid of Peer Review?" *Science* 342 (6154): 60–5.

Boland, W. 2014. *Public-Private Partnerships for the Management of Agricultural Innovation Systems*. Doctoral dissertation. University of Saskatchewan. Saskatoon Canada. Available at http://ecommons.usask.ca/handle/10388/EDT-2014-06-1579

Borins, Sandford. 2010. "Digital State 2.0." In Glen Toner, Leslie A. Pal, and Michael J. Prince (eds.), *Policy: From Ideas to Implementation*, 177–206. Montreal and Kingston: McGill-Queen's University Press.

Borins, Sandford, and David Brown. 2008. "E-consultation: Technology at the Interface between Civil Society and Government." In David Siegel and Ken Rasmussen (eds.), *Professionalism and Public Service: Essays in Honour of Kenneth Kernaghan*, 178–206. Toronto: University of Toronto Press.

Bothwell, Robert 1988. *Nucleus: The History of Atomic Energy of Canada Ltd.* Toronto: University of Toronto Press.

Bouchard, Marie J., ed. 2013. *Innovation and the Social Economy: The Quebec Experience*. Toronto: University of Toronto Press.

Boutros, Serena, Lillian Hayward, Anique Montambault, Laura Smallwood, and Glen Toner. 2010. "Growing the Children of Brundtland: The Creation and Evolution of NRTEE, IISD, CESD, and SDTC." In Glen Toner, Leslie A. Pal, and Michael J. Prince (eds.), *Policy: From Ideas to Implementation*, 257–86. Montreal and Kingston: McGill-Queen's University Press.

Boyer, Peter J. 2010. "The Covenant." *New Yorker*, 6 Sept., 60–7.

Bradford, Neil, and David A. Wolfe. 2012. "Playing against Type? Regional Economic Development Policy in the Harper Era." In G. Bruce Doern and Christopher Stoney (eds.), *How Ottawa Spends 2012–2013: The Harper Majority, Budget Cuts, and the New Majority*, 71–88. Montreal and Kingston: McGill-Queen's University Press.

Bradshaw, Tim. 2013. "Flaws and Feuds Helped Twitter Fly the Nest." *Financial Times*, 27 Oct., 15.

Braithwaite, John. 2005. *Neoliberalism or Regulatory Capitalism*. Canberra: Australian National University, Regulatory Institutions Network.

– 2008. *Regulatory Capitalism*. London: Edward Elgar.

Brassard, Daniel. 1996. *Science and Technology: The New Federal Policy*. Ottawa: Library of Parliament. http://www.parl.gc.ca/Content/LOP/ResearchPublications/bp414-e.htm

Breen, David H. 1993. *Alberta's Petroleum Industry and the Conservation Board*. Edmonton: University of Alberta Press.

Bregha, Francois. 2011. "How to Get Serious about the Strategic Environmental Assessment of Policies and Plans." In Christopher Stoney and G. Bruce Doern (eds.), *How Ottawa Spends 2011–2012: Trimming Fat or Slicing Pork?* 144–62. Montreal and Kingston: McGill-Queen's University Press.

Brenner, Neil. 2009. "Restructuring, Rescaling and the Urban Question." *Critical Planning* 16 (Summer): 61–79.

Bricker, Darrell, and John Ibbitson. 2013. *The Big Shift: The Seismic Change in Canadian Politics, Business and Culture.* Toronto: HarperCollins.

Britton, John N.H., and James M. Gilmour. 1978. *The Weakest Link: A Technological Perspective on Canadian Industrial Underdevelopment.* Ottawa: Science Council of Canada. http://artsites.uottawa.ca/sca/doc/Background-Study-No.-43-The-Weakest-Link-A-Technological-Perspective-on-....pdf

Broder, John M. 2013. "Canadian Visits U.S. to Promote Oil Pipeline." *New York Times*, 9 Apr.

Brown, G. Malcolm. 1970. "Medical Research: Servant or Master?" Lecture in Medicine, presented to the Royal College of Physicians and Surgeons of Canada, 22 Jan., Montreal.

Brundtland Commission. 1987. *Our Common Future.* Oxford: World Commission on Environment and Development.

Bulkeley, Harriet. 2010. "Cities and the Governing of Climate Change." *Annual Review of Environment and Resources* 35: 229–53.

Bush, Vannevar. 1945. *Science, the Endless Frontier: A Report to the President on a Program for Postwar Scientifc Research.* Washington, DC: US Office of Scientific Research and Development.

Cameron, Maxwell A., and Brian W. Tomlin. 2000. *The Making of NAFTA: How the Deal Was Done.* Ithaca, NY: Cornell University Press.

Campbell, Robert M., Leslie Pal, and Michael Howlett. 2004. *The Real Worlds of Canadian Politics.* Toronto: University of Toronto Press.

Canada Foundation for Innovation. 2012a. *Annual Report 2012–2013.* Ottawa: Author.

– 2012b. *CFI Strategic Roadmap 2012–17.* Ottawa: Author.

– 2014. *Our Investments.* http://www.innovation.ca/en/OurInvestments

Canadian Agency for Drugs and Technologies in Health (CADTH). 2015. *CADTH 2015–2018 Plan: Informing Choices in a New Era of Health Care in Canadà.* Ottawa: Author.

Canadian Council of Chief Executives. 2013. *Innovation and Competitiveness: Intellectual*

Property. http://www.ceocouncil.ca/en/innovation/intellectual.php.

Canadian International Council. 2011. *Rights and Rents: Why Canada Must Harness Its Intellectual Property Resources.* http://2glspd2t2a9zr20ie1z7bx8

zbb.wpengine.netdna-cdn.com/wp-content/uploads/2011/10/CIC-IP-Report-2011.pdf

Canadian Opinion Research Archive. 2013. *Canadian Opinion Research Archive*. Kingston: Queen's University. http://www.queensu.ca/cora/

CANARIE. 2014. *CANARIE at a Glance*. http://www.canarie.ca/en/about/ataglance

Carlson, Nicholas. 2011. "The Real History of Twitter." *Business Insider*, 13 Apr.

Carolan, Michael S. 2010. "The Mutability of Biotechnology Patents: From Unwieldy Products of Nature to Independent 'Objects.'" *Theory, Culture and Society* 27 (1): 110–29.

Caron, Andre H., and Ronald Cohen. 2013. *Regulating Screens: Issues in Broadcasting and Internet Governance for Children*. Montreal and Kingston: McGill-Queen's University Press.

Carpenter, Daniel. 2010. *Reputation and Power: Organizational Image and Pharmaceutical Regulation at the FDA*. Princeton, NJ: Princeton University Press.

Carson, Rachel. 1962. *Silent Spring*. Boston: Houghton Mifflin.

Castells, Manuel. 2011. *The Rise of the Network Society*. Hoboken, NJ: Wiley.

– 2012. *Networks of Outrage and Hope: Social Movements in the Internet Age*. London: Polity.

Castle, David, ed. 2009. *The Role of Intellectual Property Rights in Biotechnology Innovation*. London: Edward Elgar.

Castle, David, and Keith Culver. 2006. "Public Engagement, Public Consultation, Innovation and the Market." *Integrated Assessment* 6 (2): 137–52.

Castle, David, and Peter W.B. Phillips. 2011. "Science and Technology in Canada: Innovation Gaps and Productivity Traps." In Christopher Stoney and G. Bruce Doern (eds.), *How Ottawa Spends 2011–12: Trimming Fat or Slicing Pork?* 163–79. Montreal and Kingston: McGill-Queen's University Press.

Castle, David, Peter W.B. Phillips, Abbe Brown, Keith Culver, Daniela Castrataro, Tania Bubela, Shawn Harmon, Graham Dutfield, and Patricia Barclay. 2010. "Knowledge Management and the Contextualization of Intellectual Property Rights in Innovation Systems." *Scripted* 7 (1): 32–50.

Castle, David, and Nola Ries, eds. 2009. *Nutrition and Genomics: Issues of Ethics, Law, Regulation and Communication*. Waltham, MA: Academic Press.

Caulfield, Timothy. 2009. "Biotechnology Patents, Public Trust and Patent Pools: The Need for Governance." In David Castle (ed.), *The Role of*

Intellectual Property Rights in Biotechnology Innovation, 357–68. London: Edward Elgar.

Caulfield, Timothy, and Tania Bubela. 2007. "Why a Criminal Ban? Analyzing the Arguments against Somatic Cell Nuclear Transfer in the Canadian Parliamentary Debate." *American Journal of Bioethics* 7 (2): 51–61.

Center for Digital Democracy. 2012. *Protecting Privacy, Promoting Consumer Rights and Ensuring Corporate Accountability*. Washington, DC: Center for Digital Democracy.

Cerny, Philip G. 1995. "Globalization and the Changing Logic of Collective Action." *International Organization* 49 (4): 595–625.

Ceruzzi, Paul E. 1998. *A History of Modern Computing*. Cambridge, MA: MIT Press.

Chan, Adrienne, and Donald Fisher. 2008. "Academic Culture and the Research-Intensive University: The Impact of Commercialism and Scientism." In Adrienne Chan and Donald Fisher (eds.), *The Exchange University*, 49–69. Vancouver: UBC Press.

Chastko, Paul. 2004. *Developing Alberta's Oil Sands: From Karl Clark to Kyoto*. Calgary: University of Calgary Press.

Chesbrough, Henry William. 2005. *Open Innovation: The New Imperative for Creating and Profiting from Technology*. Cambridge, MA: Harvard Business School Press.

Chester, Jeff, and Edmund Mierzwinski. 2014. *Big Data Means Big Opportunities and Big Challenges*. Washington, DC: US Pirg Education Fund.

Clancy, Peter. 2011. *Offshore Petroleum Politics: Regulation and Risk in the Scotian Basin*. Vancouver: UBC Press.

Clarke, Amanda. 2014. "One of These Things Is Not Like the Other? Bottom-Up Reform, Open Information, Collaboration, and the Harper Government." In G. Bruce Doern and Christopher Stoney (eds.), *How Ottawa Spends 2014–2015: The Harper Government –Good to Go?* 125–38. Montreal and Kingston: McGill-Queen's University Press.

Clarke, Tony, Diana Gibson, Brendan Haley, and Jim Stanford. 2013. *The Bitumen Cliff: Lessons and Challenges of Bitumen Mega-Developments for Canada's Economy in an Age of Climate Change*. Ottawa: Canadian Centre for Policy Alternatives.

Clarkson, Stephen. 2009. "The Governance of Energy in North America: The United States and Its Continental Periphery." In Burkhard Eberlein and G. Bruce Doern (eds.), *Governing the Energy Challenge: Canada and Germany in a Multilevel Regional and Global Context*, 99–121. Toronto: University of Toronto Press.

Cleland, Michael. 2014. *From the Ground Up: Earning Public Support for Resource Development.* Calgary: Canada West Foundation.

Coglianese, Cary. 2010. "Management-Based Regulation: Implications for Public Policy." In OECD (ed.), *Risk and Regulatory Policy: Improving the Governance of Risk*, 159–84. Paris: OECD.

Coleman, Stephen, and Jay G. Blumler. 2009. *The Internet and Democratic Citizenship*. New York: Cambridge University Press.

Coleman, William. 1988. *Business and Politics: A Study of Collective Action.* Montreal and Kingston: McGill-Queen's University Press.

Collins, Harry M. 1985. *Changing Order: Replication and Induction in Scientific Practice*. London: Sage.

Committee on Publication Ethics (COPE). 2014. *About COPE.* http://publicationethics.org/about

Competition Policy Review Panel. 2008. *Competing to Win.* Industry Canada.

Connor, Steve. 2014. "UN Climate Deal in Peru Attacked by Activists as Weak and Ineffectual." *Independent*, 15 Dec.

Conway, Thomas. 1992. *The Marginalization of the Department of the Environment: Environmental Policy 1971–1978.* Doctoral dissertation, Carleton University, Ottawa.

Cooper, Charlie. 2013. "Hopes for Dementia Drug Breakthrough in Next Five years." *Independent*, 5 Dec., 16.

Corbin, Ruth M. 2010. *Intellectual Property in the 21st Century.* http://www.conferenceboard.ca/e-library/abstract.aspx?did=3452

Cosco, Amanda R. 2015. "Why Toronto Is a Hotbed of Pioneering Wearable Technology." *Globe and Mail*, 15 Jan.

Council of Canadian Academies (CCA). 2009. *Innovation and Business Strategy: Why Canada Falls Short.* Report of the Expert Panel on Business Innovation. Ottawa: Author.

– 2013a. *About Us.* http://www.scienceadvice.ca/en/about.aspx

– 2013b. *The State of Industrial R&D in Canada.* Report of the Expert Panel on Industrial R&D in Canada. Ottawa: Author.

Coyne, Andrew. 2012. "Same Old Bill, New Hysteria." *National Post*, 20 Feb.

Crawford, Adam. 2006. "Networked Governance and the Post-Regulatory State? Steering, Rowing and Anchoring the Provision of Policing and Security." *Theoretical Criminology* 10 (4): 449–79.

Creighton, Donald. 1955. *John A. Macdonald: The Old Chieftain.* Toronto: Macmillan.

Crookshanks, Rebecca, and Peter W.B. Phillips. 2013. "A Comparative Analysis of Access and Benefits-Sharing Systems." In Tania Bubela and E. Richard Gold (eds.), *Genetic Resources and Traditional Knowledge: Case Studies and Conflicting Interests*, 63–92. London: Edward Elgar.

Crow, Michael M. 1994. "Science and Technology Policy in the United States: Trading in the 1950 Mode." *Science and Public Policy* 21 (4): 202–12.

Csanady, Ashley. 2015. "CRTC Report: Nine Things to Know about Impending Changes to Your TV, Online or Off." *Ottawa Citizen*, 28 Jan.

Culnan, Mary J., Patrick McHugh, and Jesus I. Zubillaga. 2010. "How Large U.S. Companies Can Use Twitter and Other Social Media to Gain Business Value." *MIS Quarterly Executive* 9 (4): 243–59.

Curtis, Anthony. 2013. *Brief History of Social Media*. University of North Carolina, Mass Communication Department.

Dahlgren, Peter. 2013. *The Political Web*. London: Palgrave Macmillan.

Dattu, Riyaz, and Eric Morgan. 2014. "Status Report on Canada's Foreign Investment Promotion and Protection Agreement Negotiations." http://www.mondaq.com/canada/x/303750/international+trade+investment/Status+Report+On+Canadas+Foreign+Investment+Promotion+And+Protection+Agreement+Negotiations

Davenport, Thomas H., Paul Barth, and Randy Bean. 2012. "How Big Data Is Different." *MIT Sloan Management Review* 54 (1): 66–72.

Davenport, Thomas, and D.J. Patil. 2012. "Data Scientist: The Sexiest Job of the 21st Century." *Harvard Business Review* 90 (10): 70–6.

de Beer, Jeremy, Kun Fu, and Sacha Wunsch-Vincent. 2013. *The Informal Economy, Innovation, and Intellectual Property – Concepts, Metrics and Policy*. http://www.wipo.int/edocs/pubdocs/en/wipo_pub_econstat_wp_10.pdf

de Beer, Jeremy, Richard Gold, and Mauricio Guaranga. 2011. *Intellectual Property Management: Policy Issues and Options*. http://www.genomecanada.ca/medias/pdf/en/Research_Policy-Directions-Brief.pdf

Deckha, Maneesha. 2009. "Holding onto Humanity: Animals, Dignity, and Anxiety in Canada's Assisted Human Reproduction Act." *Unbound: Harvard Journal of the Legal Left* 5: 21–54.

de la Mothe, John. 2000. "Government Science and the Public Interest." In G. Bruce Doern and Ted Reed (eds.), *Risky Business: Canada's Changing Science-Based Regulatory Regime*, 31–48. Toronto: University of Toronto Press.

– 2003. "Ottawa's Imaginary Innovation Strategy: Progress or Drift?" In G. Bruce Doern (ed.), *How Ottawa Spends 2003–2004: Regime Change and Policy Shift*, 172–86. Toronto: Oxford University Press.

de la Mothe, John, and Gilles Paquet, eds. 1998. *Local and Regional Systems of Innovation*. London: Pinter.

Den Tandt, Michael. 2013. "Behold, the Revamped National Research Council – Now at Your Service." *National Post*, 7 May.

Devaney, Sarah. 2014. *Stem Cell Research and the Collaborative Regulation of Innovation*. London: Routledge.

Dewar, Denise. 2014. "Strategies for Agricultural Innovation." Presentation on behalf of CropLife International, Washington, DC, to ABIC 2014: Global Leadership in a Changing World, Agriculture Biosciences International Conference, 5–8 Oct., Saskatoon.

Di Salle, Don, Jac van Beek, and N. Bruce Baskerville. 2006. "The Innovation Challenge and the National Research Council Response." In G. Bruce Doern (ed.), *Innovation, Science, Environment: Canadian Policies and Performance, 2006–2007*, 35–58. Montreal and Kingston: McGill-Queen's University Press.

Djellal, Faridah, Dominique Francoz, Camal Gallouj, Faïz Gallouj, and Yves Jacquin. 2003."Revising the Definition of Research and Development in the Light of the Specificities of Services." *Science and Public Policy* 30 (6): 415–30.

Dobby, Christine. 2015. "Net Neutrality: CRTC Bans Bell from Subsidizing Data Usage for Mobile TV App." *Globe and Mail*, 29 Jan.

Doern, G. Bruce. 1972. *Science and Politics in Canada*. Montreal and Kingston: McGill-Queen's University Press.

– 1981. *The Peripheral Nature of Scientific and Technological Controversy in Federal Policy Formation*. Ottawa: Science Council of Canada.

– 1987. *The R&D Tax Credit Consultation Case*. Ottawa: Canadian Centre for Management Development.

– 1990. "The Department of Industry, Science and Technology: Is There Industrial Policy after Free Trade?" In Katherine Graham (ed.), *How Ottawa Spends 1990–1991: Tracking the Second Agenda*, 49–92. Ottawa: Carleton University Press.

– 1993. *Green Diplomacy: How Environmental Policy Decisions Are Made*. Toronto: C.D. Howe Institute.

– 1995a. *Institutional Aspects of R&D Tax Incentives: The SR&ED Tax Credit*. Ottawa: Industry Canada.

– 1995b. "The Regulation of Patent and Trademark Agent Qualifications: Institutional Issues and Options." Paper prepared for the Canadian Intellectual Property Office.

– 1996. "Looking for the Core: Industry Canada and Program Review." In Gene Swimmer (ed.), *How Ottawa Spends: 1996–97 – Life under the Knife*, 73–98. Ottawa: Carleton University Press.

– 1997. "Regulating on the Run: The Transformation of the CRTC as a Regulatory Institution." *Canadian Public Administration* 40 (3): 516–38.

– 1999a. "Science and Scientists in Federal Policy and Decision-Making." Paper prepared for the Policy Research Secretariat, Government of Canada.

– 1999b. *Global Change and Intellectual Property Agencies*. London: Pinter.

– 2000a. "The Therapeutics Products Program: From Traditional Science-Based Regulator to Science-Based Risk-Benefit Manager?" In G. Bruce Doern and Ted Reed (eds.), *Risky Business: Canada's Changing Science-Based Policy and Regulatory Regime*, 185–207. Toronto: University of Toronto Press.

– 2000b. "Patient Science versus Science on Demand: The Stretching of Green Science at Environment Canada." In G. Bruce Doern and Ted Reed (eds.), *Risky Business: Canada's Changing Science-Based Policy and Regulatory Regime*, 286–306. Toronto: University of Toronto Press.

– 2004. "The Agri-Food Sector and Federal Policies and Priorities: A Public Policy Framework Discussion Paper." Paper prepared for Agriculture and Agri-Food Canada.

– 2006. "The Origins and Evolution of Canada's Research Granting Agencies: NSERC, SSHRC and CIHR." Paper prepared for Industry Canada.

– 2007. *Red Tape, Red Flags: Regulation for the Innovation Age*. Ottawa: Conference Board of Canada.

– 2008. "Polytechnics in Higher Education Systems: A Comparative Review and Policy Implications for Ontario." Paper prepared for the Higher Education Quality Council of Ontario.

– 2009. "Evolving Budgetary Policies and Experiments: 1980 to 2009–2010." In Allan Maslove (ed.), *How Ottawa Spends 2009–2010: Economic Upheaval and Political Dysfunction*, 14–46. Montreal and Kingston: McGill-Queen's University Press.

– 2010. "The Relationships between Regulation and Innovation in the Transport Canada Context." Paper prepared for Transport Canada.

Doern, G. Bruce, ed. 1986. *The Politics of Economic Policy*. Toronto: University of Toronto Press.

Doern, G. Bruce, Graeme Auld, and Christopher Stoney. 2015. *Green-Lite: Complexity in 50 Years of Canadian Environmental Policy, Governance, and Democracy*. Montreal and Kingston: McGill-Queen's University Press.

Doern, G. Bruce, and Tom Conway. 1994. *The Greening of Canada*. Toronto: University of Toronto Press.

Doern, G. Bruce, and Richard Levesque. 2002. *The NRC in the Innovation Policy Era: Changing Hierarchies, Networks, and Markets*. Toronto: University of Toronto Press.

Doern, G. Bruce, and Jeffrey Kinder. 2007. *Strategic Science in the Public Interest: Canada's Government Laboratories and Science-Based Agencies*. Toronto: University of Toronto Press.

Doern, G. Bruce, and Mark MacDonald. 1999. *Free Trade Federalism: Negotiating the Agreement on Internal Trade*. Toronto: University of Toronto Press.

Doern, G. Bruce, Allan Maslove, and Michael J. Prince. 2013. *Canadian Public Budgeting in the Age of Crises: Shifting Budgetary Domains and Temporal Budgeting*. Montreal and Kingston: McGill-Queen's University Press.

Doern, G. Bruce, and Richard Phidd. 1983. *Canadian Public Policy: Ideas, Structure, Process*. Toronto: Methuen.

– 1992. *Canadian Public Policy: Ideas, Structure, Process*. 2nd ed. Toronto: Nelson Canada.

Doern, G. Bruce, and Peter W.B. Phillips. 2012. *The Genomics Regulatory-Science Regime: Issues and Options*. http://www.genomecanada.ca/medias/pdf/en/GPS-Policy-brief-October2012.pdf

Doern, G. Bruce, and Michael J. Prince. 2012. *Three Bio-Realms: Biotechnology and the Governance of Food, Health, and Life in Canada*. Toronto: University of Toronto Press.

Doern, G. Bruce, Michael J. Prince, and Richard Schultz. 2014. *Rules and Unruliness: Canadian Regulatory Democracy, Governance, Capitalism, and Welfarism*. Montreal and Kingston: McGill-Queen's University Press.

Doern, G. Bruce, and Ted Reed. 2001. "Science and Scientists in Regulatory Governance." *Science and Public Policy* 28 (3): 195–204.

Doern, G. Bruce, and Markus Sharaput. 2000. *Canadian Intellectual Property: The Politics of Innovating Institutions and Interests*. Toronto: University of Toronto Press.

Doern, G. Bruce, and Brian Tomlin. 1991. *Faith and Fear: The Free Trade Story*. Toronto: Stoddart.

Doern, G. Bruce, and Glen Toner. 1985. *The Politics of Energy*. Toronto: Methuen.

Doern, G. Bruce, and Ted Reed, eds. 2000. *Risky Business: Canada's Changing Science-Based Policy and Regulatory Regime*. Toronto: University of Toronto Press.

Doern, G. Bruce, and Chris Stoney, eds. 2009. *Research and Innovation Policy: Changing Federal Government–University Relations*. Toronto: University of Toronto Press.

Donnan, Shawn. 2014. "Data Protectionism." *Financial Times*, 5 Aug., 7.

Douglas, Heather E. 2013. "Canadian Science under Attack." *Scientist Magazine*, 2 Apr.

Drahos, Peter, and Ruth Mayne, eds. 2002. *Global Intellectual Property Rights: Knowledge, Access and Development*. London: Palgrave.

Dreyfus, Rochelle C., Harry First, and Diane L. Zimmerman, eds. 2010. *Working within the Boundaries of Intellectual Property: Innovation Policy for the Knowledge Society*. Oxford: Oxford University Press.

Dryzek, John. 2013. *The Politics of the Earth: Environmental Discourses*. 3rd ed. Oxford: Oxford University Press.

Dryzek, John, and Patrick Dunleavy. 2009. *Theories of the Democratic State*. London: Palgrave.

Dufour, Paul. 1994. "From Rhetoric to Reality: Thirty Years of Science and Technology Policy in Canada." In Industry Canada, Resource Book for Science and Technology Consultations, vol. 2. Ottawa: Industry Canada.

– 2012. *Advancing Knowledge at the Frontier: The AAAS and the Saga of Science Societies in Canada*. University of Ottawa, Institute for Science, Society and Policy.

– 2013. "A New Future for the Research Councils?" *Research Money* 27 (5): 8.

– 2014. "Do Science Policy Ministers Really Matter?" *iPolitics*, 1 Apr.

Dufour, Paul, and Yves Gingras. 1988. "The Development of Canadian Science and Technology." *Science and Public Policy* 15 (1): 13–18.

Duranton, Gilles. 2011. "California Deamin: The Feeble Case for Cluster Policies." *Review of Economic Analysis* 3: 3–45.

Dutfield, Graham. 2003. *Intellectual Property Rights and the Life Science Industries*. London: Ashgate.

Dutton, William H., Ginette Law, Gillian Bolsover, and Soumitra Dutta. 2013. *The Internet Trust Bubble: Global Values, Beliefs and Practices*. http://www3.weforum.org/docs/WEF_InternetTrustBubble_Report2_2014.pdf

Dyson, George. 2012. *Turing's Cathedral: The Origins of the Digital Universe*. London: Penguin.

Eberstadt, Nicholas. 2010. "The Demographic Future." *Foreign Affairs* 89 (6): 54–64.

Economist. 2007a. "Pharmaceuticals: A Gathering Storm." *Economist*, 9 June, 73–4.

– 2007b. "Pharmaceuticals: Beyond the Blockbuster." *Economist*, 30 June, 75–6.

– 2010. "Biology 2.0: A Special Report on the Human Genome." *Economist*, 10 June, 3–16.

– 2013b. "'Trouble in the Lab,' Briefing Unreliable Research." *Economist*, 19 Oct., 23–5.

– 2013c. "Internet Security: Besieged." *Economist*, 9 Nov., 93–4.

– 2014a. "Little Things that Mean a Lot." *Economist*, 19 July, 66.

– 2014b. "Everybody Wants to Rule the World." *Economist*, 29 Nov., 19–21.

– 2015a. "Excellence v Equity." Special Report on Universities. *Economist*, 28 Mar., 1–19.

Edelman, Murray. 1964. *The Symbolic Purposes of Politics*. Chicago: University of Illinois Press.

– 1977. *Political Language: Words that Succeed and Policies that Fail*. New York: Academic Press.

– 1988. *Constructing the Political Spectacle.* Chicago: University of Chicago Press.

Edosomwan, Simeon, Sitalaskshmi K. Prakasan, Doriane Kouame, Jonelle Watson, and Tom Seymour. 2011. "The History of Social Media and Its Impact on Business." *Journal of Applied Management and Entrepreneurship* 16 (3): 44–58.

Eggleston, Wilfred. 1978. *National Research in Canada: The NRC 1916–1966.* Toronto: Clarke Irwin.

Eissler, Rebecca, Annelise Russell, and Bryan D. Jones. 2014. "New Avenues for the Study of Agenda Setting." *Policy Studies Journal* 42 (Suppl.): S71–S86.

Elliott, Lorraine. 2004. *The Global Politics of the Environment.* 2nd ed. London: Palgrave Macmillan.

Enros, Phillip. 1991. "The Ornery Council of Scientific and Industrial Pretence: Universities and the Early NRC's Plans for Industrial Research." *Scientia Canadensis* 15 (2): 41–52.

– 2013. *Environment for Science: A History of Policy for Science in Environment Canada.* Ottawa: Author.

Etzioni, Amitai. 2012. "The Privacy Merchants: What Is to Be Done?" *University of Pennsylvania Journal of Constitutional Law* 14 (4): 929–51.

Etzkowitz, Henry, and Loet Leydesdorff, eds. 1998. *Universities and the Global Knowledge Economy.* London: Pinter.

Evans, Barbara J. 2009. "Seven Pillars of a New Evidentiary Paradigm: The Food, Drug and Cosmetic Act Enters the Genomic Era." *Notre Dame Law Review* 85: 22–36.

Ewen, Stanley W.B., and Arpád Pusztai. 1999. "Health Risks of Genetically Modified Foods." *Lancet* 354 (9179): 684–94.

Fancourt-Smith, Mark. 2015. "Apples, Patents and Trolls, Oh My!" *Western Canada Business Litigation Blog*, 2 Mar. http://www.westerncanada businesslitigationblog.com/intellectual-property/

Farazmad, Ali. 1996. *Public Enterprise Management.* New York: Greenwood.

Findlay, Scott, and Paul Dufour. 2013. "Why Canada Needs a Science Watchdog." *iPolitics*, 18 Nov.

Fisher, Donald, Kjell Rubenson, Theresa Shanahan, and Claude Trottier, eds. 2014. *The Development of Postsecondary Education Systems in Canada.* Montreal and Kingston: McGill-Queen's University Press.

Fletcher, Fred, and Andre Blais. 2012. "New Media, Old Media, Campaigns and Canadian Democracy." In Herman Bakvis and Mark D. Jarvis (eds.), *From New Public Management to New Political Governance*, 151–78. Montreal and Kingston: McGill-Queen's University Press.

Florida, Richard. 2002. *The Rise of The Creative Class*. New York: Basic Books.

Floridi, Luciano. 2011. *The Philosophy of Information*. Oxford: Oxford University Press.

– 2014. *The 4th Revolution: How the InfoSphere Is Reshaping Human Reality.* Oxford: Oxford University Press.

Fontanella-Khan, James, Alex Barker, and Jeevan Vasagar. 2014. "Google Has Fight on Its Hands in Europe." *Financial Times*, 22 May, 5.

Foot, David K. 1990. "Demographics: The Human Landscape of Public Policy." In G. Bruce Doern and Bryne B. Purchase (eds.), *Canada at Risk? Canadian Public Policy in the 1990s*, 25–45. Toronto: C.D. Howe Institute.

Foray, D. 1999. "Science, Technology and the Market." *World Social Science Report*, 246–55. Paris: UNESCO and Elsevier.

Forsdyke, Donald R. 2000. *Tomorrow's Cures Today? How to Reform the Health Research System*. Newark, NJ: Harwood Academic.

Fowke, V.C. 1973. *The National Policy and the Wheat Economy*. Toronto: University of Toronto Press.

Fox, Dov. 2008. "The Regulation of Biotechnologies: Four Recommendations." *Perspectives* 38 (2): 111–26.

Fox, Harold G. 1969. *Canadian Law and Practice Relating to Letters Patent for Inventions*. 4th ed. Toronto: Carswell.

Fox, Martha Lane. 2015. "Tech Giants Have Too Much Power: Let's Reclaim the Internet." *Guardian*, 30 Mar.

Freeman, Chris, and Luc Soete. 1997. *The Economics of Industrial Innovation*. 3rd ed. London: Pinter.

Friedman, Milton, and Anna J. Schwartz. 1971. *A Monetary History of the United States 1867–1960*. Princeton, NJ: Princeton University Press.

Fuss, Melvin, and Leonard Waverman. 1986. *The Canada-U.S. Auto Pact of 1965: An Experiment in Selective Trade Liberalism*. Cambridge, MA: National Bureau of Economic Research.

Gagnon, Marc-Andre. 2013. "Corruption of Pharmaceutical Markets: Addressing the Misalignment of Financial Incentives and Public Health." *Journal of Law, Medicine and Ethics* 41 (3): 571–80.

Gaisford, James, William Kerr, Peter Phillips, and Camille Ryan. 2010. "Canadian Agricultural Biotechnology Clusters: Innovation and Performance." In Zhiqi Chen and Marc Duhamel (eds.), *Industrial Organization in Canada: Empirical Evidence and Policy Challenges*, 284–335. Montreal and Kingston: McGill-Queen's University Press.

Galvin, Patrick. 2012. *Local Government and Cluster-Related Innovation Policy: Two Industry Clusters in the City of Toronto*. Doctoral dissertation, University of Exeter.

Gatehouse, Jonathan. 2013. "When Science Goes Silent." *Maclean's*, 3 May, 1 and 8.

Gaudreault, Valerie, and Patrick Lemire. 2003. *The Age of Public Infrastructure in Canada*. Ottawa: Statistics Canada.

Geist, Michael. 2007. "Building Confidence: Security, Privacy and User Empowerment." Paper prepared for the Canada Roundtable on the Future of the Internet Economy, 2 Oct., Ottawa.

– 2011. "Why Canada's New Copyright Law Remains Flawed." *Toronto Star*, 1 Oct., 5.

– 2014a. "Digital Canada 150: The Digital Strategy without a Strategy." *Toronto Star*, 4 Apr.

– 2014b. "Why the Government's New Digital Privacy Act Puts Your Privacy at Risk." *Toronto Star*, 11 Apr.

Genome Canada. 2000. *Guidelines and Evaluation Criteria for Genome Centres*. http://www.genomecanada.ca/medias/PDF/EN/GUIDE18-final.pdf

– 2001. *Guidelines and Evaluation Criteria for Competition II*. http://www.genomecanada.ca/medias/PDF/EN/CompIIGuidelinesfinal.pdf

– 2004a. *Annual Report 2003–2004*. http://www.genomecanada.ca/medias/PDF/EN/2003-2004AnnualReport.pdf

– 2004b. *Guidelines and Evaluation Criteria for Competition III*. http://www.genomecanada.ca/medias/PDF/EN/Guidelines3.pdf.

– 2006. *Criteria and Guidelines for the Position Paper Process*.

– 2013. "Harper Government Announces New Personalized Health Projects across Canada." 26 Mar. http://www.genomecanada.ca/en/about/news.aspx?i=427

Gershenfeld, Neil, and J.P. Vasseur. 2014. "As Objects Go On Line: The Promise (and Pitfalls) of the Internet of Things." *Foreign Affairs* (Mar./Apr.): 60–7.

Geyer, Robert, and Samie Rihani. 2010. *Complexity and Public Policy*. London: Routledge.

Gibbins, Roger, Antonia Maioni, and Janice Gross Stein. 2006. *Canada by Picasso: The Faces of Federalism*. Ottawa: Conference Board of Canada.

Gibbons, Michael. 2000. "Mode 2 Society and the Emergence of Context-Sensitive Science." *Science and Public Policy* 27 (3): 88–110.

Gibbons, Michael, Camille Limoges, Helga Nowotny, Simon Schwartzman, Peter Scott, and Martin Trow. 1994. *The New Production of Knowledge: The Dynamics of Science and Research in Contemporary Societies*. London: Sage.

Giddens, Anthony. 2011. *The Politics of Climate Change*. 2nd ed. London: Polity.

Gillette, Felix, Diane Brady, and Caroline Winter. 2013. "The Rise and Fall of BlackBerry: An Oral History." *Businessweek*, 5 Dec.

– 2014. "U.N. Panel Issues Its Starkest Warning Yet on Global Warming." *New York Times*, 2 Nov.

Global Research Council (GRC). 2012. *Statement of Principles for Scientific Merit Review*. http://www.globalresearchcouncil.org/sites/default/files/pdfs/gs_principles-English.pdf

– 2014. *About Us*. http://www.globalresearchcouncil.org/about-us

Goar, Carol. 2008. "All Politics, No Science for Harper." *Toronto Star*, 11 Feb.

Godin, Benoit. 2004. "Neglected Scientific Activities: The (Non)Measurement of Related Scientific Activities." University of Quebec at Montreal, Project on the History and Sociology of S&T Statistics.

Goel, Vindu. 2014. "Facebook's Profit Propelled by Mobile." *New York Times*, 23 July.

Goklany, Indur M. 2001. *The Precautionary Principle: A Critical Appraisal of Environmental Risk*. Washington, DC: Cato Institute.

– 2007. *The Improving State of the World*. Washington, DC: Cato Institute.

Gold, Richard, and Julia Carbone. 2010. "Myriad Genetics: In the Eye of the Policy Storm." *Genetics in Medicine* 12(4 Suppl.), S39–S70.

Goldacre, Ben. 2012. *Bad Pharma: How Drug Companies Mislead Doctors and Harm Patients*. London: Fourth Estate.

Goldsmith, Jack, and Tim Wu. 2008. *Who Controls the Internet? Illusions of a Borderless World*. Oxford: Oxford University Press.

Gollner, Andrew B., and Daniel Salee, eds. 1989. *Canada under Mulroney*. Montreal: Vehicule.

Grabosky, Peter. 1995. "Using Non-Governmental Resources to Foster Regulatory Compliance." *Governance* 8 (4): 527–50.

Graff, Gregory D., David Zilberman, and Alan B. Bennett. 2009. "The Contraction of Agbiotech Product Quality Innovation." *Nature Biotechnology* 27 (8): 702–4.

Gray, Earle. 2005. *The Great Canadian Oil Patch*. 2nd ed. Edmonton: June Warren Publishing.

Grice, Andrew. 2013. "Facebook Democracy: The Face of Things to Come?" *Independent*, 28 Nov., 1 and 5.

Grubb, Michael, Christiaan Vrolijk, and Duncan Brack. 1999. *The Kyoto Protocol: A Guide and Assessment*. London: Royal Institute of International Affairs.

Guston, David H. 2000. *Between Politics and Science: Assuring the Integrity and Productivity of Research*. Cambridge: Cambridge University Press.

Haldane, Andrew G. 2013. "Why Institutions Matter (More than Ever)." A Speech presented to the Centre for Research on Socio-Cultural Change, Annual Conference, School of Oriental and African Studies, 4 Sept.,

London. http://www.bankofengland.co.uk/publications/Documents/speeches/2013/speech676.pdf

Hale, Geoffrey. 2001. *The Politics of Taxation*. Peterborough: Broadview.

– 2002. "Innovation and Inclusion: Budgetary Policy, the Skills Agenda and the Politics of the New Economy." In G. Bruce Doern (ed.), *How Ottawa Spends 2002–2003: The Security Aftermath and National Priorities*, 20–47. Toronto: Oxford University Press.

– 2006. *Uneasy Partnership: The Politics of Business and Government in Canada*. Toronto: University of Toronto Press.

Hall, Peter, and David Soskice, eds. 2001. *Varieties of Capitalism: The Institutional Foundations of Comparative Advantage*. Oxford: Oxford University Press.

Harley, Diane. 2013. "Scholarly Communication: Cultural Contexts, Evolving Models." *Science* 342 (6154): 80–2.

Harrabin, Roger. 2015. "Fossil Fuels: The Untouchable Reserves." *BBC News*, 7 Jan.

Harris, John. 2010. "Promise and Risks from Life Not as We Know It." *Financial Times*, 27 May, 13.

Harris, Richard. 1985. *Trade, Industrial Policy and International Competition*. Toronto: University of Toronto Press.

Harris, Richard A., and Sidney M. Milkis. 1989. *The Politics of Regulatory Change: A Tale of Two Agencies*. New York: Oxford University Press.

Hawkins, Richard. 2012. *Looking at Innovation from a Uniquely Canadian Perspective: The Case for a New Alliance of Practice, Policy and Scholarship*.. University of Ottawa, Institute for Science, Society and Policy.

Hayes, F. Ronald. 1973. *The Chaining of Prometheus: Evolution of a Power Structure for Canadian Science*. Toronto: University of Toronto Press.

Hébert, Monique, Nancy Miller Chenier, and Sonya Norris. 2004. *Bill C-6: Assisted Human Reproduction Act*. Parliament of Canada. http://www.parl.gc.ca/About/Parliament/LegislativeSummaries/bills_ls.asp?ls=C6&Parl=37&Ses=3

Held, David, Anthony G. McGrew, David Goldblatt, and Jonathan. Perraton. 1999. *Global Transformations: Politics, Economics and Culture*. London: Polity.

Heller, Michael A., and Rebecca S. Eisenberg. 1998. "Can Patents Deter Innovation? The Anticommons in Biomedical Research." *Science* 280 (5364): 698–701.

Helliwell, Janet, and William Smith. 1999. "Scientific Advice in Government Decision Making: The Canadian Experience." Paper prepared for Council of Science and Technology Advisors.

Helm, Dieter. 2012. *The Carbon Crunch*. New Haven, CT: Yale University Press.

Hessing, Melody, and Michael Howlett. 2005. *Canadian Natural Resources and Environmental Policy*. 2nd ed. Vancouver: UBC Press.

Heydon, Kenneth, ed. 2012. *International Trade Policy*. London: Ashgate.

Hira, Anil, ed. 2013. *What Makes Clusters Competitive? Cases from the Global Wine Industry*. Montreal and Kingston: McGill-Queen's University Press.

Hoberg, George. 2002. "Canadian-American Environmental Relations: A Strategic Framework." In Debora L. VanNijnatten and Robert Boardman (eds.), *Canadian Environmental Policy: Context and Cases*, 2nd ed., 171–89. Toronto: Oxford University Press.

Hoberg, George, and Kathryn Harrison. 1994. "It's Not Easy Being Green: The Politics of Canada's Green Plan." *Canadian Public Policy* 20 (2): 119–37.

Hodson, Hal. 2013. "Stopping the Spooks." *New Scientist*, 23 Nov., 24.

Hong, Hannah. 2008. "Canada Abolishes Its National Science Advisor: After Just Four Years, Government Axes Post." *Nature* 451: 505.

Hood, Christopher. 1986. *Administrative Analysis*. Brighton: Wheatsheaf Books.

Hood, Christopher, Henry Rothstein, and Robert Baldwin. 2001. *The Government of Risk: Understanding Risk Regulation Regimes*. Oxford: Oxford University Press.

Houghton, John. 1997. *Global Warming: The Complete Briefing*. Cambridge: Cambridge University Press.

Howlett, Michael. 2011. *Designing Public Policies: Principles and Instruments*. London: Routledge

Howlett, Michael, Alex N. Netherton, and M.R. Ramesh. 1999. *Political Economy of Canada: An Introduction*. 2nd ed. Toronto: Oxford University Press.

Howlett, Michael, and Jeremy Rayner. 2006. "Globalization and Governance Capacity: Explaining Divergence in National Forest Programs as Instances of 'Next Generation' Regulation in Canada and Europe." *Governance* 19 (2): 251–75.

– 2013. "Patching vs. Packaging in Policy Formulation: Assessing Portfolio Design." *Politics and Governance* 1 (2): 170–82.

Huffine, Richard. 2010. "Value of Grey Literature to Scholarly Research in the Digital Age." Presented at the Elsevier customer event, Special Libraries Association Annual Conference, 15 June, New Orleans.

Hunt, Alan. 1999. *Governing Morals: A Social History of Moral Regulation*. Cambridge: Cambridge University Press.

Hunter, Lawson A.W., Edward Iacobucci, and Michael J. Trebilcock. 2010. "Scrambled Signals: Canadian Content Policies in a World of Technological Abundance." *C.D. Howe Institute Commentary* (301): 1–34.

Independent Panel of Experts (IPE): Arnold Naimark (chair), Kevin Keough, Kelvin Ogilvie, and Clive Willis. 2008. *Inter-Sectoral Partnerships for Non-Regulatory Federal Laboratories: A Report to the President of the Treasury Board of Canada.* http://www.tbs-sct.gc.ca/fedlab-labfed/docs/2008/tnfl-tglfvn/tnfl-tglfvn-eng.pdf

Information Highway Advisory Council (IHAC). 1995. *Final Report.* Ottawa: Industry Canada.

Innis, Harold. 1956. *Essays in Canadian Economic History.* Edited by Mary Q. Innis. Toronto: University of Toronto Press.

– 1977 [1930]. *The Fur Trade in Canada: An Introduction to Canadian Economic History.* Revised and reprinted. Toronto: University of Toronto Press.

Innovation Sustainability and Technology (INSITE). 2013. *The Innovation Society.* http://www.insiteproject.org/perspectives/innovation-society/

International Council for Science (ICSU). 2010. *Universality of Science in the Polar Regions.* http://www.icsu.org/publications/icsu-position-statements/universality

International Human Genome Sequencing Consortium. 2001. "Initial Sequencing and Analysis of the Human Genome." *Nature* (409): 860–921.

Ivison, John. 2013. "National Research Council Revamp." *National Post,* 7 May.

Jack, Andrew. 2009. "Remedy for a Malady." *Financial Times,* 15 Aug., 7.

– 2010. "Big Pharma Aims for Reinvention." *Financial Times,* 13 May, 14.

Jaffe, Adam B., and Manuel Trajtenberg. 2002. *Patents, Citations and Innovation: A Window on the Knowledge Economy.* Cambridge, MA: MIT Press.

Jakobi, Tobias. 2012. "Regulating Regulation? The Regulatory Policy of the OECD." Paper presented at the Conference on New Perspectives on Regulation, Governance and Learning, 27–9 June, University of Exeter.

Janisch, Hudson. 2012. "The Relationship between Governments and Independent Regulatory Agencies: Will We Ever Get It Right?" *Alberta Law Review* 40 (4): 46–58.

Jarrell, Richard, and Yves Gingras, eds. 1991. *Building Canadian Science: The Role of the National Research Council.*" Special issue. *Scientia Canadensis* 15 (2).

Jasanoff, Sheila.1990. *The Fifth Branch: Science Advisors as Policy Makers.* Boston: Harvard University Press.

– 1998. *Comparative Science and Technology Policy.* London: Edward Elgar.

Jenkins, Tom. 2013. *Canada First: Leveraging Defence Procurement through Key Industrial Capabilities.* Report of the Special Adviser to the Minister of

Public Works and Government Services, Feb. http://www.tpsgc-pwgsc.gc.ca/app-acq/documents/eam-lmp-eng.pdf

Jerome, Joseph. 2014. "Big Data: Catalyst for a Privacy Debate." *Indiana Law Review* 48 (1): 213–42.

Johnson, David, and A. Molloy. 2009. "The Quiet Crisis and the Emergence of La Relève." *Canadian Public Administration* 50 (2): 203–23.

Johnson, Steven. 2006. *The Ghost Map*. New York: Riverhead Books.

– 2009. "How Twitter Will Change the Way We Live." *Time*, 3 Apr.

Johnston, David. 1996. "Statement Speech – Information Highway Advisory Council." Presented to the 18th International Privacy and Data Protection Conference, 18 Sept., Ottawa. Office of the Privacy Commissioner of Canada. https://www.priv.gc.ca/media/sp-d/archive/02_05_a_960918_07_e.asp

Jones, Bryan D., and Frank R. Baumgartner. 2005. *The Politics of Attention: How Government Prioritizes Problems*. Chicago: University of Chicago Press.

– 2012. "From There to Here: Punctuated Equilibrium to the General Punctuation Thesis to a Theory of Government Information Processing." *Policy Studies Journal* 40 (1): 1–20.

Jones, Glen A., Patricia McCarney, and Michael Skolnik, eds. 2005. *Creating Knowledge, Strengthening Nations: The Changing Role of Higher Education*. Toronto: University of Toronto Press.

Jordana, J., and David Levi-Faur. 2004. *The Politics of Regulation: Examining Regulatory Institutions in the Governance Age*. Cheltenham: Edward Elgar.

Jorgensen, Rikke Frank. 2013. *Framing the Net: The Internet and Human Rights*. London: Edward Elgar.

Juillet, Luc, and Ken Rasmussen. 2008. *Defending a Contested Ideal: Merit and the Public Service Commission 1908–2008*. Ottawa: University of Ottawa Press.

Kalaitzandonakes, Nicholas, James Kaufman, and Douglas Miller. 2014. "Potential Economic Impacts of Zero Thresholds for Unapproved GMOs: The EU Case." *Food Policy* 45: 146–57.

Keen, Andrew. 2015. *The Internet Is Not the Answer*. London: Atlantic Books.

Kellogg, Paul. 2015. *Escape from the Staple Trap*. Toronto: University of Toronto Press.

Kietzmann, Jan H., Kristopher Hermkens, Ian P. McCarthy, and Bruno S. Silvestre. 2011. "Social Media? Getting Serious! Understanding the Functional Building Blocks of Social Media." *Business Horizons* 54 (3): 241–51.

Kilpatrick, Sean. 2013. "Stop Muzzling Scientists, Protesters Tell Tories." *Toronto Star*, 16 Sept.

Kinder, Jeffrey. 2003. "The Doubling of Government Science and Canada's Innovation Strategy." In G. Bruce Doern (ed.), *How Ottawa Spends 2003–2004: Regime Change and Policy Shift*, 204–20. Toronto: Oxford University Press.

– 2010. *Government Laboratories: Institutional Variety, Change and Design Space*. Doctoral dissertation, Carleton University, Ottawa.

– 2013. "Government Science 2020: Re-thinking Public Science in a Networked Age." Discussion paper for the Blueprint 2020 Public Service Renewal Agency.

– 2014. "GovScience 4.0: The Rise of 'Social' Science." *Canadian Government Executive* 20 (3).

Kindi, Vasso, and Theodore Arabatzis, eds. 2012. *Kuhn's* The Structure of Scientific Revolutions *Revisited*. London: Routledge.

King, M. Christine. 1989. *E.W.R. Steacie and Science in Canada*. Toronto: University of Toronto Press.

Kingdon, John W. 1995. *Agendas, Alternatives, and Public Policies*. 2nd ed. New York: HarperCollins.

Kirkpatrick, David. 2010. *The Facebook Effect*. London: Virgin Books.

Kleinberg, Jon. 2011. "What Can Huge Data Teach Us about Society and Ourselves?" In Max Brockman (ed.), *Future Science*, 72–87. Oxford: Oxford University Press.

Klinkenborg, Verlyn. 2013. "Silencing Scientists." Editorial. *New York Times*, 21 Sept.

Knoppers, Bartha M., and Rosario M. Isasi. 2004. "Regulatory Approaches to Reproductive Genetic Testing." *Human Reproduction* 19 (12): 2695–2701.

Knowles, Lori P. 2010. "Religion and Stem Cell Research." Stem Cell Network. http://www.stemcellnetwork.ca/index.php

Kotikoff, Laurence J. 2002. "Generational Policy." In Martin Feldstein and A.J. Auerbach (eds.), *Handbook of Public Economics*, vol. 4, *1873–1932*. New York: Elsevier.

KPMG. 2009. *Genome Canada Performance Audit 2009*. http://www.genomecanada.ca/medias/PDF/EN/Performance_Audit_Report09.pdf

Krashinsky, Susan. 2014. "How Canadians Are Using the Internet Differently." *Globe and Mail*, 1 Apr.

Krugman, Paul. 1996. *How the Economy Organizes Itself in Space: A Survey of the New Economic Geography*. Santa Fe Institute. http://www.santafe.edu/media/workingpapers/96-04-021.pdf

Kryder, R. David, Stanley P. Kowalski, and Anatole F. Krattiger. 2000. *The Intellectual and Technical Property Components of Pro-Vitamin A Rice (Golden Rice): A Preliminary Freedom-to-Operate Review*. https://isaaa.org/resources/publications/briefs/20/download/isaaa-brief-20-2000.pdf

Kuchler, Hannah. 2014. "Fresh Exit from Twitter as Clash Over How to Attract Users Takes Toll." *Financial Times*, 13 June, 1.

Kuhn, Thomas S. 1962. *The Structure of Scientific Revolutions*. London: Routledge.

Lafferty, William, and James Meadowcroft, eds. 2000. *Implementing Sustainable Development*. Oxford: Oxford University Press.

Lalonde, Gerard J. 1992. "1992 Proposals for SR&ED Tax Incentives: A Ten-Year Restrospective." In *1992 Ontario Tax Conference*. Toronto: Canadian Tax Foundation.

Landler, Mark. 2014. "U.S. and China Reach Climate Accord after Months of Talks." *New York Times*, 11 Nov.

Lanier, Jaron. 2013. *Who Owns the Future?* London: Penguin.

LaPointe, Russell. 2006. "The Social Sciences and Humanities Research Council: From a Granting Council to a Knowledge Council." In G. Bruce Doern (ed.), *Innovation, Science and Environment: Canadian Policies and Performance 2006–2007*, 127–48. Montreal and Kingston: McGill-Queen's University Press.

Lawrence, Amanda. 2012. *Grey Literature in Brief*. http://greylitstrategies.info/research/grey-literature-brief

Leadbeater, Charles. 2008. *We Think: Mass Innovation, Not Mass Production*. London: Profile Books.

– 2012. *Innovation in Education*. London: Bloomsbury Quatar Foundation.

Lee, Dave. 2014. "Google to Face Data Watchdogs over Right to Be Forgotten." *BBC News*, 24 July.

Lee, Ian. 2012. "The Harper Election Victory, the Liberals, and Debate over Corporate Tax Reform." In G. Bruce Doern and Christopher Stoney (eds.), *How Ottawa Spends 2012–2013: The Harper Majority, Budget Cuts, and the New Opposition*, 35–52. Montreal and Kingston: McGill-Queen's University Press.

Leiner, Barry M., Vinton G. Cerf, David D. Clark, Robert E. Kahn, Leonard Kleinrock, Daniel C. Lynch, Jon Postel, Larry G. Roberts, and Stephen Wolff. 2014. *Brief History of the Internet*. www.internetsociety.org

Leiss, William. 2000. "Between Expertise and Bureaucracy: Risk Management Trapped at the Science-Policy Interface." In G. Bruce Doern and Ted Reed (eds.), *Risky Business: Canada's Changing Science-Based Policy and Regulatory Regime*, 49–74. Toronto: University of Toronto Press.

Lemmens, Trudo, and Ron A. Bouchard. 2007. "The Regulation of Pharmaceuticals in Canada." In Jocelyn Downie, Timothy Caulfield, and Colleen M. Flood (eds.), *Canadian Health Law and Policy*, 3rd ed., 56–69. Toronto: Butterworths.

Lenard, Patti Tamara, and Richard Simeon. 2012. *Imperfect Democracies: The Democratic Deficit in Canada and the United States*. Vancouver: UBC Press.

Levasseur, Karine. 2009. "Universities and the Regulation of Research Ethics." In G. Bruce Doern and Christopher Stoney (eds.), *Research and Innovation Policy: Changing Federal Government–University Relations*, 242–64. Toronto: University of Toronto Press.

Levi-Faur, David. 2005. "The Global Diffusion of Regulatory Capitalism." *Annals of the American Academy of Political and Social Science* 598: 12–32.

Liberal Party of Canada. 1993. *Creating Opportunity: The Liberal Plan for Canada*. (The Red Book). The Platform Committee: co-chairs, Haviva Hosek and Paul Martin. Ottawa: Author.

– 2015. *Real Change: A New Plan for a Strong Middle Class*. Ottawa: Author.

Lindblom, Charles E., and Edward J. Woodhouse. 1993. *The Policy-Making Process*. 3rd ed. New York: Prentice-Hall.

Lloyd, John, and Laura Toogood. 2014. *Journalism and PR: News, Media and Public Relations in the Digital Age*. London: I.B. Tauris.

Lohr, Steve. 2015. "FCC Plans Strong Hand to Regulate the Internet." *New York Times*, 4 Feb.

Lopreite, Debora. 2006. "The Natural Sciences and Engineering Research Council as a Granting and Competitiveness-Innovation Body." In G. Bruce Doern (ed.), *Innovation, Science and Environment: Canadian Policies and Performance, 2006–2007*, 105–26. Montreal and Kingston: McGill-Queen's University Press.

Lopreite, Debora, and Joan Murphy. 2009. "The Canada Foundation for Innovation as Patron and Regulator." In G. Bruce Doern and Christopher Stoney (eds.), *Research and Innovation Policy: Changing Federal Government–University Relations*, 123–47. Toronto: University of Toronto Press.

Lorch, Mark. 2013. "Accusations of Fraud Spur a Revolution in Scientific Publishing." *Guardian*, 8 Nov., 8.

Losey, John E., Linda S. Raynor, and Maureen E. Carter. 1999. "Transgenic Pollen Harms Monarch Larvae." *Nature* 399 (6733): 214.

Lucas, Robert E. 1988. "On the Mechanics of Economic Development." *Journal of Monetary Economics* 22 (1): 3–42.

Lundvall, Bengt-Ake, ed. 1992. *National Systems of Innovation: Towards a Theory of Innovation and Interactive Learning*. London: Pinter.

Lynch, Michael P. 2014. "Privacy and the Pool of Information." *New York Times*, 27 July.

– 2016. *The Internet of Us: Knowing More and Understanding Less in the Age of Big Data*. New York: W.W. Norton.

Macdonald, Douglas. 2011. "Harper Energy and Climate Change Policy: Failing to Address the Key Problems." In Christopher Stoney and G. Bruce

Doern (eds.), *How Ottawa Spends 2011–2012: Trimming Fat or Slicing Pork?* 127–43. Montreal and Kingston: McGill-Queen's University Press.

MacDonald, Larry. 2000. *Nortel Networks: How Innovation and Vision Created a Network Giant*. Toronto: Wiley.

Mackenzie, C.J. 1964. *Report to the Prime Minister on Government Science*. Ottawa: Privy Council Office.

MacKinnon, Peter. 2014. *University Leadership and Public Policy in the Twenty-First Century: A President's Perspective*. Toronto: University of Toronto Press.

MacLean, D. 2007. "Fuelling Creativity: Innovation, Participation and Entrepreneurship." Paper prepared for the Canada Roundtable on the Future of the Internet Economy, 2 Oct., Ottawa.

MacMillan, Margaret. 2013. *The War that Ended Peace: The Road to 1914*. London: Random House.

Madgett, Paul J., and Christopher Stoney. 2009. "Universities, Commercialization, and the Entrepreneurial Process: Barriers to Industry." In G. Bruce Doern and Christopher Stoney (eds.), *Research and Innovation Policy: Changing Federal Government–University Relations*, 148–71. Toronto: University of Toronto Press.

Maeder, Thomas. 2003. "The Orphan Drug Backlash." *Scientific American* 288 (5): 70–7.

Magnuson-Ford, Karen, and Katie Gibbs. 2014. *Can Scientists Speak?* https://evidencefordemocracy.ca/sites/default/files/reports/Can%20Scientists%20Speak_.pdf

Malloy, Jonathan. 2004. "The Executive and Parliament in Canada." *Journal of Legislative Studies* 10 (2/3): 206–17.

Marcellin, Sherry S. 2010. *The Political Economy of Pharmaceutical Patents*. London: Ashgate.

Marchant, Gary E., Kenneth W. Abbott, and Braden Allenby, eds. 2013. *Innovative Governance Models for Emerging Technologies*. London: Edward Elgar.

Marland, Alex, Thierry Glasson, and Tamara A. Small. 2015. *Political Communication in Canada: Meet the Press and Tweet the Rest*. Vancouver: UBC Press.

Marshall, Alfred. 1890. *Principles of Economics*. London: Macmillan.

Martin, Lawrence. 2010. *Harperland: The Politics of Control*. Toronto: Viking.

Martin, Roger, and Karen Christensen, eds. 2013. Rotman *on Design: The Best on Design Thinking from* Rotman *Magazine*. Toronto: University of Toronto Press.

Maybee, Gareth E., and Robert E. Mitchell. 1985. *History of the Patent and Trademark Profession in Canada*. Ottawa: Patent and Trademark Institute of Canada.

Mayer-Schönberger, Viktor, and Kenneth Cukier. 2013. *Big Data: A Revolution that Will Transform How We Live, Work and Think*. New York: Houghton Mifflin Harcourt.

Mazzacato, Mariana. 2013. *The Entrepreneurial State*. London: Anthern Press.

McCabe, Christopher, and Don Husereau. 2014. *Personalized Medicine and Health Care Policy: From Science to Value*. http://www.genomecanada.ca/medias/pdf/en/PersonalizedMedecine_Policy-Directions-Brief.pdf

McCarthy, Shawn. 2013. "Alberta's Bold Plan to Cut Emissions Stuns Ottawa and Oil Industry." *Globe and Mail*, 13 Apr.

McDonald, David. 2014. "Big Data and Big Problems." *Carleton University Magazine* (Spring): 28–33.

McPhee Knowles, Sara, and Peter W.B. Phillips. 2014. "New Methods in Science, Technology and Policy Analysis". Presented to EU-SPRI Conference, Manchester, UK, 30 June.

McFetridge, Donald. 1981. "Research and Development Expenditures." In G. Bruce Doern (ed.), *How Ottawa Spends Your Tax Dollars: Federal Priorities 1981*, 255–79. Toronto: James Lorimer.

McFetridge, Donald, ed. 1985. *Canadian Industrial Policy in Action*. Toronto: University of Toronto Press.

McGlade, Christofe, and Paul Ekins. 2015. "The Geographical Distribution of Fossil Fuels Unused When Limiting Global Warming to 2 Degrees Centigrade." *Nature* 517: 187–90.

McGrane, David. 2013. "National Unity through Disengagement: The Harper Government's One-Off Federalism." In Christopher Stoney and G. Bruce Doern (eds.), *How Ottawa Spends 2013–2014: The Harper Government – Mid-Term Blues and Long-Term Plans*, 114–26. Montreal and Kingston: McGill-Queen's University Press.

McGrath, Matt. 2014. "Fossil Fuels Should Be Phased Out by 2100, Says IPCC." *BBC News*, 2 Nov.

McKenna, Barrie. 2011. "Flawed R&D Scheme Costs Taxpayers Billions." *Globe and Mail*, 11 Mar.

McKenzie-Brown, Peter. (in press). *Powerhouse: A Biography of Canada's Oil Sands*. Edmonton: University of Alberta Press.

Meadowcroft, James, and Oluf Langhelle, eds. 2011. *Caching the Carbon: The Politics and Policy of Carbon Capture and Storage*. London: Edward Elgar.

Merton, Robert K. 1973. *The Sociology of Science: Theoretical and Empirical Investigations*. Chicago: University of Chicago Press.

Meslin, Eric M. 2010. *What to Expect from the New Bioethics Commission*. http//www.scienceprogress.org/2010/05/problemsolvers

Meulien, Pierre. 2014. "Genomics and Personalized Heath." Presentation to the NHGRI Global Leaders in Genomic Medicine Meeting, 8–9 Jan., Washington, DC.

Migone, Andrea, and Michael Howlett. 2009. "Classifying Biotechnology-Related Policy, Regulatory and Innovation Regimes: A Framework for the Comparative Analysis of Genomics Policy-Making." *Policy and Society* 28 (4): 267–78.

Milke, Mark. 2013. *Corporate Welfare at Industry Canada since John Diefenbaker*. https://www.fraserinstitute.org/sites/default/files/corporate-welfare-at-industry-canada-since-john-diefenbaker-rev.pdf

Miller Chenier, Nancy. 1994. *Reproductive Technologies: Royal Commission Final Report*. Ottawa: Library of Parliament, Political and Social Affairs Division.

– 2002. *Intergovernmental Consultations on Health: Toward a National Framework on Reproductive Technologies*. http://dsp-psd.pwgsc.gc.ca/Collection-R/LoPBdP/PRB-e/PRB0234-e.pdf

Miller, Claire Cain. 2013. "Angry Over U.S. Surveillance, Tech Giants Bolster Defences." *New York Times*, 31 Oct.

Miller, Jeff. 2012. "Net Neutrality Regulation in Canada: Assessing the CRTC's Statutory Competency to Regulate the Internet." *Appeal Review of Current Law and Law Reform* 17 (1): 47–62.

Miller, James R. 2006. *A Short History of SSHRC*. Unpublished. University of Saskatchewan.

Mills, Lisa, and Ashley Weber. 2006. "Access to Emergency Medicines: How Ottawa Amends the Patent Act." In G. Bruce Doern (ed.), *How Ottawa Spends 2007–2007: In From the Cold – The Tory Rise and the Liberal Demise*, 229–46. Montreal and Kingston: McGill-Queen's University Press.

Mitchell, Ronald B. 2013. *2002-2013 International Environmental Agreements Database Project*. (Version 2013.1). http://iea.uoregon.edu/

Moir, Hazel. 2013. *Patent Policy and Innovation*. London: Edward Elgar.

Mokyr, J. 2002. "Innovation in an Historical Perspective." In B. Steil, G. Victor, and R. Nelson (eds.), *Technological Innovation and Economic Performance*, 23–46. Princeton, NJ: Princeton University Press.

Mook, Laurie, Jack Quarter, and Sherida Ryan, eds. 2012. *Businesses with a Difference: Balancing the Social and the Economic*. Toronto: University of Toronto Press.

Mooney, Chris. 2005. *The Republican War on Science*. New York: Basic Books.

Morcol, Goktug. 2012. *A Complexity Theory for Public Policy*. London: Routledge.

Morgan, Clara. 2009. "Higher Education Funding and Policy Trade-offs: The AUCC and Federal Research in the Chrétien-Martin Era." In G. Bruce Doern and Christopher Stoney (eds.), *Research and Innovation Policy: Changing Federal Government–University Relations*, 59–88. Toronto: University of Toronto Press.

Morris, Stephen G. 2007. "Canada's Assisted Human Reproduction Act: A Chimera of Religion and Politics." *American Journal of Bioethics* 7 (2): 69–70.

Mowat Centre. 2014. *A Federal Economic Agenda for Ontario: Discussion Paper*. University of Toronto, School of Public Policy and Governance.

Mueller, Lisa L. 2014. *Patent Trolls: A Global Perspective*. http://www.lexology.com/library/detail.aspx?g=e6a23668-b328-42b3-bfff-06d54b2d33ee

Mueller, Milton L., and Hadi Asghari. 2012. "Deep Packet Inspection and Bandwidth Management: The Battle Over Bittorrent in Canada and the United States." *Telecommunication Policy* 36 (6): 462–75.

Mundie, Craig. 2014. "Privacy Pragmatism: Focus on Data Use, Not Data Collection." *Foreign Affairs* 93 (2): 28–38.

Murphy, David. 2014. "44 Percent of Twitter Accounts Have Never Tweeted." *PC Magazine*, 3 Apr.

Murphy, Joan. 2006. "Multilevel Regulatory Governance in the Health Sector." In G. Bruce Doern and Robert Johnson (eds.), *Rules, Rules, Rules, Rules: Multilevel Regulatory Governance*, 305–24. Toronto: University of Toronto Press.

– 2007. "Transforming Health Sciences Research: From the Medical Research Council to the Canadian Institutes of Health Research." In G. Bruce Doern (ed.), *Innovation, Science, Environment: Canadian Policies and Performance 2007–2008*, 240–61. Montreal and Kingston: McGill-Queen's University Press.

Naughton, John. 2012. *From Guttenberg to Zuckerberg: What You Really Need to Know about the Internet*. London: Quercus.

Nelson, Richard R., ed., 1993. *National Innovation Systems: A Comparative Analysis*. Oxford: Oxford University Press.

Nielsen, Michael. 2011. *Reinventing Discovery: The New Era of Networked Science*. Princeton, NJ: Princeton University Press.

Niosi, Jorge. 2000. *Canada's National System of Innovation*. Montreal and Kingston: McGill-Queen's University Press.

– 2005. *Canada's Regional Innovation System*. Montreal and Kingston: McGill-Queen's University Press.

Norberg, Jon, and Graeme S. Cumming, eds. 2012. *Complexity Theory for a Sustainable Future*. New York: Columbia University Press.

Nordhaus, William D. 2007. "A Review of the Stern Review on the Economics of Climate Change." *Journal of Economic Literature* 65 (3): 686–702.

Nordmann, Alfred. 2004. *Converging Technologies: Shaping the Future of European Societies*. http://www.ntnu.no/2020/pdf/final_report_en.pdf

Nye, David E. 2006. *Technology Matters: Questions to Live With*. Cambridge, MA: MIT Press.

O'Doherty, Kieran, and Edna Einsiedel, eds. 2013. *Public Engagement and Emerging Technologies*. Vancouver: UBC Press.

O'Hara, Kathryn, and Paul Dufour. 2014. "How Accurate is the Harper Government's Misinformation? What Role Does Evidence Play in Federal Policy Making?" In Christopher Stoney and Bruce Doern (eds.), *How Ottawa Spends 2013–14*, 178–91. Montreal and Kingston: McGill-Queen's University Press.

Okereke, Chukwumerije , Harriet Bulkeley, and Heike Schroeder. 2009. "Conceptualizing Climate Governance beyond the International Regime." *Global Environmental Politics* 9 (1): 58–78.

O'Malley, Kady. 2013. "Research Agency Faced Existential Confusion before Rebrand, Data Reveals." *CBC News*, 13 May.

Oreskes, Naomi, and Erik M. Conway. 2010. *Merchants of Doubt*. London: Bloomsbury Press.

Organization for Economic Cooperation and Development (OECD). 2002. *OECD Reviews of Regulatory Reform: Canada 2002 – Maintaining Leadership through Innovation*. Paris: Author.

– 2010. *The OECD Innovation Strategy: Getting a Head Start on Tomorrow*. Paris: Author.

– 2013. *Main Science and Technology Indicators*. Paris: Author.

– 2014. *Main Science and Technology Indicators*. Issue 2. Paris: Author

– 2015a. *Science, Technology and Innovation*. http://www.oecd-library.org/science-and-technology/

– 2015b. *Defining Innovation*. http://www.oecd-library.org/site/innovationstrategy/defining/innovation.html

– 2015c. *Oslo Manual Guidelines for Collecting and Interpreting Innovation Data*. 3rd ed. http://www.oecd-ilibrary.org/science-and-technology/oslo-manual_9789264013100-en

Osborne, David, and Ted Gaebler. 1992. *Reinventing Government*. Reading, MA: Addison-Wesley.

Ostrom, Elinor. 1990. *Governing the Commons: The Evolution of Institutions for Collective Action*. New York: Cambridge University Press.

– 2012. *The Future of the Commons: Beyond Market Failure and Government Regulation.* London: Institute of Economic Affairs.

Ostry, Sylvia. 1997. *The Post–Cold War Trading System: Who's on First?* Chicago: University of Chicago Press.

Owens, Brian. 2013. "Does Parliament Need a Science Watchdog?" http://www.scibri.com/2013/11/26/does-parliament-need-a-science-watchdog/

Pal, Leslie A. 1998. "Balancing Act: The Post-Deficit Mandate." In Leslie A. Pal (ed.), *How Ottawa Spends 1998–99: Balancing Act – The Post-Deficit Mandate*, 1–30. Toronto: Oxford University Press.

– 2000. "Past Imperfect, Future Tense." In Leslie A. Pal (ed.), *How Ottawa Spends 2000–2001: Past Imperfect, Future Tense*, 1–32. Toronto: Oxford University Press.

– 2011. "Assessing Incrementalism: Formative Assumptions, Contemporary Realities." *Policy and Society* 30 (1): 29–39.

– 2012. *The Frontiers of Governance: The OECD and Global Public Management Reform*. London: Palgrave Macmillan.

– 2013. *Beyond Policy Analysis: Public Issue Management in Turbulent Times.* 5th ed. Toronto: Nelson.

Panitch, Leo, ed.1977. *The Canadian State*. Toronto: University of Toronto Press.

Papadopoulos, Stavroula. 2009. *History of the Creation of the Council of Canadian Academies*. www.scienceadvice.ca/uploads/eng/workshop report and history/(2009) history of the council of canadian academies.pdf

Parsons, Mark, and Nicholas Phillips. 2007. *An Evaluation of the Federal Tax Credit for Research and Experimental Development*. Ottawa: Department of Finance.

Pavitt, Keith. 1991. "What Makes Basic Research Economically Useful?" *Research Policy* 20: 109–19.

Peekhaus, Wilhelm. 2013. *Resistance Is Fertile: Canadian Struggles on the BioCommons*. Vancouver: UBC Press.

Pembina Institute. 2011. "Pembina's Work on Natural Capital." Edmonton: Pembina Institute.

Performance Management Network Inc. 2013. "Review of Relevance and Effectiveness of the Networks of Centres of Excellence (NCE) Program." Prepared for NCE.

Personalized Medicine Coalition. 2014. *The Case For Personalized Medicine*. 4th ed. Washington, DC: Author.

Pew Research Center. 2014. Pew Research Internet Project. *World Wide Web Timeline*. March. http://www.pewinternet.org

Phillips, Peter W.B. 2003. "The Economic Impact of Herbicide Tolerant Canola in Canada." In N. Kalaitzanondakes (ed.), *The Economic and Environmental Impacts of Agbiotech: A Global Perspective*, 119–40. Dordrecht: Kluwer.

– 2007. *Governing Transformative Technological Innovation: Who's in Charge?* London: Edward Elgar.

– 2012. "Democracy, Governance and Public Engagement: A Critical Assessment." In K.O'Doherty and E. Einsiedel (eds.), *Public Participation and Emerging Technologies*, 45–65. Vancouver: UBC Press.

Phillips, Peter W.B., and David Castle. 2010. "Science and Technology Spending: Still No Viable Federal Innovation Agenda." In Bruce Doern and Christopher Stoney (eds.), *How Ottawa Spends 2010–2011: Recession, Realignment, and the New Deficit Era*, 168–86. Montreal and Kingston: McGill-Queen's University Press.

– 2012. "Science and Technology in Canada: Government Investment at a Crossroad?" In G. Bruce Doern and Christopher Stoney (eds.), *How Ottawa Spends 2012–2013: The Harper Majority, Budget Cuts, and the New Opposition*, 89–105. Montreal and Kingston: McGill-Queen's University Press.

– 2013. "Is the Budget Action Plan and Related Harper Agenda a Coherent Plan for Economic Growth?" In Christopher Stoney and Bruce Doern (eds.), *How Ottawa Spends 2013–2014: The Harper Government – Mid-Term Blues and Long-Term Plans*, 19–30. Montreal and Kingston: McGill-Queen's University Press.

Phillips, Peter, and G.G. Khachatourians. 2001. *The Biotechnology Revolution in Global Agriculture: Invention, Innovation and Investment in the Canola Sector*. Oxfordshire, UK: CAB International.

Phillips, Peter , and H. McNeill. 2001. "Labelling for GM Foods: Theory and Practice." In V. Santaniello, R.E. Evenson and D. Zilberman (eds.), *Market Development for Genetically Modified Foods*, 245–61. Oxfordshire, UK: CAB International.

Phillips, Peter W.B., A. Boland, and C. Ryan, 2013. *Public-Private-Producer Partnerships (P4s) in Canada*. Report submitted to Agriculture and Agri-Food Canada.

Phillips, Peter, and S. Smyth. 2004. "Managing the Value of New-Trait Varieties in the Canola Supply Chain in Canada." *Supply Chain Management* 9 (4): 313–22.

Phillips, Peter W.B., G. Webb, J. Karwandy, and C. Ryan. 2012. *Innovation in Agri-food Research Systems: Theory and Case Studies*. Oxfordshire, UK: CAB International.

Phillips, Peter W.B., Sidi Zhang, Tara Williams, and Laural DeBusschere. 2013. "Canada's First Nations' Policies and Practices Related to Managing Traditional Knowledge." In T. Bubela and E.R. Gold (eds.), *Genetic Resources and Traditional Knowledge: Case Studies and Conflicting Interests*, chapter 8. London: Edward Elgar.

Phillips, Peter, and G.G. Khachatourians. 2001. *The Biotechnology Revolution in Global Agriculture: Invention, Innovation and Investment in the Canola Sector*. Oxfordshire, UK: CAB International.

Phillips, Peter, S. Smyth, and W. Kerr, eds. 2006. *Governing Risk in the 21st Century: Lessons from the World of Biotechnology*. Huntington, NY: Nova Science.

Phillips, Peter W.B., and R. Wolfe, eds. 2001. *Governing Food: Science, Safety and Trade*. Montreal and Kingston: McGill-Queen's University Press and Queen's University, School of Public Policy.

Phillips, Sarah. 2007. "A Brief History of Facebook." *Guardian*, 25 July.

Phillips McDougall. 2011. *The Cost and Time Involved in the Discovery, Development and Authorization of a New Plant Biotechnology Derived Trait*. Consultancy Study for Crop Life International. https://croplife.org/wp-content/uploads/2014/04/Getting-a-Biotech-Crop-to-Market-Phillips-McDougall-Study.pdf

Phillipson, Donald C. 1991. "The National Research Council of Canada: Its Historiography, Its Chronology, Its Biography." *Scientia Canadensis* 15 (2): 177–97.

Phoenix Strategic Perspectives Inc. 2013. *Final Report: Focus Group Research Assessing Communication Approaches for the NSERC Strategy for Partnerships and Innovation*. Prepared for Natural Sciences and Engineering Research Council of Canada. http://www.nserc-crsng.gc.ca/_doc/Reports-Rapports/StudyFinalReport.pdf

Picard, Andre. 2015. "Governments Need to Plan for Expensive Orphan Drugs." *Globe and Mail*, 6 Mar.

Polanyi, Michael. 1967. *The Tacit Dimension*. New York: Doubleday Anchor Books.

Porter, Jene, and Peter W.B. Phillips. 2007. *Public Science in Liberal Democracy*. Toronto: University of Toronto Press.

Porter, Michael. 1990. *The Competitive Advantage of Nations*. New York: Free Press.

Preston, Alex. 2014. "The Death of Privacy." *Observer*, 3 Aug.

Preston, Jennifer. 2011. "Social Media History Becomes a New Job Hurdle." *New York Times*, 20 July.

Price, Don K. 1965. *The Scientific Estate*. Boston: Harvard University Press.
PricewaterhouseCoopers. 2011. "SR&ED Tax Clips: 2010 Provincial and Territorial R&D Tax Credits." 15 July.
Professional Institute of the Public Service of Canada (PIPSC). 2013. *Who We Are*. http://www.pipsc.ca/portal/page/portal/website/aboutinstitute
– 2014. *Vanishing Science*. http://www.pipsc.ca/portal/page/portal/website/news/magazine/autumn2014/6
Public Policy Forum. 1998. *Blood, Fish and Tears: A Roundtable Discussion on the Credibility and Acceptability of Science Advice for Decision Making*. https://www.ppforum.ca/sites/default/files/Blood_Fish_Tears.pdf
– 2014. *Canada's Evolving Internal Market: An Agenda for a More Cohesive Economic Union*. http://www.ppforum.ca/publications/canada%E2%80%99s-evolving-internal-market-agenda-more-cohesive-economic-union
Rainie, Lee, and Barry Wellman. 2012. *Networked: The New Social Operating System*. Cambridge, MA: MIT Press.
Rao, S., and Andrew Sharpe, eds. 2002. *Productivity Issues in Canada*. Calgary: University of Calgary Press.
Rayner, Steve. 2012. "Uncomfortable Knowledge: The Social Construction of Ignorance in Science and Environmental Policy Discourses." *Economy and Society* 41 (1): 107–23.
Regier, Dean A., Stephanie J. Peacock, Reka Palaky, Kimberley van der Hoek, Gail P. Jarvik, Jeffrey Hoch, and David Veenstra. 2015. "Societal Preferences for the Return of Incidental Findings from Clinical Genomic Sequencing: A Discrete-Choice Experiment." *Canadian Medical Association Journal* 187 (6): E190–7.
Reguly, Eric. 2013. "The Decline and Fall of Canada's Global Corporate Superstars." *Globe and Mail*, 6 Aug.
Revkin, Andrew C. 2014. "A Darker View of the Age of Us – The Anthropocene," *New York Times*, 18 June.
Rhodes, Roderick A.W. 1997. *Understanding Governance*. London: Open University Press.
Rice, James J., and Michael J. Prince. 2013. *Changing Politics of Canadian Social Policy*. 2nd ed. Toronto: University of Toronto Press.
Richards, Gregory. 2014. "Big Data and the Public Service." *Ottawa Citizen*, 1 Aug.
Riddell, W. Craig. 2008. *Investing in Human Capital: Policy Priorities for Canada*. Montreal: Institute for Research on Public Policy.
Rivers, Nic, and Mark Jaccard. 2009. "Talking without Walking: Canada's Ineffective Climate Effort." In Burkard Eberlein and G. Bruce Doern (eds.),

Governing the Energy Challenge: Canada and Germany in a Multilevel Regional and Global Context, 285–313.Toronto: University of Toronto Press.

Rixen, Thomas. 2013. "Why Regulation after the Crisis Is Feeble: Shadow Banking, Offshore Financial Centers, and Jurisdictional Competition." *Regulation and Governance* 7 (4): 435–59.

Romer, Paul M. 1990. "Endogenous Technological Change." *Journal of Political Economy* 98 (5): 71–102.

Royal Society of Canada (RSC). 2013. *History*. https://www.rsc-src.ca/en/about-us/our-purpose/history RPX Corporation. 2015. *2014 NPE Litigation Report*. https://www.rpxcorp.com/wp-content/uploads/2014/12/Compressed-RPX_Litigation-Report-2014_FNL_031215.indd_.pdf

Ruiz, Rebecca. 2015. "F.C.C. Sets Net Neutrality Rules." *New York Times*, 12 Mar., 1.

Russell, Peter A. 2012. *How Agriculture Made Canada*. Montreal and Kingston: McGill-Queen's University Press.

Salutin, Rick. 2014. "Dismantling of DFO Libraries a Blow to Democracy." *Toronto Star*, 9 Jan.

Saner, Marc A. 2007. *A Map of the Interface between Science and Policy*. University of Ottawa, Institute for Science, Society and Policy.

– 2014. "Evaluating Science-Policy Interfaces." Presentation to Gordon Research Conference, 10–15 Aug., Waterville.

Savoie, Donald. 1987. *Regional Economic Development: Canada's Search for Solutions*. Toronto: University of Toronto Press.

– 1999. *Governing from the Centre: The Concentration of Power in Canadian Politics*. Toronto: University of Toronto Press.

– 2003a. *Breaking the Bargain: Ministers, Public Servants and Parliament*. Toronto: University of Toronto Press.

– 2003b. "Reviewing Canada's Regional Development Efforts." In Victor L. Young, vol. 4 of *Collected Research Papers of the Royal Commission on Renewing and Strengthening Our Place in Canada*, 147–84. St. John's, NL: Queen's Printer. http://www.gov.nf.ca/publicat/royalcomm/research/Savoie.pdf

– 2010. *Power: Where Is It?* Montreal and Kingston: McGill-Queen's University Press.

– 2013. *Whatever Happened to the Music Teacher? How Government Decides and Why*. Montreal and Kingston: McGill-Queen's University Press.

Scala, F. 2007. "Scientists, Government, and 'Boundary Work': The Case of Reproductive Technologies and Genetic Engineering." In Michael Orsini and Miriam Smith (eds.), *Critical Policy Studies*, 211–31. Vancouver: UBC Press.

Schooler, Dean. 1971. *Science, Scientists and Public Policy*. New York: Free Press.

Schultz, Richard. 2000. *The Consumers' Association of Canada and the Federal Regulatory System, 1973–1992*.Vancouver: SFU-UBC Centre for the Study of Government and Business.

– 2003. "Dancing around the Digital Divide: The Fight for a Federal Broadband Policy." In G. Bruce Doern (ed.), *How Ottawa Spends 2003–2004: Regime Change and Policy Shift*, 187–203. Toronto: Oxford University Press.

– 2008. "Telecommunications: What a Difference a Minister Can Make." In Allan M. Maslove (ed.), *How Ottawa Spends 2008–2009*, 155–71. Montreal and Kingston: McGill Queens University Press.

– 2011. "Industry Canada as Economic Regulator: Globealive and the Lessons of Political Licensing." In Christopher Stoney and G. Bruce Doern (eds.), *How Ottawa Spends 2011–2012: Trimming Fat or Slicing Pork?* 198–217. Montreal and Kingston: McGill-Queen's University Press.

Schultz, Richard, and Alan Alexandroff. 1985. *Economic Regulation and the Federal System*. Toronto: University of Toronto Press.

Science Council of Canada (SCC). 1968. *Towards a National Science Policy for Canada*. Ottawa: Queen's Printer. http://artsites.uottawa.ca/sca/doc/Report-no.-4-Towards-a-National-Science-Policy-for-Canada.pdf

– 1971. *A Trans-Canada Computer Communications Network: Part I of a Major Program on Computers*. Ottawa: Information Canada. http://artsites.uottawa.ca/sca/doc/Report-no.-13-A-Trans-Canada-Computer-Communications-Network.pdf

– 1973. *Strategies of Development for the Canadian Computer Industry*. Ottawa: Information Canada. http://artsites.uottawa.ca/sca/doc/Report-no.-21-Strategies-of-Development-for-the-Canadian-Computer-Industry.pdf

– 1982. *Planning Now for an Information Society – Tomorrow Is Too Late*. Ottawa: Supply and Services Canada. http://artsites.uottawa.ca/sca/doc/Report-no.-33-Planning-Now-for-an-Information-Society-Tomorrow-is-too-late.pdf

Science-Metrix. 2010. *Tenth-Year Evaluation of the Canada Research Chairs Program Final Evaluation Report*. Montreal: Author.

Science, Technology and Innovation Council (STIC). 2012. *State of the Nation Report*. Ottawa: Author.

– 2014. *Mandate*. http://www.stic-csti.ca/eic/site/stic-csti.nsf/eng/h_00005.html

Schneier, Bruce. 2015. *Data and Goliath: The Hidden Battles to Collect Your Data and Control Your World*. New York: W.W. Norton.

Schumpeter, Joseph A. 1934. *The Theory of Economic Development*. Cambridge, MA: Harvard University Press.

– 1954. *History of Economic Analysis*. London: Allen and Unwin.

Scott, Mark. 2014a. "One Court Ruling on Privacy in Europe, and 28 Regulators." *New York Times*, 14 May.

– 2014b. "Principles Are No Match for Europe's Love of U.S. Web Titans." *New York Times*, 6 July.

SECOR. 2012. "CADTH Patient Input Process Review." Prepared for the Canadian Agency for Drugs and Technologies in Health. https://www.cadth.ca/media/pdf/2012_SECOR_Patient-Input-Review_e.pdf

Sell, Susan. 1998. *Power and Ideas: North-South Politics of Intellectual Property and Antitrust*. New York: State University of New York Press.

– 2010. "The Rise and Rule of Trade-Based Strategy: Historical Institutionalism and the International Regulation of Intellectual Property." *Review of International Political Economy* 17 (4): 762–90.

Selley, Chris. 2012. "We Won't Let C-30 Limit Our Freedom without a Fight." *National Post*, 22 Feb.

Séralini, Gilles-Eric, Emilie Clair, Robin Mesnage, Steve Gress, Nicolas Defarge, Manuela Malatesta, Didier Hennequin, and Joël Spiroux De Vendômois. 2012. "Long-Term Toxicity of a Roundup Herbicide and a Roundup-Tolerant Genetically Modified Maize." *Food and Chemical Toxicology* 50 (11): 4221–31.

Shade, Leslie Regan. 1994. "Computer Networking in Canada: From CA*net to CANARIE." *Canadian Journal of Communication* 19 (1): 22–34.

Shapin, Steven, and Simon Shaffer. 1985. *Leviathan and the Air-Pump: Hobbes, Boyle and the Experimental Life*. Princeton, NJ: Princeton University Press.

Shepherd, Robert P. 2013. "How Ottawa Controls: Harper Era Strategic Reviews in the Context of the 1993–1996 Liberal Program Review." In Christopher Stoney and G. Bruce Doern (eds.), *How Ottawa Spends 2013–2014:The Harper Government – Mid-Term Blues and Long-Term Plans*, 101–13. Montreal and Kingston: McGill-Queen's University Press.

Shreeve, James. 2004. *The Genome War*. New York: Knopf.

Silcoff, Sean. 2014. "Rescue in Motion." *Report on Business*, Mar., 30–5.

Skinner, Mel. 1978. "The Federal Make or Buy R&D Policy: A Preliminary Analysis of Policy and Implementation." MA thesis, Carleton University, Ottawa.

Slaughter, Sheila, and Gary Rhoades. 2004. *Academic Capitalism and the New Economy*. Baltimore, MD: Johns Hopkins University Press.

Smarden, Bruce. 2015. *Asleep at the Switch: The Political Economy of Federal Research and Development Policy Since 1960*. Montreal and Kingston: McGill-Queen's University Press.

Smith, Alex. 2006. *The Accountability of Deputy Ministers Before Parliaments*. Ottawa: Library of Parliament. http://www.parl.gc.ca/Content/LOP/ResearchPublications/prb0548-e.htm

Smyth, Stuart, and Peter W.B. Phillips. 2002. "Competitors Co-operating: Establishing a Supply Chain to Manage Genetically Modified Canola." *International Food and Agribusiness Management Review* 4 (1): 51–66.

Sommer, Jeff. 2014. "Defending the Open Internet." *New York Times*, 10 May.

Soroka, Stuart N. 2002. *Agenda-Setting Dynamics in Canada*. Vancouver: UBC Press.

Spears, Tom. 2015. "Why Canada Now Has Two Science Ministers." *Ottawa Citizen*. 5 Nov. 2015. http://ottawacitizen.com/news/national/why-canadian-science-now-has-two-science-ministers

Spence, Jennifer. 2014. "The National Shipbuilding Model for Government Procurement: Separating the Wheat from the Chaff." In G. Bruce Doern and Christopher Stoney (eds.), *How Ottawa Spends 2014–2015: The Harper Government – Good to Go?* 165–77. Montreal and Kingston: McGill-Queen's University Press.

Spencer, Gregory M. 2014. *Cluster Atlas of Canada*. University of Toronto, Munk School of Global Affairs. https://localideas.files.wordpress.com/2014/05/cluster-atlas.pdf

Stern, Nicholas. 2007. *Stern Review on the Economics of Climate Change*. Cambridge: Cambridge University Press.

Stilborn, Jack. 1998. *Federal Public Service Renewal: The La Relève Initiative*. Ottawa: Library of Parliament. http://www.parl.gc.ca/Content/LOP/researchpublications/prb987-e.htm

Stirling, Andy. 2013. "Why the Precautionary Principle Matters." *Guardian*, 8 July.

Stone, Richard, and Barbara Jasny. 2013. "Scientific Discourse: Buckling at the Seams." *Science* 342 (4): 57.

Stoney, Christopher, and G. Bruce Doern, eds. 2015. *How Ottawa Spends 2015–2016: The Liberal Rise and the Tory Demise*, 1–20. Ottawa: School of Public Policy and Administration, Carleton University.

Storer, Norman W. 1966. *The Social System of Science*. New York: Holt, Rinehart and Winston.

Strangelove. Michael. 2015. *Post-TV: Piracy, Cord-Cutting, and the Future of Television*. Toronto: University of Toronto Press.

Streitfeld, David. 2014. "Amazon, a Friendly Giant as Long as It's Fed." *New York Times*, 12 July.

Stross, Randall. 2008. *Planet Google*. New York: Free Press.

Swann, Peter. 2014. *Common Innovation: How We Create the Wealth of Nations*. London: Edward Elgar.

Sweeny, Alastair. 2009. *Blackberry Planet: The Story of Research in Motion and the Little Device that Took the World by Storm*. Toronto: Wiley.

Swimmer, Gene, ed. 1996. *How Ottawa Spends 1996–97: Life Under The Knife*. Ottawa: Carleton University Press.

Taylor, Matthew, Nick Hopkins, and Jenima Kiss. 2013. "NSA Surveillance May Cause Break-up of the Internet, Warn Experts." *Guardian*, 1 Nov.

Teichler, Ulrich. 2007. *Higher Education Systems, Conceptual Frameworks, Comparative Perspectives*. New York: Sense-Publishers.

Thierer, Adam. 2014. *Permissionless Innovation: The Continuing Case for Comprehensive Technological Freedom*. George Mason University, Mercatus Center.

Thistle, Mel. 1965. *The Inner Ring*. Toronto: University of Toronto Press.

Thomas, Kerry, and Janet Chan, eds. 2013. *Handbook of Research on Creativity*. London: Edward Elgar.

Thurow, Lester. 1997. "Needed: A New System of Intellectual Property Rights." *Harvard Business Review* 75 (3): 94–103.

Toner, Glen. 2008. "The Harper Minority Government and ISE: Second Year – Second Thoughts." In Glen Toner (ed.), *Innovation Science and Environment: Canadian Policies and Performance 2008–2009*, 1–29. Montreal and Kingston: McGill-Queen's University Press.

Toner, Glen, and Carey Frey. 2004. "Governance for Sustainable Development: Next Stage Institutional and Policy Development." In G. Bruce Doern (ed.), *How Ottawa Spends 2004-2005: Mandate Change in the Paul Martin Era*, 198–221. Montreal and Kingston: McGill-Queen's University Press.

Toner, Glen, and Jennifer McKee. 2014. "Harper's Wedge Politics: Bad Environmental Policy and Bad Energy Policy." In G. Bruce Doern and Christopher Stoney (eds.), *How Ottawa Spends 2014–2015*, 108–24. Montreal and Kingston: McGill-Queen's University Press.

Trebilcock, Michael J. 2011. *Understanding Trade Law*. London: Edward Elgar.

Trebilcock, Michael, Robert Howse, and Antonia Eliason. 2013. *The Regulation of International Trade*. 4th ed. London: Routledge.

Tumin, Zachary. 2000. *Connecting Canadians: Canada's Community Access Program. A Case Study of Government Strategic Investment on the Internet*. Harvard Information Infrastructure Project. Harvard University, John F. Kennedy School of Government. http://thepublicvoice.org/events/CAP_Case.pdf

Tuohy, Caroline, and Alan Wolfson. 1978. "Self-Regulation: Who Qualifies?" In Phillip Slayton and Michael Trebilcock (eds.), *The Professions and Public Policy*, 111–22. Toronto: University of Toronto Press.

Tupper, Allan. 2009. "Pushing Federalism to the Limits: Post-Secondary Education Policy in the Millennium." In G. Bruce Doern and Christopher

Stoney (eds.), *Research and Innovation Policy: Changing Federal Government–University Relations*, 35–58. Toronto: University of Toronto Press.

Turk, Eli. 1999. *Canada's Strategy for the Information Highway*. Harvard Information Infrastructure Project. Harvard University, John F. Kennedy School of Government. http://thepublicvoice.org/events/IHAC_Case.pdf

Ugur, Mehmet, ed. 2013. *Governance, Regulation and Innovation: Theory and Evidence from Firms and Nations*. London: Edward Elgar.

United States. Food and Drug Administration (FDA). 2013. *Paving the Way for Personalized Medicine*. Washington, DC: Food and Drug Administration.

– National Human Genome Research Institute. 2015a. *About NHGRI*. https://www.genome.gov/About/

– 2015b. *A Brief Guide to Genomics*. http://www.genome.gov/18016863

– National Institutes of Health. 2014. *Peer Review Process*. http://grants.nih.gov/grants/peer_review_process.htm

– National Research Council. 1983. *Risk Assessment in the Federal Government: Managing the Process*. Washington, DC: National Academy Press.

– 1994. *Science and Judgment in Risk Assessment*. Washington, DC: National Academy Press.

– 1996. *Understanding Risk: Informing Decisions in a Democratic Society*. Washington DC: National Academy Press.

– National Science Foundation. 2014. *Merit Review*. http://www.nsf.gov/bfa/dias/policy/merit_review/

Usborne, Simon. 2014. "Route Cause." *Independent*, 18 June, 25–7.

VALGEN. 2010. *Genome Canada ABC Integration Workshop Report*. http://www.valgen.ca/wp-content/uploads/2009/09/VALGEN-Banff-Workshop-Final-Report.pdf

van den Hoonaard, Will C. 2013a. "The Ethics Rupture Summit, Fredericton, New Brunswick, October 25–28, 2012." *Journal of Empirical Research on Human Research Ethics* 8 (1): 3–7.

– 2013b. "The Social and Policy Contexts of the New Brunswick Declaration on Research Ethics, Integrity, and Governance: A Commentary." *Journal of Empirical Research on Human Research Ethics* 8 (2): 104–9.

van den Hoonard, Will C., and Ann Hamilton, eds. 2016. *The Ethics Rupture: Exploring Alternatives to Formal Research-Ethics Review*. Toronto: University of Toronto Press.

van Dijck, Jose. 2013. *The Culture of Connectivity: A Critical History of Social Media*. Oxford: Oxford University Press.

Ventose, Eddy D. 2011. *Medical Patent Law: The Challenges of Medical Treatment*. London: Edward Elgar.

Vincent, James. 2014. "Facebook Will Use Drones and Lasers to Deliver Internet Access to the Developing World." *Independent*, 28 Mar.

Vodden, Christy. 1992. *No Stone Unturned: The First 150 Years of the Geological Survey of Canada*. Ottawa: Energy Mines and Resources Canada.

Vogel, David. 2012. *The Politics of Precaution: Regulating Health, Safety, and Environmental Risks in Europe and the United States*. Princeton, NJ: Princeton University Press.

Volger, John. 2013. "Environmental Issues." In John Baylis, Steve Smith, and Patricia Owens (eds.), *The Globalization of World Politics*, 222–34. Oxford: Oxford University Press.

Voyer, Roger. 1999. "Thirty Years of Canadian Science Policy: From 1.5 to 1.5." *Science and Public Policy* 26 (4): 277–82.

Waddell, Christopher. 2010. "The Auto Industry Bailout: Industrial Policy or Job-Saving Social Policy?" In G. Bruce Doern and Christopher Stoney (eds.), *How Ottawa Spends 2010– 2011: Recession, Realignment, and the New Deficit Era*, 150–67. Montreal and Kingston: McGill-Queen's University Press.

Wade, Nicholas. 2010. "Researchers Say They Created a 'Synthetic Cell.'" *New York Times*, 20 May.

Walby, Sylvia. 2007. "Complexity Theory, Systems Theory, and Multiple Intersecting Social Inequalities." *Philosophy of the Social Sciences* 37 (4): 449–70.

Warrian, Peter. 2010. *The Importance of Steel Manufacturing to Canada – A Research Study*. Munk School Briefing. http://munkschool.utoronto.ca/wpcontent/uploads/2013/05/CSPA_report_web.pdf

Waters, Richard. 2014. "Google Bows to EU Privacy Ruling." *Financial Times*, 30 May, 1.

– 2015. "Culture of Innovation Faces Squeeze." *Financial Times*, 15 Apr., 5.

Waverman, Leonard. 2007. "Benefiting from Convergenve: Access, Mobility and Ubiquity." Paper prepared for the Canada Roundtable on the Future of the Internet Economy, 2 Oct., Ottawa.

Weir, Lorna. 2010. *Global Public Health Vigilance: Creating a World on Alert*. London: Routledge.

Wellcome Trust. 1999. "From Genome to Health," *Wellcome New* (20, Q 3): 10–27.

Whitley, Richard, Jochen Glaser, and Lars Engwall, eds. 2010. *Reconfiguring Knowledge Production*. Oxford: Oxford University Press.

Whittington, Michael. 1974. "Environmental Policy." In G. Bruce Doern and V. Seymour Wilson (eds.), *Issues in Canadian Public Policy*, 203–27. Toronto: Macmillan.

– 1980. "The Department of the Environment." In G. Bruce Doern (ed.), *How Ottawa Spends Your Tax Dollars: Federal Priorities 1981*, 99–118. Toronto: James Lorimer.

Wijen, Frank, Kees Zoeteman, Jan Peters, and Paul van Seters, eds. 2013. *The Handbook of Globalization and Environmental Policy*. 2nd ed. London: Edward Elgar.

Wilks, Stephen. 2013. *The Political Power of the Business Corporation*. London: Edward Elgar.

Wilks, Stephen, and G. Bruce Doern. 2007. "Accountability and Multilevel Governance in UK Regulation." In Peter Vass (ed.), *Regulatory Review 2006–2007*, 341–71. University of Bath, Centre for the Study of Regulated Industries.

Williams, Glen. 2009. *Canadian Politics in the 21st Century*. 7th ed. Toronto: Nelson.

Wilson, Barry K. 1990. *Farming the System: How Politicians and Producers Shape Canadian Agricultural Policy*. Saskatoon: Western Producer Prairie Books.

Wolfe, David. 2002. "Innovation for the Knowledge-Based Economy: From the Red Book to the White Paper." In G. Bruce Doern (ed.), *How Ottawa Spends 2002–2003: The Security Aftermath and National Priorities*, 137–50. Toronto: Oxford University Press.

– 2009. "Universities and Knowledge Transfer: Powering Local Economic and Cluster Development." In G. Bruce Doern and Christopher Stoney (eds.), *Research and Innovation Policy: Changing Federal Government–University Relations*, 265–87. Toronto: University of Toronto Press.

Wolfe, David, and Meric S. Gertler. 2004. "Clusters from the Inside and Out: Local Dynamics and Global Linkages." *Urban Studies* 41 (5–6): 1071–93.

Wolfe, David, and Matthew Lucas. 2003. *Clusters in a Cold Climate*. Montreal and Kingston: McGill-Queen's University Press.

Wolfe, Michelle, Bryan D. Jones, and Frank R. Baumgartner. 2013. "A Failure to Communicate: Agenda Setting in Media and Policy." *Political Communication* 30 (2): 175–92.

Woods, Michael. 2013. "*New York Times* Criticizes Harper Government's Alleged Muzzling of Scientists to Protect Oil Sands." *National Post*, 22 Sept.

World Health Organization (WHO). 2003. *Diet, Nutrition and the Prevention of Chronic Diseases: Report of a Joint WHO/FAO Expert Consultation*. http://www.who.int/dietphysicalactivity/publications/trs916/en/

– 2013. *About WHO*. http://www.who.int/about/role/en/
World Intellectual Property Office (WIPO). 2013. *WIPO IP Facts and Figures*. http://www.wipo.int/edocs/pubdocs/en/statistics/943/wipo_pub_943_2013.pdf
World Wide Web (WWW) Foundation. 2014. *History of the Web*. http://webfoundation.org/about/vision%20/history-of-the-web/
Wyatt, Edward. 2014a. "F.C.C., in a Shift, Backs Fast Lanes for Web Traffic." *New York* Times, 23 Apr.
– 2014b. "F.C.C. Backs Opening Net Neutrality Rules for Debate." *New York Times*, 15 May.
Yakabuski, Konrad. 2013. "BlackBerry Left Behind? That's so Canadian." *Globe and Mail*, 7 Nov.
Zaslow, Morris. 1975. *Reading the Rocks: The Story of the Geological Survey of Canada*. Toronto: Macmillan.
Zhang, Lucy Chen. 2013. *Policy Evaluation: A Case Study of Genome Canada Programming 2000–2011*. Master's thesis, University of Saskatchewan.
Zucker, Lynne G., Michael R. Darby, and Marilynn B. Brewer. 1998. "Intellectual Human Capital and the Birth of U.S. Biotechnology Enterprises." *American Journal of Economics* 88 (1): 290–306.
Zussman, David. 2010. "The Precarious State of the Federal Public Service." In G. Bruce Doern and Christopher Stoney (eds.), *How Ottawa Spends 2010–2011: Recession, Realignment, and the New Deficit Era*, 219–42. Montreal and Kingston: McGill-Queen's University Press.
– 2014. "Preparing for the Impact of Disruptive Technologies." *Canadian Government Executive* 20 (9).

Index